NEW MICROBIAL TECHNOLOGIES FOR ADVANCED BIOFUELS

Toward More Sustainable Production Methods

NEW MICROBIAL TECHNOLOGIES FOR ADVANCED BIOFUELS
Toward More Sustainable Production Methods

Edited by
Juan Carlos Serrano-Ruiz, PhD

APPLE ACADEMIC PRESS

Apple Academic Press Inc. | Apple Academic Press Inc.
3333 Mistwell Crescent | 9 Spinnaker Way
Oakville, ON L6L 0A2 | Waretown, NJ 08758
Canada | USA

©2016 by Apple Academic Press, Inc.

First issued in paperback 2021

Exclusive worldwide distribution by CRC Press, a member of Taylor & Francis Group
No claim to original U.S. Government works

ISBN 13: 978-1-77463-551-3 (pbk)
ISBN 13: 978-1-77188-130-2 (hbk)

Library and Archives Canada Cataloguing in Publication

New microbial technologies for advanced biofuels: toward more sustainable production methods/edited by Juan Carlos Serrano-Ruiz, PhD.

Includes bibliographical references and index.
ISBN 978-1-77188-130-2 (bound)
1. Biomass energy. 2. Microbial biotechnology. I. Serrano-Ruiz, Juan Carlos, editor

TP339.N49 2015   662'.88   C2015-900156-0

Library of Congress Cataloging-in-Publication Data

New microbial technologies for advanced biofuels : toward more sustainable production methods/editor, Juan Carlos Serrano-Ruiz.

pages cm
Includes bibliographical references and index.
ISBN 978-1-77188-130-2 (alk. paper)
1. Microbial biotechnology. 2. Biomass conversion. 3. Biomass energy. I. Serrano-Ruiz, Juan Carlos, editor.

TP248.27.M53N49 2015   662'.88--dc23   2014049259

ABOUT THE EDITOR

JUAN CARLOS SERRANO-RUIZ, PhD

Juan Carlos Serrano-Ruiz studied Chemistry at the University of Granada (Spain). In 2001 he moved to the University of Alicante (Spain) where he received a PhD in Chemistry and Materials Science in 2006. In January 2008, he was awarded a MEC/Fulbright Fellowship to conduct studies on catalytic conversion of biomass in James Dumesic's research group at the University of Wisconsin-Madison (USA). He is (co)author of over 50 manuscripts and book chapters on biomass conversion and catalysis. He is currently Senior Researcher at Abengoa Research, the research and development division of the Spanish company, Abengoa (Seville, Spain).

CONTENTS

ACKNOWLEDGMENT AND HOW TO CITE

The editor and publisher thank each of the authors who contributed to this book. The chapters in this book were previously published in various places in various formats. To cite the work contained in this book and to view the individual permissions, please refer to the citation at the beginning of each chapter. Each chapter was read individually and carefully selected by the editor; the result is a book that provides a nuanced look at new microbial technologies for sustainable production of advanced biofuels. The chapters included are broken into three sections, which describe the following topics:

- The articles in chapter 1 were all chosen to provide a well-rounded overview of microbial technology's applicability to biofuels.
- Chapter 2 offers a review of which microbes are appropriate to use for generating electricity, and what standards might be used to determine this, both now and in the future.
- In chapter 3 we turn to microalgae and their usefulness within biorefineries. These microorganisms have the advantage as a feedstock because they do not compete with food production, and they have additional environmental benefits.
- Chapter 4 picks up yet another useful microbe: yeast cells. This article examines the role that metabolic engineering plays in yeast's future in the world's use of biofuels.
- Chapter 5 adds to our understanding of this topic with a discussion of bacteria that are useful for microbial fuel cells.
- Finally, in chapter 6, we complete our investigation of specific types of microbes useful in biofuel, focusing here on bacteria-fungal flora found in forest soil.
- Chapter 7 begins our investigation of microbial pretreatment technologies. This article introduces the topic with research into the advantages and limitations of various approaches to microbial pretreatment.
- The authors of chapter 8 provide us with genomic research into the use of *E. coli* as a pretreatment.
- Chapter 9 uses ultrasonic technology as a pretreatment for microalgae and finds that it enhances bioavailability.

- In chapter 10 we investigate oxidative pretreatments of woody biomass. The authors present compelling evidence that it improves efficiency significantly, thereby reducing costs as well.
- Chapter 11 begins our focus on metabolic engineering (a topic introduced in chapter 4). The authors of this article investigate cloning as a useful strategy for developing more efficient microbial technologies for biofuels.
- The authors of Chapter 12 isolate mutant *E. coli* genes beneficial for producing hydrocarbons.
- In chapter 13 we turn out attention to generating phenotypic diversity, using ATMT with alcohol stress.
- The authors of chapter 14 describe an in vivo mutagenesis mechanism in microbial cells by introducing a group of genetically modified proofreading elements of the DNA polymerase, thus accelerating the evolutionary process under stressful conditions. This chapter concludes our investigation, laying a groundwork for further fruitful research in the future.

LIST OF CONTRIBUTORS

Reda A. I. Abou-Shanab
Department of Environmental Engineering, Yonsei University, Wonju, Gangwon-do 220-710, South Korea and Department of Environmental Biotechnology City of Scientific Research and Technology Applications, New Borg El Arab City, Alexandria 21934, Egypt

Jessica Bergmann
Department of Genomics Science and Biotechnology, Universidade Católica de Brasília, Brasília DF 70790-160, Brazil

Zhen Cai
CAS Key Laboratory of Microbial Physiological and Metabolic Engineering, Institute of Microbiology, Chinese Academy of Sciences, No. 1 West Beichen Road, Chaoyang District, Beijing 100101, China

Thomas Canam
Department of Biological Sciences, Eastern Illinois University, Charleston, IL, USA

Charles H. Chen
Department of Chemical Engineering and Materials Science, Michigan State University, East Lansing, USA and DOE Great Lakes Bioenergy Research Center, Michigan State University, East Lansing, USA

Chansoo Choi
Department of Applied Chemistry, Daejeon University, Daejeon, South Korea

Jeong-A Choi
Department of Environmental Engineering, Yonsei University, Wonju, Gangwon-do 220-710, South Korea

Brian A. Dempsey
Department of Civil and Environmental Engineering, Pennsylvania State University, 212 Sackett Building, University Park, PA 16802, USA

Patrik D'haeseleer
Physical and Life Sciences Directorate, Lawrence Livermore National Laboratory, Livermore, CA, USA

Yaser Dhaman
Department of Chemical Engineering, Ryerson University, Ontario, Canada

Francisco Dini-Andreote
Department of Microbial Ecology, Center for Ecological and Evolutionary Studies (CEES), University of Groningen (RUG), Nijenborgh 7, 9747AG Groningen, The Netherlands

Fiona Doohan
Molecular Plant-Microbe Interactions Laboratory, School of Biology and Environmental Science, University College Dublin, Dublin, Ireland

Tim Dumonceaux
Agriculture and Agri-Food Canada Saskatoon Research Centre, Saskatoon, Canada and Department of Veterinary Microbiology, University of Saskatchewan, Saskatoon, Canada

Jean Escudero
Department of Biological science, Texas A&M University-Kingsville, Texas 78363, USA

Jee Loon Foo
School of Chemical and Biomedical Engineering, Nanyang Technological University, 62 Nanyang Drive, Singapore 637459, Singapore

Brigitte Gasser
Department of Biotechnology, University of Natural Resources and Life Sciences, Muthgasse 18, 1190 Vienna, Austria and Austrian Centre of Industrial Biotechnology (ACIB GmbH), Vienna, Austria

John M. Gladden
Physical Biosciences Division, Lawrence Berkeley National Laboratory, Joint BioEnergy Institute (JBEI), 1 Cyclotron Road, Berkeley, CA 94720, USA and Biological and Materials Science Center, Sandia National Laboratories, Livermore, CA, USA

Luisa Gouveia
LNEG - National Laboratory of Energy and Geology, Bioenergy Unit, Portugal

Eric L. Hegg
DOE Great Lakes Bioenergy Research Center, Michigan State University, East Lansing, USA and Department of Biochemistry & Molecular Biology, Michigan State University, East Lansing, USA

Rosanna C. Hennessy
Department of Crop Science, Teagasc Research Centre, Oak Park, Carlow, Ireland and Molecular Plant-Microbe Interactions Laboratory, School of Biology and Environmental Science, University College Dublin, Dublin, Ireland

David B. Hodge
Department of Chemical Engineering and Materials Science, Michigan State University, East Lansing, USA, DOE Great Lakes Bioenergy Research Center, Michigan State University, East Lansing, USA, Department of Biosystems & Agricultural Engineering, Michigan State University, East Lansing, USA, and Department of Civil, Environmental and Natural Resources Engineering, Luleå University of Technology, Luleå, Sweden

Ebtesam N. Hosseny
Department of Botany and Microbiology, Al-Azhar University, Nasr City, Cairo 11884, Egypt

Jae-Hoon Hwang
Department of Environmental Engineering, Yonsei University, Wonju, Gangwon-do 220-710, South Korea

Eman Ibrahim
Faculty of Science, Nasr City, Cairo, Egypt and Department of Botany and Microbiology, Al-Azhar University, Nasr City, Cairo 11884, Egypt

Kingsley Iroba
Department of Chemical and Biological Engineering, University of Saskatchewan, Saskatoon, Canada

Byong-Hun Jeon
Department of Environmental Engineering, Yonsei University, Wonju, Gangwon-do 220-710, South Korea

Diego Javier Jiménez
Department of Microbial Ecology, Center for Ecological and Evolutionary Studies (CEES), University of Groningen (RUG), Nijenborgh 7, 9747AG Groningen, The Netherlands

Kim D. Jones
Department of Environmental Engineering, Texas A&M University-Kingsville, Texas 78363, USA

Hyun-Chul Kim
Department of Environmental Engineering, Yonsei University, Wonju, Gangwon-do 220-710, South Korea, Current address: Research Institute for Sustainable Environments, Ilshin Environmental Engineering Co., Ltd., Reclean Building 3rd Fl., 692-2 Jangji-dongSongpa-gu, Seoul 138-871, South Korea

Jung Rae Kim
School of Chemical and Biomolecular Engineering, Pusan National University, Busan 609-735, South Korea

Taekwon Lee
School of Civil and Environmental Engineering, Yonsei University, Seoul, South Korea and Department of Microbiology and Ecosystem Science, University of Vienna, Vienna, Austria

Susanna Su Jan Leong
School of Chemical and Biomedical Engineering, Nanyang Technological University, 62 Nanyang Drive, Singapore 637459, Singapore

Yin Li
CAS Key Laboratory of Microbial Physiological and Metabolic Engineering, Institute of Microbiology, Chinese Academy of Sciences, No. 1 West Beichen Road, Chaoyang District, Beijing 100101, China

Zhenglun Li
Department of Chemical Engineering and Materials Science, Michigan State University, East Lansing, USA and DOE Great Lakes Bioenergy Research Center, Michigan State University, East Lansing, USA

Bongsu Lim
Department of Environmental Engineering, Daejeon University, Daejeon, South Korea

Guodong Luan
CAS Key Laboratory of Microbial Physiological and Metabolic Engineering, Institute of Microbiology, Chinese Academy of Sciences, No. 1 West Beichen Road, Chaoyang District, Beijing 100101, China and University of Chinese Academy of Sciences, Beijing 100049, China

Yanhe Ma
State Key Laboratory of Microbial Resources, Institute of Microbiology, Chinese Academy of Sciences, Beijing 100101, China

Longfei Mao
Centre for Advanced Computational Solutions, Wine, Food and Molecular Bioscience Department, Lincoln University, Ellesmere Junction Road, Lincoln, 7647, New Zealand

Diethard Mattanovich
Department of Biotechnology, University of Natural Resources and Life Sciences, Muthgasse 18, 1190 Vienna, Austria and Austrian Centre of Industrial Biotechnology (ACIB GmbH), Vienna, Austria

Ewen Mullins
Department of Crop Science, Teagasc Research Centre, Oak Park, Carlow, Ireland

Joonhong Park
School of Civil and Environmental Engineering, Yonsei University, Seoul, South Korea

Joshua I. Park
Physical Biosciences Division, Lawrence Berkeley National Laboratory, Joint BioEnergy Institute (JBEI), 1 Cyclotron Road, Berkeley, CA 94720, USA, Biological and Materials Science Center, Sandia National Laboratories, Livermore, CA, USA, Current address: Department of Biological Sciences, Takeda California, Inc., San Diego, CA, USA

Betania F. Quirino
Department of Genomics Science and Biotechnology, Universidade Católica de Brasília, Brasília DF 70790-160, Brazil and Embrapa-Agroenergy, Brasília DF 70770-901, Brazil

Pattanathu K.S.M. Rahman
School of Science & Engineering, Technology Futures Institute, Teesside University, Middlesbrough, UK

John M. Regan
Department of Civil and Environmental Engineering, Pennsylvania State University, 212 Sackett Building, University Park, PA 16802, USA

Vimalier Reyes-Ortiz
Physical Biosciences Division, Lawrence Berkeley National Laboratory, Joint BioEnergy Institute (JBEI), 1 Cyclotron Road, Berkeley, CA 94720, USA

Pallavi Roy
Culture Link Canada, Ryerson University, Canada

Kenneth L. Sale
Physical Biosciences Division, Lawrence Berkeley National Laboratory, Joint BioEnergy Institute (JBEI), 1 Cyclotron Road, Berkeley, CA 94720, USA and Biological and Materials Science Center, Sandia National Laboratories, Livermore, CA, USA

Michael Sauer
Department of Biotechnology, University of Natural Resources and Life Sciences, Muthgasse 18, 1190 Vienna, Austria and Austrian Centre of Industrial Biotechnology (ACIB GmbH), Vienna, Austria

Kamaljeet Kaur Sekhon
School of Science & Engineering, Technology Futures Institute, Teesside University, Middlesbrough, UK

Blake A. Simmons
Physical Biosciences Division, Lawrence Berkeley National Laboratory, Joint BioEnergy Institute (JBEI), 1 Cyclotron Road, Berkeley, CA 94720, USA and Biological and Materials Science Center, Sandia National Laboratories, Livermore, CA, USA

Steven W. Singer
Physical Biosciences Division, Lawrence Berkeley National Laboratory, Joint BioEnergy Institute (JBEI), 1 Cyclotron Road, Berkeley, CA 94720, USA and Department of Geochemistry & Department of Ecology, Earth Sciences Division, Lawrence Berkeley National Laboratory, Berkeley, CA, USA

Mariam B. Sticklen
Department of Plant, Soil and Microbial Sciences, Michigan State University, East Lansing, USA

Lope Tabil
Department of Chemical and Biological Engineering, University of Saskatchewan, Saskatoon, Canada

Jennifer Town
Agriculture and Agri-Food Canada Saskatoon Research Centre, Saskatoon, Canada and Department of Veterinary Microbiology, University of Saskatchewan, Saskatoon, Canada

Jan Dirk van Elsas
Department of Microbial Ecology, Center for Ecological and Evolutionary Studies (CEES), University of Groningen (RUG), Nijenborgh 7, 9747AG Groningen, The Netherlands

Wynand S. Verwoerd
Centre for Advanced Computational Solutions, Wine, Food and Molecular Bioscience Department, Lincoln University, Ellesmere Junction Road, Lincoln, 7647, New Zealand

Zejie Wang
Department of Environmental Engineering, Daejeon University, Daejeon, South Korea and Institute of Urban Environment, Chinese Academy of Sciences, Urban, China

INTRODUCTION

The world needs renewable and clean forms of energy. Thanks to our dependency on fossil fuels, our planet faces the crisis of climate change, making the search for viable fuel alternatives even more urgent. Biofuels offer one of these alternatives, one that the media initially promoted as a glowing answer to our energy problems. It has become evident, however, that first-generation biofuels had many challenges to be overcome before biofuel could make a practical contribution to the world's energy crisis.

One strategy that second-generation biofuels are employing is microbial technology. Microorganisms such as yeast, bacteria, fungi, and microalgae can be effective bioreactors for the conversion of both solar energy and waste carbon dioxide into biodiesel, biohydrogen, and bioethanol, all renewable and biodegradable alternatives to fossil fuel. Even better, the use of microbial technology for biofuel production means that we can turn to sources of biomass other than food and feed crops.

This compendium gathers together recent research within this vital field of investigation. It offers first an overview of the topic and the varieties of microorganisms useful for this technology, and then goes on to examine pretreatment methodologies and the role of genetic engineering to further this technology. The collection offers a platform from which to build future investigations that will create even more advanced biofuels, furthering the world's search for viable answers to its energy crisis.

Juan Carlos Serrano-Ruiz, PhD

Chapter 1, a collection of three articles by Sticklen, Sekhon and Rahman, and Dhaman and Roy, is designed to provide an overall introduction to the topic of second-generation biofuels. The collection provides a well-rounded overview of microbial technology's applicability to biofuels.

A microbial fuel cell (MFC) is a device that uses microorganisms as biocatalysts to transform chemical energy or light energy into electricity. However, the commercial applications of MFCs are limited by their performance. Chapter 2, by Mao and Verwoerd, presents the perspective that in silico metabolic modelling based on genome-scale metabolic networks can be used for understanding the metabolisms of the anodic microorganisms and optimizes the performance of their metabolic networks for MFCs. This is in contrast to conventional research that focuses on engineering designs and study of biological aspects of MFCs to improve interactions of anode and microorganisms. Four categories of biocatalysts—microalgae, cyanobacteria, geobacteria and yeast—are nominated for future in silico constraint-based modelling of MFCs after taking into account the cell type, operation mode, electron source and the availability of metabolic network specifications. In addition, the advantages and disadvantages of each organism for MFCs are discussed and compared.

The study of microalgae is an emerging research field, due to their high potential as a source of several biofuels in addition to the fact that they have a high-nutritional value and contain compounds that have health benefits. They are also highly used for water stream bioremediation and carbon dioxide mitigation. Therefore, the tiny microalgae could lead to a huge source of compounds and products, giving a good example of a real biorefinery approach. In Chapter 3, Gouveia shows and presents examples of experimental microalgae-based biorefineries grown in an autotrophic mode at a laboratory scale.

Yeasts are regarded as the first microorganisms used by humans to process food and alcoholic beverages. The technology developed out of these ancient processes has been the basis for modern industrial biotechnology. Yeast biotechnology has gained great interest again in the last decades. Joining the potentials of genomics, metabolic engineering, systems and synthetic biology enables the production of numerous valuable products of primary and secondary metabolism, technical enzymes and biopharmaceutical proteins. An overview of emerging and established substrates and products of yeast biotechnology is provided in Chapter 4, and Mattanovich and colleagues discuss this technology in the light of the recent literature.

The microbial fuel cell represents a novel technology to simultaneously generate electric power and treat wastewater. Both pure organic matter and real wastewater can be used as fuel to generate electric power and the substrate type can influence the microbial community structure. Chapter 5, Wang and colleagues use rice straw, an important feedstock source in the world, as fuel after pretreatment with diluted acid method for a microbial fuel cell to obtain electric power. Moreover, the microbial community structures of anodic and cathodic biofilm and planktonic culture were analyzed and compared to reveal the effect of niche on microbial community structure. The microbial fuel cell produced a maximum power density of 137.6 ± 15.5 mW/m^2 at a COD concentration of 400 mg/L, which was further increased to 293.33 ± 7.89 mW/m^2 through adjusting the electrolyte conductivity from 5.6 mS/cm to 17 mS/cm. Microbial community analysis showed reduction of the microbial diversities of the anodic biofilm and planktonic culture, whereas diversity of the cathodic biofilm was increased. Planktonic microbial communities were clustered closer to the anodic microbial communities compared to the cathodic biofilm. The differentiation in microbial community structure of the samples was caused by minor portion of the genus. The three samples shared the same predominant phylum of *Proteobacteria*. The abundance of exoelectrogenic genus was increased with *Desulfobulbus* as the shared most abundant genus; while the most abundant exoelectrogenic genus of *Clostridium* in the inoculum was reduced. Sulfate reducing bacteria accounted for large relative abundance in all the samples, whereas the relative abundance varied in different samples. The results demonstrated that rice straw hydrolysate can be used as fuel for microbial fuel cells; microbial community structure differentiated depending on niches after microbial fuel cell operation; exoelectrogens were enriched; sulfate from rice straw hydrolysate might be responsible for the large relative abundance of sulfate reducing bacteria.

Mixed microbial cultures, in which bacteria and fungi interact, have been proposed as an efficient way to deconstruct plant waste. The characterization of specific microbial consortia could be the starting point for novel biotechnological applications related to the efficient conversion of lignocellulose to cello-oligosaccharides, plastics and/or biofuels. In Chapter 6, the diversity, composition and predicted functional profiles of novel bacterial-fungal consortia are reported by Jiménez and colleagues, on the basis of replicated aerobic wheat straw enrichment cultures. In order to set up biodegradative microcosms, microbial communities were retrieved from a forest soil and introduced into a mineral salt medium containing 1% of (un)treated wheat straw. Following each incubation step, sequential transfers were carried out using 1 to 1,000 dilutions. The microbial source next to three sequential batch cultures (transfers 1, 3 and 10) were analyzed by bacterial 16S rRNA gene and fungal ITS1 pyrosequencing. Faith's phylogenetic diversity values became progressively smaller from the inoculum to the sequential batch cultures. Moreover, increases in the relative abundances of *Enterobacteriales, Pseudomonadales, Flavobacteriales* and *Sphingobacteriales* were noted along the enrichment process. Operational taxonomic units affiliated with *Acinetobacter johnsonii, Pseudomonas putida* and *Sphingobacterium faecium* were abundant and the underlying strains were successfully isolated. Interestingly, *Klebsiella variicola* (OTU1062) was found to dominate in both consortia, whereas *K. variicola*-affiliated strains retrieved from untreated wheat straw consortia showed endoglucanase/xylanase activities. Among the fungal players with high biotechnological relevance, we recovered members of the genera- *Penicillium, Acremonium, Coniochaeta* and *Trichosporon*. Remarkably, the presence of peroxidases, alpha-L-fucosidases, beta-xylosidases, beta-mannases and beta-glucosidases, involved in lignocellulose degradation, was indicated by predictive bacterial metagenome reconstruction. Reassuringly, tests for specific (hemi)cellulolytic enzymatic activities, performed on the consortial secretomes, confirmed the presence of such gene functions. In an in-depth characterization of two wheat straw degrading microbial consortia, we revealed the enrichment and selection of specific bacterial and fungal taxa that were presumably involved in (hemi) cellulose degradation. Interestingly, the microbial community composition was strongly influenced by the wheat straw pretreatment. Finally, the functional bacterial-metagenome prediction and the evaluation of enzymatic activities (at the consortial secretomes) revealed the presence and enrichment of proteins involved in the deconstruction of plant biomass.

The purpose of Chapter 7, by Canam and colleagues, is to review the various pretreatment options available for lignocellulosic biomass, with particular emphasis on agricultural residues and on strategies that exploit the natural metabolic activity of microbes to increase the processability of the biomass. These microbial-based strategies can be effective pretreatments on their own or, more probably, can be used in combination with thermomechanical pretreatments in order to provide a cost-effective means to make lignocellulosic substrates available for conversion to bio-fuels by microorganisms. The key advantages and disadvantages of this strategy will be presented along with a vision for how microbial pretreat-ment can be integrated into an economical biorefinery process for biofuels and co-product production.

The development of advanced biofuels from lignocellulosic biomass will require the use of both efficient pretreatment methods and new bio-mass-deconstructing enzyme cocktails to generate sugars from lignocel-lulosic substrates. Certain ionic liquids (ILs) have emerged as a promising class of compounds for biomass pretreatment and have been demonstrated to reduce the recalcitrance of biomass for enzymatic hydrolysis. However, current commercial cellulase cocktails are strongly inhibited by most of the ILs that are effective biomass pretreatment solvents. Fortunately, re-cent research has shown that IL-tolerant cocktails can be formulated and are functional on lignocellulosic biomass. Chapter 8, by Gladden and col-leagues sought to expand the list of known IL-tolerant cellulases to further enable IL-tolerant cocktail development by developing a combined in vi-tro/in vivo screening pipeline for metagenome-derived genes. Thirty-sev-en predicted cellulases derived from a thermophilic switchgrass-adapted microbial community were screened in this study. Eighteen of the twenty-one enzymes that expressed well in *E. coli* were active in the presence of the IL 1-ethyl-3-methylimidazolium acetate ([C2mim][OAc]) concentra-tions of at least 10% (v/v), with several retaining activity in the presence of 40% (v/v), which is currently the highest reported tolerance to [C2mim][OAc] for any cellulase. In addition, the optimum temperatures of the en-zymes ranged from 45 to 95°C and the pH optimum ranged from 5.5 to 7.5, indicating these enzymes can be used to construct cellulase cocktails that function under a broad range of temperature, pH and IL concentra-tions. This study characterized in detail twenty-one cellulose-degrading enzymes derived from a thermophilic microbial community and found that 70% of them were [C2mim][OAc]-tolerant. A comparison of opti-mum temperature and [C2mim][OAc]-tolerance demonstrates that a posi-tive correlation exists between these properties for those enzymes with a optimum temperature >70°C, further strengthening the link between ther-motolerance and IL-tolerance for lignocelluolytic glycoside hydrolases.

Microalgal biomass contains a high level of carbohydrates which can be biochemically converted to biofuels using state-of-the-art strategies that are almost always needed to employ a robust pretreatment on the biomass for enhanced energy production. In Chapter 9, Jeon and colleagues used an ultrasonic pretreatment to convert microalgal biomass (*Scenedesmus obliquus* YSW15) into feasible feedstock for microbial fermentation to produce ethanol and hydrogen. The effect of sonication condition was quantitatively evaluated with emphases on the characterization of carbohydrate components in microalgal suspension and on subsequent production of fermentative bioenergy. *Scenedesmus obliquus* YSW15 was isolated from the effluent of a municipal wastewater treatment plant. The sonication durations of 0, 10, 15, and 60 min were examined under different temperatures at a fixed frequency and acoustic power resulted in morphologically different states of microalgal biomass lysis. Fermentation was performed to evaluate the bioenergy production from the non-sonicated and sonicated algal biomasses after pretreatment stage under both mesophilic (35°C) and thermophilic (55°C) conditions. A 15 min sonication treatment significantly increased the concentration of dissolved carbohydrates (0.12 g g^{-1}), which resulted in an increase of hydrogen/ethanol production through microbial fermentation. The bioconvertibility of microalgal biomass sonicated for 15 min or longer was comparable to starch as a control, indicating a high feasibility of using microalgae for fermentative bioenergy production. Increasing the sonication duration resulted in increases in both algal surface hydrophilicity and electrostatic repulsion among algal debris dispersed in aqueous solution. Scanning electron microscope images supported that ruptured algal cell allowed fermentative bacteria to access the inner space of the cell, evidencing an enhanced bioaccessibility. Sonication for 15 min was the best for fermentative bioenergy (hydrogen/ethanol) production from microalga, and the productivity was relatively higher for thermophilic (55°C) than mesophilic (35°C) condition. These results demonstrate that more bioavailable carbohydrate components are produced through the ultrasonic degradation of microalgal biomass, and thus the process can provide a high quality source for fermentative bioenergy production.

One route for producing cellulosic biofuels is by the fermentation of lignocellulose-derived sugars generated from a pretreatment that can be effectively coupled with an enzymatic hydrolysis of the plant cell wall. While woody biomass exhibits a number of positive agronomic and logistical attributes, these feedstocks are significantly more recalcitrant to chemical pretreatments than herbaceous feedstocks, requiring higher chemical and energy inputs to achieve high sugar yields from enzymatic hydrolysis. Li and colleagues previously discovered that alkaline hydrogen peroxide (AHP) pretreatment catalyzed by copper(II) 2,2′-bipyridine complexes significantly improves subsequent enzymatic glucose and xylose release from hybrid poplar heartwood and sapwood relative to uncatalyzed AHP pretreatment at modest reaction conditions (room temperature and atmospheric pressure). In Chapter 10, the reaction conditions for this catalyzed AHP pretreatment were investigated in more detail with the aim of better characterizing the relationship between pretreatment conditions and subsequent enzymatic sugar release. The authors found that for a wide range of pretreatment conditions, the catalyzed pretreatment resulted in significantly higher glucose and xylose enzymatic hydrolysis yields (as high as 80% for both glucose and xylose) relative to uncatalyzed pretreatment (up to 40% for glucose and 50% for xylose). They identified that the extent of improvement in glucan and xylan yield using this catalyzed pretreatment approach was a function of pretreatment conditions that included H_2O_2 loading on biomass, catalyst concentration, solids concentration, and pretreatment duration. Based on these results, several important improvements in pretreatment and hydrolysis conditions were identified that may have a positive economic impact for a process employing a catalyzed oxidative pretreatment. These improvements include identifying that: (1) substantially lower H_2O_2 loadings can be used that may result in up to a 50-65% decrease in H_2O_2 application (from 100 mg H_2O_2/g biomass to 35–50 mg/g) with only minor losses in glucose and xylose yield, (2) a 60% decrease in the catalyst concentration from 5.0 mM to 2.0 mM (corresponding to a catalyst loading of 25 μmol/g biomass to 10 μmol/g biomass) can be achieved without a subsequent loss in glucose yield, (3) an order of magnitude improvement in the time required for pretreatment (minutes versus hours or days) can be realized using the catalyzed pretreatment approach, and (4) enzyme dosage can be reduced to less than 30 mg protein/g glucan and potentially further with only minor losses in glucose and xylose yields. In addition, the authors established that the reaction rate is improved in both catalyzed and uncatalyzed AHP pretreatment by increased solids concentrations. This work explored the relationship between reaction conditions impacting a catalyzed oxidative pretreatment of woody biomass and identified that significant decreases in the H_2O_2, catalyst, and enzyme loading on the biomass as well as decreases in the pretreatment time could be realized with only minor losses in the subsequent sugar released enzymatically. Together these changes would have positive implications for the economics of a process based on this pretreatment approach.

Lignocellulosic biomass has potential for bioethanol, a renewable fuel. A limitation is that bioconversion of the complex lignocellulosic material to simple sugars and then to bioethanol is a challenging process. Recent work has focused on the genetic engineering of a biocatalyst that may play a critical role in biofuel production. *Escherichia coli* have been considered a convenient host for biocatalysts in biofuel production for its fermentation of glucose into a wide range of short-chain alcohols and production of highly deoxygenated hydrocarbon. The bacterium *Pectobacterium carotovorum* subsp. *carotovorum* (*P. carotovorum*) is notorious for its maceration of the plant cell wall causing soft rot. The ability to destroy plants is due to the expression and secretion of a wide range of hydrolytic enzymes that include cellulases and polygalacturonases. In Chapter 11, Ibrahim and colleagues used *P. carotovorum* ATCC™ no. 15359 as a source of DNA for the amplification of celB, celC and peh. These genes encode 2 cellulases and a polygalacturonase, respectively. Primers were designed based on published gene sequences and used to amplify the open reading frames from the genomic DNA of *P. carotovorum*. The individual PCR products were cloned into the pTAC-MAT-2 expression vector and transformed into *Escherichia coli*. The deduced amino acid sequences of the cloned genes have been analyzed for their catalytically active domains. Estimation of the molecular weights of the expressed proteins was performed using SDS-PAGE analysis and celB, celC and peh products were approximately 29.5 kDa, 40 kDa 41.5 kDa, respectively. Qualitative determination of the cellulase and polygalacturonase activities of the cloned genes was carried out using agar diffusion assays.

The depletion of fossil fuels and the rising need to meet global energy demands have led to a growing interest in microbial biofuel synthesis, particularly in *Escherichia coli*, due to its tractable characteristics. Besides engineering more efficient metabolic pathways for synthesizing biofuels, efforts to improve production yield by engineering efflux systems to overcome toxicity problems is also crucial. Chapter 12, by Foo and Leong, aims to enhance hydrocarbon efflux capability in *E. coli* by engineering a native inner membrane transporter, AcrB, using the directed evolution approach. The authors developed a selection platform based on competitive growth using a toxic substrate surrogate, which allowed rapid selection of AcrB variants showing enhanced efflux of linear and cyclic fuel molecule candidates, n-octane and α-pinene. Two mutants exhibiting increased efflux efficiency for n-octane and α-pinene by up to 47% and 400%, respectively, were isolated. Single-site mutants based on the mutations found in the isolated variants were synthesized and the amino acid substitutions N189H, T678S, Q737L and M844L were identified to have conferred improvement in efflux efficiency. The locations of beneficial mutations in AcrB suggest their contributions in widening the substrate channel, altering the dynamics of substrate efflux and promoting the assembly of AcrB with the outer membrane channel protein TolC for more efficient substrate export. It is interesting to note that three of the four beneficial mutations were located relatively distant from the known substrate channels, thus exemplifying the advantage of directed evolution over rational design. Using directed evolution, the authors have isolated AcrB mutants with improved efflux efficiency for n-octane and α-pinene. The utilization of such optimized native efflux pumps will increase productivity of biofuels synthesis and alleviate toxicity and difficulties in production scale-up in current microbial platforms.

Consolidated bioprocessing (CBP) of lignocellulosic biomass offers an alternative route to renewable energy. The crop pathogen *Fusarium oxysporum* is a promising fungal biocatalyst because of its broad host range and innate ability to co-saccharify and ferment lignocellulose to bioethanol. A major challenge for cellulolytic CBP-enabling microbes is alcohol inhibition. Chapter 13, by Hennessy and colleagues, tested the hypothesis that *Agrobacterium tumefaciens*-mediated transformation (ATMT) could be exploited as a tool to generate phenotypic diversity in *F. oxysporum* to investigate alcohol stress tolerance encountered during CBP. A random mutagenesis library of gene disruption transformants (n=1,563) was

constructed and screened for alcohol tolerance in order to isolate alcohol sensitive or tolerant phenotypes. Following three rounds of screening, exposure of select transformants to 6% ethanol and 0.75% n-butanol
resulted respectively in increased ($\geq$11.74%) and decreased ($\leq$43.01%)
growth compared to the wild –type (WT). Principal component analysis
(PCA) quantified the level of phenotypic diversity across the population
of genetically transformed individuals and isolated candidate strains for
analysis. Characterisation of one strain, Tr. 259, ascertained a reduced
growth phenotype under alcohol stress relative to WT and indicated the
disruption of a coding region homologous to a putative sugar transporter
(FOXG_09625). Quantitative PCR (RT-PCR) showed FOXG_09625 was
differentially expressed in Tr. 259 compared to WT during alcohol-induced
stress ($P<0.05$). Phylogenetic analysis of putative sugar transporters suggests diverse functional roles in *F. oxysporum* and other filamentous fungi
compared to yeast for which sugar transporters form part of a relatively
conserved family. This study has confirmed the potential of ATMT coupled with a phenotypic screening program to select for genetic variation
induced in response to alcohol stress. This research represents a first step
in the investigation of alcohol tolerance in *F. oxysporum* and has resulted
in the identification of several novel strains, which will be of benefit to
future biofuel research.

Microbial production of biofuels requires robust cell growth and metabolism under tough conditions. Conventionally, such tolerance phenotypes were engineered through evolutionary engineering using the principle of "Mutagenesis followed-by Selection". The iterative rounds of
mutagenesis-selection and frequent manual interventions resulted in discontinuous and inefficient strain improvement processes. Chapter 14, by
Luan and colleagues, aimed to develop a more continuous and efficient
evolutionary engineering method termed as "Genome Replication Engineering Assisted Continuous Evolution" (GREACE) using "Mutagenesis
coupled-with Selection" as its core principle. The core design of GREACE
is to introduce an in vivo continuous mutagenesis mechanism into microbial cells by introducing a group of genetically modified proofreading elements of the DNA polymerase complex to accelerate the evolution process
under stressful conditions. The genotype stability and phenotype heritability can be stably maintained once the genetically modified proofreading

element is removed, thus scarless mutants with desired phenotypes can be obtained. Kanamycin resistance of *E. coli* was rapidly improved to confirm the concept and feasibility of GREACE. Intrinsic mechanism analysis revealed that during the continuous evolution process, the accumulation of genetically modified proofreading elements with mutator activities endowed the host cells with enhanced adaptation advantages. They further showed that GREACE can also be applied to engineer n-butanol and acetate tolerances. In less than a month, an *E. coli* strain capable of growing under an n-butanol concentration of 1.25% was isolated. As for acetate tolerance, cell growth of the evolved E. coli strain increased by 8-fold under 0.1% of acetate. In addition, the authors discovered that adaptation to specific stresses prefers accumulation of genetically modified elements with specific mutator strengths. The authors developed a novel GREACE method using "Mutagenesis coupled-with Selection" as core principle. Successful isolation of *E. coli* strains with improved n-butanol and acetate tolerances demonstrated the potential of GREACE as a promising method for strain improvement in biofuels production.

CHAPTER 1

OVERVIEW

CO-PRODUCTION OF HIGH-VALUE RECOMBINANT BIOBASED MATTER IN BIOENERGY CROPS FOR EXPEDITING THE CELLULOSIC BIOFUELS AGENDA

Mariam B. Sticklen

At present, food crops such as sugarcane sugar and corn seed starch are used to commercially produce ethanol. The goal of the biofuels industry is to produce biofuels from crop waste matter, that are non-food residues whose cellulosic matter are converted into hydrocarbon liquid fuels (mostly ethanol) after transportation and storage, pretreatment and enzymatic hydrolysis of cellulose via the use of microbial cellulases for production of fermentable sugars, followed by fermentation of sugars. Despite recent improvements in pretreatment technology and modifications of genomes of bioenergy crops via anti-sense methods to lower their lignin contents for an enhanced hydrolysis [1], and improvements made in genomes of cellulase-producing microbes and even considering the synthetic biology [2], the costs associated with cellulosic biofuel production remains to be the inhibitory factor to the non-subsidized commercialization and sustainable economy of cellulosic biofuel industry.

Co-Production of High-Value Recombinant Biobased Matter in Bioenergy Crops for Expediting the Cellulosic Biofuels Agenda. © *Sticklen MB.* Advances in Crop Science and Technology, *1 (2013), doi: 10.4172/2329-8863.1000e101. Licensed under Creative Commons License, http://creativecommons. org/licenses/by/3.0.*

The author proposes the use of the petroleum industry model to expedite biofuels market agenda. Petroleum industry makes its profit not only from petro-fuels (gasoline or petrol, petrodiesel, ethane, kerosene, liquefied petroleum gas and natural gas), but also from over 6,000 petroleum-derivative co-products such as alkenes (olefins), lubricants, wax, sulfuric acid, bulk tar, asphalt, the solid fuel called petroleum coke, paraffin wax, and aromatic petrochemicals that are used for production of hydrocarbon fuels and hydrocarbon chemicals. Using the petroindustry model is not only co-producing of plant-based matters such as for example lactic acid and ascorbic acid, but also pioneering systems on changing the genetic structure of bioenergy crops for production of higher value biobased co-products in crop waste matter. For example, all three microbial cellulases have been produced in maize stover, not in plant seeds, flowers or roots. Such recombinant co-products were accumulated in maize plant sub-cellular compartments such as apoplast (cell wall areas), chloroplast, endoplasmic reticulum and mitochondria [3]. Sub-cellular compartment accumulation of cellulases has the advantage of keeping the heterologous enzyme away from cytoplasm to avid harm to transgenic plant growth and development, and to keep these heterologous cellulases in compartments that can accumulate during the stages of plant growth and development under ideal pH conditions [4,5]. A report indicates that it is possible to co-produce recombinant heterologous cellulases in the cellulosic biomass of maize at over 1 kg/ton of its Stover waste [3].

A few examples of other high-value matter that have been coproduced in recombinant plants either via nuclear of chloroplast transgenesis include, but not limited to silk and silk-like structures [6], Vaccines [7] Therapeutics [8,9], biopharmaceuticals such as interferon [10] and interleukin (Matakas, 2011), A biodegradable plastics [11], and fatty acid commodities [12-15]. Also, advances have also been made to match the glycosylation of plant-produced heterologous proteins to match those of humans when used for human treatments of diseases [16,17].

For example, the global market for cellulases will be enormous. Also, the global market for biotech drugs of 2007 was worth $86.8 billion and is predicted to double by 2013. Co-production of cellulases and high value biopharmaceuticals in bioenergy crops will not only expedite cellulosic biofiels agenda and be lucrative to biotech drugs industry, but will also help to make these drugs available to financially less fortunates at the global level.

After produced, the extraction of the bioenergy crop-produced recombinant high-value co-products fits well in the biofuel industry processes, and would add value to the bioenergy crops allowing the farmers to sell their crop residues at a higher price. Furthermore, the co-production of high-value recombinant commodities in crop waste can boost the cellulosic biofuels industry revenues through the sales of those co-products.

The co-production and use of high-value co-products is a necessary step towards the sustained economy of cellulosic biofuels. The coproduction of high-value recombinant molecules in cellulosic wastes of bioenergy crops can revolutionize the economic sustainability and unsubsidized commercialization of cellulosic biofuels including ethanol and butanol.

REFERENCES

1. Park SH, Mei C, Pauly M, Ong RG, Dale BE, et al. (2012) Downregulation of Maize Cinnamoyl-Coenzyme A Reductase via RNA Interference Technology Causes Brown Midrib and Improves Ammonia Fiber Expansion-Pretreated Conversion into Fermentable Sugars for Biofuels. Crop Sci 52: 2687-2701.

2. Witze A (2013) Factory of Life: Synthetic biologists reinvent nature with parts, circuits. The Science News 183: 22

3. Park SH, Ransom C, Mei C, Sabzikar R, Qi C, et al. (2011) The quest for alternatives to microbial cellulase mix production: corn stover-produced heterologous multi-cellulases readily deconstruct lignocellulosic biomass into fermentable sugars. J Chem Technol Biotechnol 86: 633-641

4. Sticklen MB (2007) Feedstock genetic engineering for alcohol fules. Crop Science 47: 2238-2248.

5. Sticklen M (2009) Expediting biofuels agenda via genetic manipulations of BioEnergy crops. Biofuels, Bioproducts and Biorefinery 3: 448-455.

6. Yang J, Barr LA, Fahnestock SR, Liu Z (2005) High yield recombinant silk-like protein production in transgenic plants through protein targeting. Transgenic Res 14: 313-324.

7. Kamarajugadda S, Daniell H (2006) Chloroplast-derived anthrax and other vaccine antigens: their immunogenic and immunoprotective properties. Expert Review of Vaccines 5: 839-849.

8. Staub JM, Garcia B, Graves J, Hajdukiewicz PT, Hunter P, et al. (2000) High-yield production of a human therapeutic protein in tobacco chloroplasts. Nat Biotechnol 18: 333-338.

9. Gomord V, Fitchette AC, Menu-Bouaouiche L, Saint-Jore-Dupas C, Plasson C, et al. (2010) Plant-specific glycosylation patterns in the context of therapeutic protein production. Plant Biotechnol J 8: 564-87

10. Arlen PA, Falconer R, Cherukumilli S, Cole A, Cole AM, et al. (2007) Field production and functional evaluation of chloroplast-derived interferon-alpha2b. Plant Biotechnol J 5: 511-525.

11. Mooney BP (2009) The second green revolution? Production of plant-based biodegradable plastics. Biochem J 418: 219-32.

12. Napier JA, Beaudoin F, Michaelson LV, Sayanova O (2004) The production of long chain polyunsaturated fatty acids in transgenic plants by reverse-engineering. Biochimie 86: 785-792.

13. Ohlrogge JB, Chapman K (2011) The seeds of green energy - expanding the contribution of plant oils as biofuels. The Biochemist 33: 34-38.

14. Troncoso-Ponce MA, Kilaru A, Cao X, Durrett TP, Fan J, et al. (2011) Comparative deep transcriptional profiling of four developing oilseeds. Plant J 68: 1014-1027.

15. Chapman KD, Ohlrogge JB (2012) Compartmentation of triacylglycerol accumulation in plants. J Biol Chem 287: 2288-2294.

16. Faye L, Boulaflous A, Benchabane M, Gomord V, Michaud D (2005) Protein modifications in the plant secretory pathway: current status and practical implications in molecular pharming. Vaccine 23: 1770-1778.

17. Obembe OO, Popoola JO, Leelavathi S, Reddy SV (2011) Advances in plant molecular farming. Biotechnol Adv 29: 210-222.

SYNTHETIC BIOLOGY: A PROMISING TECHNOLOGY FOR BIOFUEL PRODUCTION

Kamaljeet Kaur Sekhon and Pattanathu K.S.M. Rahman

With the increasing awareness among the masses and the depleting natural resources and oil reservoirs, a replacement for the fossil fuels is urgently required. There are rising global concerns about climate change and energy security. The current biofuel production trends are no doubt promising and increasing steadily. The biofuel markets are getting bigger and better in the European Union, USA, Brazil, India, China and Argentina and con-

Synthetic Biology: A Promising Technology for Biofuel Production. © *Sekhon KK and Rahman PKSM.*
Journal of Petroleum & Environmental Biotechnology, *4 (2013), doi: 10.4172/2157-7463.1000e121.*
Licensed under Creative Commons License, http://creativecommons.org/licenses/by/3.0.

tributing to their bio-economies, respectively. In the US, biodiesel production exceeded 1 billion gallons in 2012 and reports claim that the global biofuels market will touch the figure of $185 billion in 2021. However the big question is: will the current biofuel production rate be able to meet the escalating transportation fuel demands?

The total estimated generation of biomass in the world is 150 billion tons annually. Increase in the production of biofuels in the recent years and the usage of edible commodities like maize, sugarcane and vegetable oil has led to the worldwide apprehension towards the future of biofuels and to the 'food vs fuel' debate. The second generation biofuels, however, are produced from renewable, cheap and sustainable feed-stocks for example citrus peel, corn stover, sawdust, bagasse, straw, rice peel and are attracting ever-increasing attention. A great deal of research, by the scientific community, is carried out in various parts of the world in order to improve the yield of second generation biofuels to meet the future demands but hasn't achieved any remarkable success.

Sustainable, economic production of second generation biofuels is of global importance. However, major technological hurdles remain before widespread conversion of non-food biomass into biofuel. Various multidisciplinary teams of scientists, technologists and engineers work together collaboratively in integrated teams to carry out research that underpins the generation and implementation of sustainable second generation biofuels from algal biomass using biological processes. The advantages of algal biomass from both micro and macroalgae as a raw material for producing biofuels have been well recognised for decades. Billions of tonnes of algal biomass are enzymatically converted into food energy by marine and freshwater animals and microbes every day, in a sustainable manner. However, the industrial, enzyme driven conversion of such biomass for bioenergy applications is still in its infancy.

The production of commercially attractive biofuels using enzymatic methods, all the same, is not as easy as it appears. The various polysaccharides viz. cellulose, starch, lignin, hemicellulose, or lignocelluloses need to be enzymatically degraded for their transformation into glucose or sugar molecules which in turn are fermented into biofuels (bioethanol or biobutanol). In case of cellulose, the process of cellulolysis involves enzymes like cellulases and glucosidases. Cellulases are expensive, unstable and slow in action; therefore they increase the overall economics of the process of cellulolysis and hence biofuel production. The bulk production of cellulases at industrial level seems to be the relevant solution. The microbes that produce cellulases include symbiotic anaerobic bacteria (e.g. *Cellulomonasfimi, Clostridium thermocellum, Clostridium phytofermentans, Thermobifidafusca*) found in ruminants such as cow and sheep, flagellate protozoa present in hindguts of termites, and filamentous fungi isolated from decaying plants (e.g. *Hypocreajecorina, Thermoascusaurantiacus, Phanerochaetechrysosporium, Neurosporacrassa, Tricodermareesei, Asperigillusniger, Fusariumoxysporum*). The gene(s) responsible for cellulase production are characterized, isolated and recombinantly introduced into *Escherichia coli* for the enhancedcellulase expression levels.

Apart from the conventional biotechnology methods for biofuel production, synthetic biology has shown promising results lately. Understanding the DNA sequences, precisely measuring the gene behaviour paves way for fabricating or synthesizing the cellulase gene de novo. To put it in simple words, synthetic biology is a science of designing and constructing new biological parts, devices and systems for programming cells and organisms and endowing them with novel functions. It is a technique of writing the DNA / genetic code base by base using several computational tools and software like Gene designer, GenoCAD, Eugene and Athena to name a few. Gene designer is a DNA design tool for de novo assembly of genetic constructs, GenoCAD is a computer-assisted-design application for synthetic biology for designing complex gene constructs and artificial gene networks, Eugene is a language designed to develop novel biological devices and Athena is a CAD / CAM software for constructing biological models as modules. These synthetic biology approaches can be useful in bringing down the cost of cellulases and, thereby, of biofuels. Several companies are spending a fortune on the production of bioethanol for example; Amyris Biotechnologies, Verenium, Iogen, Bioethanol Japan, Mascoma, POET, SolixBiofuels, Pacific Ethanol, NextGen Fuel Inc. and Jatro Diesel. However, the cost-effective production of the second generation biofuels is still a cherished desire of the scientific community.

Synthetic biology is an evolving field still dealing with the inherent complexity of biological systems and overcoming the biosafety issues involved with engineering the living systems. Indeed the proliferation of the computer modelling tools is leading to the revolution of this discipline which might write the success story of some of the present and future scientific challenges.

CHALLENGES AND GENERATIONS OF BIOFUELS: WILL ALGAE FUEL THE WORLD?

Yaser Dhaman and Pallavi Roy

Utilisation of biomass started with the discovery of fire, when early men burnt branches to cook their food and stay warm. Although we are still utilising biomass directly through burning them, we have come far in utilising those sources of energy with improved conversions. We have been forever looking for different and more efficient sources of energy. Biofuels and oil generated from peanuts and legumes were used in the 1800s to run farm machinery and more. The discovery of petroleum entirely revolutionised our way of life. Nowadays, it is hard to imagine life without the comforts that fossil fuels bring. However, this boon is slowly turning into one of the major problems of our time. Clearly, by burning fossil fuel we are releasing immense amounts of stored carbon back into the environment, polluting it and causing problems on a global scale. As a result, there has been a recent push to develop solutions to this problem yet maintaining our way of life.

Biofuels theoretically are one of the best approaches for finding solutions to this problem. Many nations are also bound by various international treaties to reduce emissions. All these have been pushing for increased production and utilization of biofuels. It started with converting mostly edible foods being converted to ethanol and biodiesel, which was called 1st generation biofuels. Ethanol from corn (US) or sugarcane (Brazil) is

Challenges and Generations of Biofuels: Will Algae Fuel the World? © *Dhaman Y and Roy P.* Fermentation Technology, 2,119 (2013), doi: 10.4172/2167-7972.1000e119. *Licensed under Creative Commons License, http://creativecommons.org/licenses/by/3.0.*

the most common among first generation biofuels. Ethanol was first mixed with gasoline in the US in the 1990s as a replacement for Methyl-ter-butyl ether (MTBE) as an additive to oxygenate gasoline, since MTBE had issues of water contamination. Brazil and USA started investing in Ethanol following an energy crisis in the 1970s. Both countries are the major producers and consumers of ethanol biofuel generated from food crops. However, major issue with first generation fuels is the food vs. fuel debate. This led to focus being shifted to cellulosic biofuels in the 2000s. These are the 2nd generation biofuels where the inedible part of food crops, agricultural wastes, wood, grasses etc. are being used to produce biofuels. Agricultural residue is very attractive due to the large amount of carbon which can be converted to fuel, and the cheap cost of the feedstock is lucrative. But the 2nd generation biofuels are still plagued by lower yields, stopping the product from being commercially viable.

This is where the new categories of biofuels come in, often referred to as the 3rd generation biofuels or algal biofuel. Algae produce triglycerides that can be extracted and converted in a refinery to biodiesel. Producing biodiesel from algae is widely regarded as one of the most efficient ways of generating biofuels and has great potential to replace gasoline/diesel demand in transport [1]. Algal biofuel is very attractive because [2].

- Algal fuel can be produced using freshwater, saltwater and wastewater,
- The oil is biodegradable so it is harmless to the environment if spilled.
- The bio-oil production is around 60% of the biomass much higher than the 2-3% produced from soybean.

Not only biodiesel can be produced from Algae, but also the remaining biomass of algae after oil extraction can be utilized as a renewable and sustainable resource for carbon. This can replace agriculture residues to produce biofuels such as ethanol and butanol [3]. Algae are known to produce biomass faster and on reduced land surface as compared with lignocellulosic biomass [4].

Recently, a 4th generation biofuel has been proposed where the biofuels are created using petroleum-like hydroprocessing or advanced biochemistry. The 4th generation of biofuels has introduced the concept of "cell factory". Studies on metabolic engineering of algae to increase the photosynthetic ability of the cell to produce higher yield of fuel has shown lots of potentials [5]. Joule Biotechnologies—the folks behind the unusual hybrid solar-biofuel technology—uses sunlight, waste CO_2 and engineered microorganisms in a "solar converter" to create fuel [6].

Now, what future is holding for us? It is being concluded that by 2022, eight percent of the global fuel volumes consumed by transportation will be biofuels. Right now, 1st and 2nd generations of biofuels account for 99% of the world's biofuel production. Interestingly, 1st generation ethanol and biodiesel are the only commercially viable product. While lots of interest is heading towards algae, technological advances promise to be helpful in making the 3rd and 4th generation fuels commercially competitive and hopefully the future of transportation fuels.

REFERENCES

1. Pabbi, S., Dhar, D. W., Bhatnagar, S. K., Saxena, A., &Kraan, S. (2011). Feasibility of algal biomass for biodiesel production. Algae biofuel.
2. Lundquist, T. J., Woertz, I. C., Quinn, N. W. T., &Benemann, J. R. (2010). A realistic technology and engineering assessment of algae biofuel production. Energy Biosciences Institute 1.
3. Potts, T., Du, J., Paul, M., May, P., Beitle, R., &Hestekin, J. (2012). The production of butanol from Jamaica bay macro algae. EnvironmentalProgress & Sustainable Energy31(1).
4. Lee, R. A., & Lavoie, J. M. (2013). From first-to third-generation biofuels: Challenges of producing a commodity from a biomass of increasing complexity. Animal Frontiers3(2).
5. Lü, J., Sheahan, C., & Fu, P. (2011). Metabolic engineering of algae for fourth generation biofuels production. Energy & Environmental Science4(7).

PART I

MICROORGANISMS

CHAPTER 2

SELECTION OF ORGANISMS FOR SYSTEMS BIOLOGY STUDY OF MICROBIAL ELECTRICITY GENERATION: A REVIEW

LONGFEI MAO and WYNAND S. VERWOERD

2.1 INTRODUCTION

The technology for extracting an electrical current for use in external circuits from the metabolic processes of living microbes has been in development for more than a century [1]. The resulting devices, termed microbial fuel cells (MFCs), have several potential advantages over more prominent sustainable energy technologies such as solar or wind power. For example, they can directly convert organic waste into electricity [2] without pollution or inefficient intermediate steps that involve mechanical generators. This feature, energy recovery from solid wastes, has been exploited in proposed national strategies for many Asian countries [3]. It may be possible to achieve the same goal by inorganic catalysts or enzymes, but using living cells makes it possible to exploit their adaptability to environmental conditions and avoids the high capital cost of installation for other waste-to-energy systems reviewed by Eddine and Salah [4]. The whole organ-

Selection of Organisms for Systems Biology Study of Microbial Electricity Generation: A Review. © *Mao L and Verwoerd WS.* International Journal of Energy and Environmental Engineering, *4,17 (2013). doi:10.1186/2251-6832-4-17 Originally distributed under the Creative Commons Attribution 2.0 Generic License, http://creativecommons.org/licenses/by/2.0/.*

isms used in MFCs contain various enzymes and therefore allow different substrates (or mixed substrates) to be used. The organisms in the fuel cell system can be considered as micro-reactors and provide optimal conditions for different enzymes. Because the organisms are self-replicating, the organic matter oxidation implemented by these bio-catalysts are self-sustaining [5] and not subject to catalytic poisoning like metallic catalysts or degradation of enzyme catalysts. By selecting photosynthetic microbes, solar energy could be converted at the same time. One can envisage portable electronics powered by MFCs that are 'charged' by feeding them nutrients rather than electric current, or medical implants that derive their power directly from nutrients circulating in the bloodstream. Perhaps the process can be reversed, and external electrical power supplied to an MFC converted into biomass, as a temporary storage, to overcome the intermittent nature of many other sustainable energy sources—a possibility currently under serious consideration [6–8].

However, these future possibilities are still severely hampered by the low energy yields per mass or volume that are currently achieved. Generally MFC energy output is reported in milliwatts per square metre of electrode area or per cubic metre of electrolyte volume [9]. Scientific research has increased the densities of MFCs to over 1 kW m^{-3} (reactor volume) and to 6.9 W m^{-2} (anode area) under optimal conditions in the laboratory [10]. However, these values still cannot meet the needs of many applications, which require a power output larger than 100 kW m^{-2}[11]. For this reason large-scale waste water treatment is the application closest to industrial realisation and is the domain of much current MFC research.

A variety of designs are under development to improve the efficiency and potential application of MFCs to industry. Areas under investigation include the selection of electrode materials for optimal electrochemical performance and maximising electrode surface to volume ratios; improving charge transfer between microbes and electrodes either chemically or by mechanical design; and finding and maintaining optimal living conditions for microbe colonies, efficient supply of nutrients and removal of effluent. Different configurations are being investigated for extracting current, sometimes in combination with production of hydrogen or other metabolites of further use in energy generation and with or without exploitation of photosynthesis. The choice of process configuration and engineering design is also closely linked with the selection of the most suitable organism for a particular design or for whether overall priority is given to energy generation, waste disposal or some other objective.

A schematic representation of MFC research activity is shown in Figure 1. While there is a large volume of biochemical research literature on, for example, electron transfer chains and redox processes in cell metabolism that is relevant to MFC, relatively few studies focus specifically on MFC. This is exacerbated by the fact that ongoing research continues to identify new mechanisms for electron exchange between microbes and electrodes, new design strategies to exploit these and consequently new candidate organisms. Such organisms have not necessarily been well studied experimentally before.

In silico modelling is well suited to bridge this gap and extend knowledge in the biochemical interface between MFC biology and engineering design. Externally, electron flow (in the external circuit) and the counterflow of protons in the electrolyte make up the current that carries useful electrical power. Internally, both of these are comprehensively woven into the fabric of metabolism: electrons being transported by redox carrier molecules such as reduced nicotinamide adenine dinucleotide (NADH) that participate in a large fraction of all biochemical reactions, and protons that, for example, drive adenosine triphosphate (ATP) synthesis needed for energy transport are also ubiquitously involved in a great many reac-

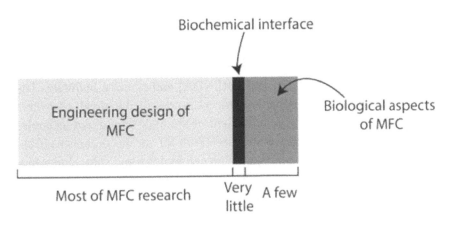

FIGURE 1: Areas covered in current microbial fuel cell research.

tions. This clearly calls for a systems-level approach rather than the reductionist strategy of pathway-oriented, conventional biochemistry.

This is the domain of systems biology [12]. Systems biology provides in silico models that incorporate biological data, metabolic flux data and different physico-chemical constraints such as the conservation of mass and energy, thermodynamics, redox balance, etc. [12] and thus provides an opportunity to identify the bottlenecks hidden in a complex network of interactions and cellular compartmentation [13].

The kinetic behaviour of a metabolic network at a whole-genome level can be constructed and analysed through a mathematical model [14, 15]. However, the characterization of metabolic networks is still far from comprehensive in databases [16] and even in the best-understood organisms, the majority of kinetic parameters are undetermined.

The development of new computational methods allows for the whole-network modelling of metabolism and conduction of compelling and testable predictions even without many parameters. The key idea is to incorporate stoichiometry and other fundamental principles as mathematical constraints, which separate feasible and infeasible metabolic behaviours. Compared with kinetic parameters, these constraints are much easier to identify and make it possible to build a large-scale model [17]. These constraint-based modelling approaches allow integration of high-throughput post-genomic data but describe steady states and generally offer no information about metabolite concentrations or the temporal dynamics of the system [18–20].

Genome-scale metabolic modelling requires high-quality metabolic network reconstructions [14]. The reconstructions are based on sequenced genomes and are generally built manually using information from metabolic databases (e.g. KEGG and BRENDA) and primary literature. The metabolic network reconstruction process is described in detail elsewhere [21]. Recently, due to the development of the high-throughput technologies, the reconstructions have now been built for various organisms [22].

Flux analysis can be combined with cell biology and sub-cellular biochemistry to reveal the functionality and efficiency of the enzymes associated with cell biological components or structures [23]. Metabolic regulation can be deeply understood only when multiple system components are examined simultaneously. This kind of analysis has been conducted in microbial and medical research in recent years.

Flux determinations can produce results that are hardly predictable from observed changes in transcript or protein levels because most of metabolic control takes place at post-translational levels and enzyme activities are often not correlated with changes in transcript or protein levels [23]. The incorporation of data from enzyme platforms should make the functionality of genomics strategies more clear. System-wide metabolic flux characterization is an important part of metabolic engineering [23].

Nevertheless, there is no published literature that uses genome-scale flux models to study the metabolic behaviour of biocatalysts in MFCs. The only attempt to use a genome-scale model to study biocatalyst behaviour in an MFC was presented as a conference abstract paper in the 17th European Symposium on Computer Aided Process Engineering (2007) [24]. However, this paper is an immature work that did not provide the source of the central metabolic network, describe the gene knockout methodology or discuss the results in relation to other MFC experimental work. Therefore, future research activities are urgently needed to fill the research gap indicated in Figure 1, with recently advanced constraint-based modelling approaches.

An essential first step in applying constraint-based analysis to microbial fuel cells is to choose appropriate organisms for further study. Due to the varied strategies and designs alluded to above, no single organism can serve as a suitable model, and a major advantage of in silico modelling is that different organisms can be studied in the same framework to facilitate mutual comparisons. This paper reviews the background against which such choices can be made and proposes a set of four organisms for the purpose.

The 'Microbial fuel cells' section explains the construction, operation and classification of current MFC designs, and the 'Current directions of MFC research' section reviews the issues being addressed in current MFC research. Based on this, the 'Microorganisms for in silico study of MFC functioning' section discusses a selection of organisms that are representative of various combinations of biological aspects that can be exploited in MFCs, while also featuring well-established metabolic network reconstructions, suitable for the computational analysis.

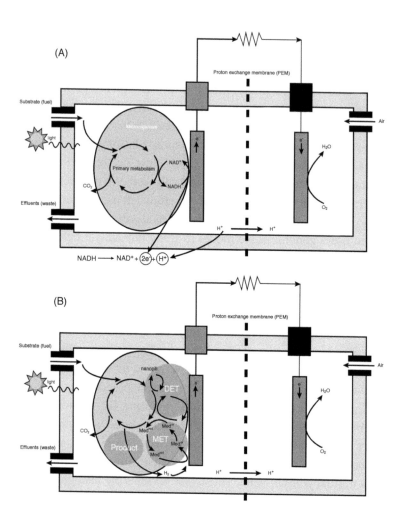

FIGURE 2: The working principle of a microbial fuel cell. (A) A bacterium in the anode compartment transfers electrons obtained from an electron donor (glucose or light in the case of photosynthetic organisms) to the anode electrode. Protons are also produced in excess during electron production. These protons flow through the proton exchange membrane (PEM) into the cathode chamber. The electrons flow from the anode through an external resistance (or load) to the cathode where they react with the final electron acceptor (oxygen) and protons. (B) Three electron transfer modes: (1) directly via membrane-associated components (DET), (2) mediated by soluble electron shuttles (MET) or (3) primary product (Product), e.g. H2 can act as a fuel to be oxidised to provide electrons for the electricity circuit. Med, redox mediator; Red oval, terminal electron shuttle in or on the bacterium.

2.2 MICROBIAL FUEL CELLS

MFCs are unique devices that can use microorganisms as catalysts for transforming chemical energy directly into electricity. The biggest advantage of an MFC is that it can generate combustion-less, pollution-free bio-electricity directly from the organic matter in biomass [2]. In an MFC the energy stored in chemical bonds in organic compounds is converted to electrical energy through enzymatic reactions by microorganisms. Thus, the electricity production by MFC is associated with the normal living processes of bacteria capturing and processing energy.

In a typical MFC configuration (Figure 2), microorganisms are situated in the anodic compartment and use the biomass for growth while forming electrons and protons [25]. The electrons are transported out of cells to an electrode using redox mediators or directly expelled by some micro-organisms for reducing the substrate. The protons or H^+ ions are diffused through the electrolyte to the cathode where it is oxidised to water. The cathode can be in a separate chamber (i.e. double-chambered MFCs) or in the same chamber (i.e. single-chambered MFCs). A single-chambered MFC eliminates the need for the cathodic chamber by exposing the cathode directly to the air. The only by-product released by MFCs is carbon dioxide, which can be fixed by plants for photosynthesis.

MFCs require running under conditions predefined by the optimum growth and living conditions of the used microorganisms. Thus, factors affecting the MFC's efficiency include electrode material, pH buffer and electrolyte, proton exchange system and operating conditions in both the anodic chamber and the cathodic chamber. MFCs are usually operated at ambient temperature, at atmospheric pressure and at pH conditions that are neutral or only slightly acidic [26].

MFCs harness the electrons from these systems in three main operation modes: mediated electron transfer (MET), direct electron transfer (DET) and product mode. Photosynthetic MFCs use photosynthesis as the electron source and can also be operated in the same modes.

2.2.1 MEDIATED ELECTRON TRANSFER

MET is defined as where a mediator molecule acts an electron relay that repeatedly cycles between the reaction sites and the electrode [11]. MET is the most common electron transfer mode used in MFCs and can be classified into two sub-types [11]:

- Indirect transfer systems that involve freely diffusing mediator molecules (i.e. diffusive MET)
- Indirect transfer systems in which the mediator is integrated into the electrode or the cell membrane (i.e. non-diffusive MET)

In diffusive MET, the mediators enter the cell membrane and exchange electrons between cellular metabolism inside the cell and the electrode outside it. In the non-diffusive MET, the mediator can collect the electrons from the cell membrane without penetrating the cell.

Based on the type of mediators, diffusive MET can also be classified into three sub-categories:

- MET via exogenous (artificial) redox mediators
- MET via secondary metabolites
- MET via primary metabolites

The detailed mechanisms in those three classes are discussed in [27]. Because the terminal electron transfer to or from the electrode determines the overall cell potential, potential (voltage) losses can be minimised by using a mediator that has a reaction potential near that of the biological component.

2.2.2 DIRECT ELECTRON TRANSFER

DET is defined as the case where electrons cycle directly between a microorganism and an electrode. DET can be achieved through two naturally occurring mechanisms:

- Membrane-bound c-type cytochromes, which exist in the cell membrane in some organisms [28, 29] to provide electron transfer capacity. For example, multi-heme proteins have especially evolved in sediment-inhabiting metal-reducing microorganisms such as *Geobacter[30], Rhodoferax*[31] and *Shewanella*[32]. In their natural environment, iron(III) oxides act as the solid terminal electron acceptors, but in the case of MFC, the anode is used as the solid electron acceptor.

- Electronically conducting nanowires. The DET via outer membrane cytochromes requires the cytochrome (the bacterial cell) to be physically adhered to the fuel cell anode. When a biofilm is formed, only bacteria in the first monolayer at the anode surface are electrochemically active [30]. Thus, the maximum cell density in this bacterial monolayer usually influences the MFC performance. However, it has been shown that some *Geobacter* and the *Shewanella* strains can evolve electronically conducting molecular pili (nanowires of 2 to 3 µm long, made of fibrous protein structures [33]) that make the microorganism able to reach and use more distant solid electron acceptors [34, 35]. The pili are connected to the membrane-bound cytochromes and allow transference of the electrons to the distant electron acceptors without cellular contact (Figure 2B). Thus, thicker electroactive biofilms can be formed to increase anode performance. It was shown that fuel cell performance can be increased up to tenfold upon nanowire formation of *Geobacter sulfurreducens*[35].

2.2.3 PRODUCT TYPE

In product-type MFCs, microbes metabolize the substrate, releasing a secondary fuel product such as hydrogen that then diffuses to the electrode and is oxidised or reduced (as appropriate) to form a final waste product, which is discharged [11]. The product operation is similar to conventional fermentation processes, in that products of the microbial metabolism are used as the fuel at the electrode.

The product system is made up of two independent stages: one is storage of the microbial reaction product, and the other one is the product being fed to a conventional fuel cell process driven by non-biological catalysis, such as in the case of a proton exchange membrane fuel cell, where H_2 is converted into electricity [36]. These stages may also be physically separated in different containment vessels. However, a product system only truly belongs to a biofuel cell system when the microbes and the electrode are together in the same anode compartment [11]. Nowadays, the fermentation (mostly to hydrogen gas) usually takes place in the fuel cell itself [37, 38].

Product systems have two main drawbacks, one is that the efficiency of the conversion of the biological substrate to hydrogen is quite low, and the other is that hydrogen oxidation requires high fuel cell temperatures. Also, the produced biofuel gas is always contaminated with other by-products such as CO, H_2S and (poly)siloxanes making it not sufficiently pure for direct use in a fuel cell [39].

2.2.4 PHOTOSYNTHETIC MFCS

Photosynthetic MFCs are MFCs that generate electricity from a light source rather than a fuel substrate and require the mediator involved to be light stable [40]. Conventionally two operating modes exist for photosynthetic MFCs:

- Energy is produced and stored by the microorganism during illumination and then released and processed in the same way as in a non-photosynthetic biofuel cell.
- The energy produced during illumination may be directly extracted in the form of electrons for creating an external electrical circuit.

A single photosynthetic MFC may possess both of these two modes of action. However, it is recently thought better to classify photosynthetic MFCs into categories based on seven approaches that integrate photosynthesis with MFCs—photosynthetic MFCs [40]:

1. Photosynthetic bacteria at the anode with artificial mediators
2. Hydrogen-generating photosynthetic bacteria with an electrocatalytic anode
3. A mixed culture, with photosynthetic bacteria supplying organic matter to heterotrophic electroactive bacteria at the anode
4. Photosynthesis in plants, supplying organic matter via rhizodeposits to heterotrophic electroactive bacteria at the anode
5. An external photosynthetic bioreactor, where only biomass or metabolic products are transferred to the anode compartment to feed heterotrophic electroactive bacteria
6. Direct electron transfer between photosynthetic bacteria and electrodes

7. Photosynthesis at the cathode to provide oxygen

These sub-types have been discussed in detail by Rosenbaum et al. [40].

2.2.5 MICROORGANISMS SUITABLE FOR MFCS

Most microorganisms are unable to donate sufficient electrons outside of cells to produce usable currents, because the outer layers of most microbial species are made up of non-conductive lipid membrane, peptididoglycans and lipopolysaccharides which restrain electron transfer to the anode [41]. Since the 1980s, it has been found that artificial water-soluble electron shuttles (i.e. methylene blue, thionine, neutral red and 2-hydroxy-1,4-naphthoquinone) can be used as mediators that transport the electrons from electron carrier molecules inside the cell (e.g. NADH, NADPH or reduced cytochromes) to the anode surface [41]. For example, an MFC based on *Proteus vulgaris* used thionine as a mediator to generate electricity from sucrose [42].

Since the 1990s, some bacterial species such as *Pseudomonas aeruginosa*[43] and *Clostridium butyricumcan*[44] have been found to be able to self-mediate extracellular electron transfer using their own metabolic products. Meanwhile, direct transfer of electrons (DET) that involves use of electrochemically active redox enzymes (i.e. cytochromes) has been discovered in a number of bacterial species such as *Shewanella putrefaciens*[28, 29, 45], *Shewanella oneidensis*[46], *Geobacter sulfurreducens*a[30, 47], *Rhodoferax ferrireducens*[31], and the oxygenic phototrophic cyanobacterium *Synechocystis sp.* PCC 6803a[34]. These microorganisms are termed as exoelectrogens, and among them *S. oneidensis* and *G. sulfurreducens* have evolved electronically conducting molecular pili (nanowires) to further facilitate the DET [34, 35]. Besides DET mode, *S. oneidensis* can conduct MET using a self-produced mediator [48]. The exoelectrogens in MFCs are thought to actively use electrodes to conserve electrochemical energy required for their growth and thus ensure high rates of fuel oxidation and electron transfer for the production of electrical energy [5].

In most of the previous MFC studies, bacteria have been used for electricity generation. On the other hand, since 2000, eukaryotes such as microalgae (e.g. *Chlamydomonas reinhardtii*a) and yeast (e.g. *Saccharomyces* *cerevisiae*[a]) have also emerged as good choices for MFC use, because they have been studied as model microorganisms in the lab and have been widely used in the industry for a long time.

2.3 CURRENT DIRECTIONS OF MFC RESEARCH

2.3.1 ENGINEERING DESIGN AND BIOLOGICAL ASPECTS

Most previous studies tended to improve power densities of MFCs by optimizing the reactor configuration and operation parameters [49, 50], such as modifications of the electrode materials to incorporate metals that contain current collectors [51, 52], use of metals highly optimized for bacterial adhesion and metals possessing high electrical conductivity to minimise ohmic losses [10], and application of a biocathode that can increase MFC performance by improved oxidation of hydrogen at the cathode [53]. Applications of chemical treatments and precious metals to electrodes in order to increase power production in the laboratory have also been investigated [54, 55]. However, these modifications, like the use of larger laboratory reactors, may increase the cost and lead to compromises on performance based on material costs. Many bottlenecks also exist for improving those physical and chemical properties.

Since the fundamental source of electrons is the cellular metabolism, it is particularly important to focus on biological processes that take place in the microbial cells. In this regard, further clarification is needed of the factors relevant to the anodic catalysis process, such as the diversity of the electrochemically active microorganisms [56] and especially the electricity generation mechanisms in relation to normal metabolic states.

2.3.2 MEDIATOR-LESS, MEDIATOR-SELF-PRODUCING AND ARTIFICIAL MEDIATOR-BASED MFCS

Mediator-less MFCs are a more recent development relying on evolved ability of exoelectrogens for disposal of electrons originating from substrate oxidation. This type of MFCs has been increasingly preferred, because use of mediators complicates the cell design and these mediators are usually toxic, costly and unsustainable, limiting MFC development [27, 57–60]. In addition, mediator-involved MFCs usually produce low current densities (0.1 to 1 A m^{-2}) [27]. Unfortunately, mediator-less MFCs may not yet find a wide range of applications since the discovered exoelectrogens are still few in number and it is non-exoelectrogens that are largely used in the agricultural and industrial areas [61]. Thus, an important direction of MFC research is development of MFCs using non-exoelectrogens without exogenous mediators [61].

However, problems arise from the fact that redox molecules used in electron transfer reactions are not situated on the outer membrane, but in the cytoplasmic membrane. One way is to develop direct electron transfer using carbon nanoparticles that can contact the redox centres that are incorporated in the interior cell membrane [61]. Another way is to identify and develop self-produced mediators (e.g. in the case of *Shewanella* species mentioned above) through engineering methods [56].

2.3.3 CONVENTIONAL PHOTOSYNTHETIC MFCS

It is appealing to study whether phototrophic microbes can be used in an MFC generating electricity from sunlight, because sunlight is an unlimited energy resource and more solar energy reaches the Earth in 1 h (4.1 × 10^{20} J) than the energy consumed on the planet in a year [56, 62, 63]. In addition, the development of a self-sustainable photosynthetic MFC is important to meeting energy requirements at remote locations, where routine addition of fuel would be technically difficult and expensive [56].

Photosynthetic MFCs can generate electricity indirectly or directly. For example, in the indirect way, *Rhodobacter sphaeroides* in the MFC can produce H$_2$ that is oxidised at a platinum coated anode to generate electricity [64, 65], whereas in the direct way, *Rhodopseudomonas palustris*, a photosynthetic purple non-sulphur bacterium, can generate electricity in a biofilm anode MFC by direct electron transfer [66].

There are also more traditional photosynthetic MFC configurations where photosynthetic organisms live with other microbes and supply products to heterotrophs [67]. Photosynthetic microorganisms (e.g.

cyanobacteria or microalgae) and heterotrophic bacteria exhibit synergistic interactions [68] that can be used in self-sustained phototrophic MFCs [62]. An indirect synergistic relationship between photosynthetic organisms and electricigens has been exemplified in a recent study, in which algal photobioreactors were used to supply organic matter produced via photosynthesis to an MFC for electricity generation [69]. The operation of this type of photosynthetic MFCs is CO_2 neutral and does not need buffers or exogenous electron transfer mediators [67]. However, the photosynthetic MFC power densities obtained are quite low when compared with those that are currently reported for conventional MFCs, e.g. 0.95 mW m^{-2} for polyaniline-coated and 1.3 mW m^{-2} for polypyrrole-coated anodes [70] versus values in the watt per square metre range for conventional cells.

2.3.4 MFCS BASED ON THE PHOTOSYNTHETIC ELECTRON TRANSFER CHAIN

Recently, the photosynthetic electron transfer chain is considered as a source of the electrons harvested on the anode surface, which is different from the previously designed anaerobic MFCs, sediment MFCs or anaerobic photosynthetic MFCs [70]. A single-chamber photosynthetic MFC based on two photosynthetic cultures, planktonic cyanobacteria *Synechocystis sp.* PCC 6803 and a natural freshwater biofilm, has shown a positive light response (i.e. immediate increase in current upon illumination) [70]. This phenomenon proves that it is possible to extract electrons directly from the photosynthetic electron transfer chain, and not only from the respiratory transfer chain or through oxidation of hydrogen [71].

2.4 MICROORGANISMS FOR IN SILICO STUDY OF MFC FUNCTIONING

2.4.1 CATEGORIES AND REPRESENTATIVES

Because the electricity generation in MFCs is based on the metabolic activity of living microorganisms, experimental screening of different microorganisms for better anodic activity has long been recognised as a fundamental way to improve MFC performance. It is also possible to improve the performance by culturing microorganisms under selective pressure for enhanced power production [30, 72].

Compared to experimental studies, in silico modelling is less constrained to a particular MFC design and operating mode. Clearly no single organism is likely to be optimal for all of the varied designs discussed before. To date, every microorganism used in previous MFC studies has advantages and disadvantages. Selection of the microorganism depends on a variety of factors such as types of application, the capability of power generation, the availability of types of energy source for bacterial survival and the ability of extracellular electron transfer, in that electrodes are not natural electron acceptors.

From the modelling perspective, a broader view is possible. Categories of microbes can be identified to cover the range of operating modes and, within these, individual organisms selected that will allow different modes to be individually studied and also compared quantitatively.

MFC microbial communities can be divided into three groups: heterotrophic cells, photoheterotrophic cells and sedimentary cells [9]. The distinction between phototrophic and heterotrophic metabolism is fundamentally important in determining the operating mode. Another key distinction is between prokaryotes and eukaryotes. Compared to prokaryotic species and mixed cultures that have been mostly studied for different MFC applications, fewer studies involve eukaryotes as biocatalysts in MFC operations [73]. This is because the metabolic processes of eukaryotic cells take place in the membrane surrounded cell organelles (e.g. chloroplasts) and is thus putative to be difficult for some commonly used redox mediators such as 2-hydroxy-1,4-nepthoquinone to get access to [57, 74]. While prokaryotes have the advantage that their simpler cell membranes and internal structure are more amenable for physical electron extraction, the more complex metabolism of eukaryotes may be more efficient and be able to support a larger diversion of redox carrier flux without undue harmful effects on the organism.

The four anodic microorganisms: *C. reinhardtii*, *Synechocystis sp.* PCC 6803, *S. cerevisiae* and *G. sulfurreducens*, each combine a different pair of key features and are proposed as good candidates for MFC in silico characterization. As illustrated in Figure 3, *C. reinhardtii* and *S. cerevisiae* are eukaryotes, whereas *Synechocystis sp.* PCC 6803 and *G. sulfurreducens* are prokaryotes. The four organisms also cover the three groups of the MFC microbial communities mentioned above, i.e. *C. reinhardtii* and *Synechocystis sp.* PCC 6803 are photoheterotrophic cells, *S. cerevisiae* belongs to heterotrophic cells and *G. sulfurreducens* is a typical sedimentary cell.

  C. reinhardtii and *Synechocystis sp.* PCC 6803 are photosynthetic organisms that are also capable of producing hydrogen. The comparison of organ-

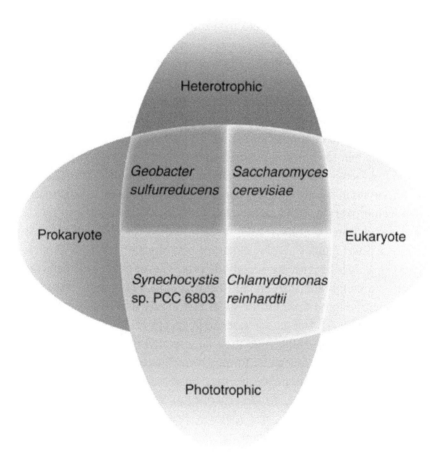

FIGURE 3: Classification of proposed microbes.

isms with/without photosynthesis can be used to study exploitation of photosynthetic and respiratory electron transport chains to supply MFC current.

A further consideration in a modelling study is whether the data are available and computationally manageable. All four microorganisms have been studied extensively as model organisms and used in various industries for a long time, and thus, related molecular tools and biological mechanisms are abundant. In particular, genome-scale metabolic networks have been reconstructed and are regularly updated for these four organisms. Based on a literature review, the most updated models are shown in Table 1. These models include the natural redox mediators (i.e. NADH) that are well balanced for the cellular energy metabolisms (e.g. oxidative phosphorylation, glycolysis, Calvin cycles and tricarboxylic acid (TCA) cycles) and are thus practicable for MFC modelling.

TABLE 1: The scope of the genome-scale models of the four selected organisms

	Chlamydomonas reinhardtii		*Synechocysti ssp. PCC 6083*		*Geobacter sulfurreducens*	*Saccharomyces cerevisiae*
Gene	2,249	1,080	1,811	678	617	918
Metabolite	1,862	1,068	465	795	644	1,655
Reaction	1,725	2,190	493	863	709	2,110
Compartment	4	10	3	4	2	17
Date	December 2011	August 2011	October 2011	January 2012	2009	2012
Reference	[75]	[76]	[77]	[78]	[79]	[80]

The number indicates the counts of relative items in the network models.

It is noted that two models were published in the same period for *C. reinhardtii*[75, 76] and *Synechocystis*[77, 78]. These models for the same organisms are not compared with each other by the authors, and thus, their limitations can only be revealed during the MFC modelling.

The biological features of the four microorganisms are summarized in Table 2, and the relevance of each of these microbes is reviewed in more detail in the following sections.

TABLE 2: Comparison of facts of four selected microorganisms

Name	Chlamydomonas reinhardtii	Synechocystis sp. PCC 6083	Saccharomyces cerevisiae	Geobacter sulfurreducens
Domain	Eukaryote	Prokaryote (Gram-negative)	Eukaryote	Prokaryote (Gram-negative)
Mitochondria	Multiple	N/A	Multiple	N/A
Chloroplast	Single chloroplast occupies two thirds of the cell	Chloroplast analogy	N/A	N/A
Hydrogen synthesis enzyme	Fe hydrogenase	NiFe hydrogenase	N/A	N/A
MFC mode	Product mode	Photosynthetic DET	MET DET (output extremely low)	DET
MFC performance	0.4 W m^{-2}	0.02 W m^{-2}	1.5 W m^{-2} (MET)	1.88 W m^{-2}
	3.3 W m^{-3}	0.007 W m^{-3}	90 W m^{-3} (MET)	43 W m^{-3}
			0.003 W m^{-2} (DET)	
Optimum growth temperature	20°C to 25°C [81]	30°C to 33°C [82]	25°C to 35°C [83]	30°C to 35°C [84]
Growth mode	Autotrophic	Autotrophic	Heterotrophic	Heterotrophic and sedimentary (soil inhabitant)
	Heterotrophic	Heterotrophic		
	Mixotrophic	Mixotrophic		

N/A, not applicable.

2.4.2 CHLAMYDOMONAS REINHARDTII

C. reinhardtii is a unicellular green alga that belongs to the Chlorophytes division, which diverged from the Streptophytes division (including land plants) more than one billion years ago [85]. *C. reinhardtii* is an approximately 10-µm, unicellular, soil-dwelling green alga with an eyespot, a nucleus, multiple mitochondria, two anterior flagella for motility and mating, and a single cup-shaped chloroplast that accommodates the photosynthetic apparatus [86, 87].

Like plants, *C. reinhardtii* has a cell wall and can grow in a medium lacking carbon and energy sources when illuminated [87]. Unlike angiosperms (flowering plants), this microorganism has functional photosynthetic apparatus even when in dark conditions and with an organic carbon source [87]. In the dark, acetate is the sole carbon source used by wild-type *C. reinhardtii* in vivo[88].

Because of the relative adaptability and quick generation time, *C. reinhardtii* has been used as a model to study eukaryotic photosynthesis, eukaryotic flagella and basal body functions and the pathological effects of their dysfunction [89, 90], and investigated for water bioremediation and biofuel generation [87, 91–93]. The cDNA, genomic sequence and mutant strains of *C. reinhardtii* are publicly available through the *Chlamydomonas* Center [94].

2.4.2.1 ADVANTAGES OF ALGAE

C. reinhardtii inherits all potential advantages of algae for industrial use and scientific study, including [95–97] the following: (1) Algae biomass can potentially be produced at extremely high volumes, and this biomass can yield a much higher oil (1,000 to 4,000 gal/acre/year) than soybeans and other oil crops [97]. (2) Algae do not compete with traditional agriculture because they are a non-food-based resource which can be cultivated in large open ponds or in closed photobioreactors located on low-productive or non-arable land. (3) Algae have a good adaption to different climate and water conditions and can be grown in a wide range of water sources, such as brackish, saline, or fresh and waste water. (4) Algae can make use of resources that would otherwise be considered waste as substrate for growth [97]. (5) Algae can use and sequester CO_2 from many sources such as flue gases of fossil fuel power plants and other waste streams. (6) Algae can be processed into a wide range of products such as biodiesel, bioethanol, methane, bio-oil and biochar, and high-protein animal feed. (7) The 'simple' photosynthetic alga *C. reinhardtii* is an excellent model organism for a systems biology approach compared to a complex vascular plant [12].

2.4.2.2 BIOFUEL AND ELECTRICITY PRODUCTION BY ALGAE

Due to the advantages listed above, algae have been examined in many studies for the generation of energy products, such as bio-oil, methane, methanol and hydrogen [98]. Nevertheless, these technologies have one disadvantage in that the fuel produced must be stored, transported and further processed to produce electricity. To circumvent these problems, MFC is used as an alternative way to directly generate electricity in only one process unit by means of hydrolysis and fermentation of algae and makes use of energy originating from sunlight.

However, algae are not exoelectrogens and the conventional mediators do not perform well in the extraction of the redox potential for algae-based MFCs because the redox species are produced through the metabolic mechanisms that take place in membrane-surrounded cell organelles in algae such as mitochondria and chloroplasts [74]. Thus, neither DET nor MET mode has been applied in MFCs using algae.

Previous studies tend to use algae in MFCs of the product mode that depend on the production of hydrogen molecules, which is then oxidised at the anode for electron transfer to the MFC circuit. Another mechanism is where algae produce organic matter that is used as a substrate for electrochemically active bacteria, which then supply electrons for MFCs from oxidization of the organic matter [69].

In one landmark study of MFC using algae, C. reinhardtii was used in a product-mode MFC to produce hydrogen for oxidation at the anode. A maximum hydrogen production rate of 7.6 ml/l culture h^{-1}[99] was achieved, at a current yield of 9 mA at a constant electrode potential of 0.2 V. Using the culture volume and electrode dimensions, this corresponds to power densities of 0.4 W m^{-2} and 3.3 W m^{-3}. In another study [98], *Chlorella vulgaris* microalgae were used as a biomass source to feed a mixed microbial culture, producing a maximum power density of 0.98 W m^{-2} (277 W m^{-3}).

Furthermore, algae have been inoculated into a bioelectrode to generate oxygen as the electron acceptor [100]. Under illumination, algae produced oxygen as the electron acceptor for the MFC cathodic reactions, changing the bioelectrode into biocathode mode, while in darkness, the algal oxygen production stops and the bioelectrode mainly functioned as the bioanode. The reversible bioelectrode can relieve the pH membrane gradient generated by the acidification at the anode and the alkalisation at the cathode during normal MFC operation [100, 101].

2.4.2.3 HYDROGEN PRODUCTION BY C. REINHARDTII

Only a specific group of green microalgae and cyanobacteria, e.g. microalga *C. reinhardtii*, have evolved the additional ability to harness the huge solar energy resource to drive molecular H_2 production [102–108]. The release of hydrogen by *C. reinhardtii* under light exposure was first reported in 1942 [105]. In 2000, sustained hydrogen production was achieved using induced sulphur depletion in a culture medium containing acetate, a carbon source that is used to cause the shift from aerobic to anaerobic state [107].

C. reinhardtii is one of the best eukaryotes for H_2 production [109]. The available experimental information, including genomics, indicates that *C. reinhardtii* possesses a complex metabolic network containing aerobic respiration and molecular flexibility associated with fermentative metabolism. The molecular flexibility is accomplished with adjustments in the rates of accumulation of organic acids, ethanol, CO_2 and H_2 [86, 110–115] and underlies the cell's adaptive ability for hypoxic and anoxic conditions.

Compared to other H_2-producing organisms such as chemotrophic and phototrophic bacteria, *C. reinhardtii* is more practical for H_2 production as it can be easily and efficiently grown in bioreactors using solar light, grows rapidly (doubling times of the order 6 h or less) and has a flexible metabolism [116]. The genome of this model microorganism was fully sequenced in 2007 [86], which makes it possible to increase production yields of H_2 from water by optimization of cell metabolism.

2.4.2.4 LIMITATION OF HYDROGEN PRODUCTION BY C. REINHARDTII

In fact, hydrogen production by *C. reinhardtii* can still not meet the commercial requirement because of several biochemical and engineering shortcomings, for example, hydrogen production demands anoxia because oxygen can suppress transcription and function of hydrogenase(s). However, the anoxia is constrained by the function of the photosystem II (PSII), which provides electron and protons from water and conducts oxygen evolution in the photosynthetic electron transport chain. Economic assessments have suggested that microalgae should achieve an efficiency of 10% in the conversion of solar energy to bioenergy to be competitive with other H_2 production methods, such as biomass gasification or photovoltaic electrolysis [117]. This is a more than a fivefold increase in efficiency from current levels. Exploiting hydrogen production directly in a product-type MFC may help to bridge this gap.

2.4.3 SYNECHOCYSTIS SP. PCC 6803

Synechocystis sp. PCC 6803 is a unicellular cyanobacterium, one of the earliest groups of microbes to evolve on earth. The first primitive bacteria on Earth are dated at 3.8 to 3.6 billion years ago [118]. It is thought that cyanobacteria flourished during the period from 3.5 to 1.8 billion years ago, consuming CO_2 and providing Earth with oxygen, making possible the development of the different forms of aerobic life. At present, cyanobacteria deliver amounts of oxygen to the atmosphere similar to those that are produced by higher plants [118]. Moreover, cyanobacteria harness 0.2% to 0.3% of the total solar energy (178,000 TW) that reaches the Earth [106] and convert the solar energy into biomass-stored chemical energy at the rate of approximately 450 TW, contributing to 20% to 30% of Earth's primary photosynthetic productivity [119].

Until 1982, the cyanobacteria were called blue-green algae because they can photosynthesize and look like chloroplasts. Since then, cyanobacteria were re-classified as prokaryotes [120]. It is suggested that cyanobacteria entered into a symbiosis with cells, which were not capable of absorbing CO_2 and releasing oxygen, and later became photosynthetic organelles of plants [121]. Nowadays many species of cyanobacteria, e.g. *Synechocystis sp.* PCC 6803, are widely distributed in nature.

Synechocystis sp. PCC 6803 and plants have similar oxygen-evolving apparatus and are thus used for studying photosynthesis in plant cells. The difference is that *Synechocystis sp.* PCC 6803 grow much faster than plants, and they are relatively easy organisms for genetic manipulations [118]. Also, plants are fixed at places where they grow and they have less adaptation abilities for their growth and propagation than cyanobacteria.

Synechocystis sp. PCC 6803 grow photoautotrophically on carbon dioxide and light, as well as heterotrophically on glucose. Like *C. reinhardtii*, *Synechocystis sp.* PCC 6803 is one of several hydrogen-yielding species of cyanobacteria [122]. After its genome was fully sequenced in the 1990s [123, 124], this cyanobacteria species has become a popular model photosynthetic organism studied by many researchers.

2.4.3.1 ADVANTAGES OF CYANOBACTERIA

Cyanobacteria, besides other photosynthetic microorganisms such as microalgae, can establish synergistic relationships with heterotrophic bacteria, for instance, in a microbial mat [68]. Thus, they could be potentially manipulated to establish an indirect synergistic relationship with electricigens in phototrophic MFC [69]. However, phototrophic MFCs usually have low conversion efficiency [62], and the study of phototrophic MFCs is in its nascent stages [62].

Since *Synechocystis sp.* PCC 6803 is a photoautotroph that divides rapidly, it has been enlisted as a platform for production of biofuels by using sunlight as an inexpensive energy source [125, 126]. This feature makes this species suitable as a candidate for MFCs of product mode.

2.4.3.2 ELECTROGENIC ACTIVITY OF CYANOBACTERIA

Unlike other exoelectrogens, such as *G. sulfurreducens*, in which the electrons are derived from biochemical oxidation of organic compounds via the respiratory electron transfer chain [127], cyanobacterial electrogenic activity does not need exogenous organic fuel and is entirely dependent on the energy of light, which drives the biophotolysis of water through the photosynthetic electron transfer chain in the cyanobacteria, releasing electrons [128, 129]. The electrogenic activity of cyanobacteria may represent a form of overflow metabolism to protect cells under high-intensity light

[70, 129]. This light-driven electrogenic activity is conserved in diverse genera of cyanobacteria and is an important microbiological channel of solar energy into the biosphere [129].

The electrogenic activity of *Synechocystis sp.* PCC 6803 has been captured in an MFC for electricity generation. The MFC can achieve a steady power density of 6.7 mW m^{-3} (peaking at 7.5 mW m^{-3}) [130, 131]. These power densities are still much lower than the values achieved by the other microbes under discussion. Despite that, it is included in the selection list because it offers a unique combination of photosynthetic activity that is plausibly accessible to direct-mode electron transfer. The quoted measurements are quite recent, and it is worth exploring if this organism has the potential to deliver competitive power densities in the future.

2.4.3.3 HYDROGEN PRODUCTION BY CYANOBACTERIA

Cyanobacteria have a similar process for hydrogen production as algae, except that they use NiFe hydrogenases rather than Fe hydrogenases in microalgae and the hydrogenase of cyanobacteria is 100 times less active than those of the green algae *C. reinhardtii*[106]. These hydrogenases contain [Ni-Fe] catalytic centres that are extremely sensitive to inactivation by O_2, one of the major barriers to hydrogen production. Natural mechanisms such as consumption by respiration, chemical reduction via PSI and reversible inactivation of PSII O_2 evolution can reduce intracellular O_2 content and thereby increase H_2 production.

2.4.4 SACCHAROMYCES CEREVISIAE

The *Saccharomyces* genus currently contains eight species [132]. *Saccharomyces cerevisiae*, *Saccharomyces bayanus* and *Saccharomyces pastorianus* are associated with anthropic environments, whereas *Saccharomyces paradoxus*, *Saccharomyces kudriavzevii*, *Saccharomyces cariocanus*, *Sac-*

charomyces mikatae and the recently described *Saccharomyces arboricolus* are mostly isolated from natural environments [133, 134]. These *Saccharomyces* species can play a major role in food or beverage fermentation. However, the ale yeasts involved in alcoholic fermentation mostly belong to the species *S. cerevisiae*[132]. Besides its important role in baking and brewing, this yeast species has been used as a eukaryotic model organism in molecular and cell biology, for example, the characteristic of many proteins can be discovered by studying their homologs in *S. cerevisiae*.

S. cerevisiae cells are round to ovoid, are 5 to 10 μm in diameter, reproduce by a budding process and can grow aerobically on glucose, maltose and trehalose but not on lactose or cellobiose. In the presence of oxygen, it is even able to operate in a mixed fermentation/respiration mode. The ratio of fermentation to respiration varies slightly among strains but is approximately 80:20 [135]. Furthermore, *S. cerevisiae* can be processed to produce potential advanced biofuels such as long-chain alcohols and isoprenoid- and fatty acid-based biofuels, which have physical properties that more closely resemble petroleum-derived fuels [136].

2.4.4.1 ADVANTAGES OF YEAST FOR MFC

Yeast is sometimes thought to be impractical as a biocatalyst, due to difficulties with transferring electrons out of cellular organelles [137]. However, since yeasts are robust, are easily handled, are mostly non-pathogenic, have high catabolic rates and grow on substrate spectrum, they are well worth considering as promising biocatalysts for MFCs [138]. In addition, several other merits may exist for using *S. cerevisiae* in MFCs. First, *S. cerevisiae* can survive and function in an anaerobic condition that is required for the anode compartment of traditional MFC. Second, the optimal growth temperature for *S. cerevisiae* is around 30°C, which is a convenient ambient temperature. Third, the metabolism of this species is well understood, which helps locate mechanisms responsible for electricity generation in MFCs. Lastly, yeast-based fuel cells could be retrofitted into ethanol plants for in situ power generation [139].

2.4.4.2 YEAST FOR IN SITU POWER GENERATION

In an anaerobic condition, yeasts usually switch to fermentation reactions where one glucose molecule is consumed for the production of two molecules of pyruvates. Pyruvate is further transformed into alcohol or organic acid by recycling NADH to NAD^+, which is a key step to sustain the glycolysis process [140]. This glycolysis reaction takes place in the cytosol of the cell rather than in the mitochondria, so NADH could be easily accessed by the mediator molecule present in the cell membrane of the yeast [73]. The glycolysis and the oxidation of NADH to NAD^+ are not influenced by the energy extraction process in the MFCs. Based on these characteristics, MFCs using yeast can be directly applied in fermenters for in situ power generation [139].

2.4.4.3 LIMITATION OF S. CEREVISIAE FOR MFC USE

Limitations exist for *S. cerevisiae* to be used in MFCs. First, *S. cerevisiae* has a weak ability to oxidise the substrate to supply the maximum number of electrons available for yeast-based MFC. In the mitochondrial process of *S. cerevisiae*, there is a total of only 14 ATP per glucose molecule produced, which is much less than a net of 28 to 30 ATP typically achieved by most aerobes [138]. Also, mediators are commonly required to facilitate the transfer of electrons to the anode, which makes exogenous mediators necessary to MFCs based on *S. cerevisiae* because this yeast is thought incapable of producing such mediators indigenously [139].

2.4.4.4 THE OUTPUT OF S. CEREVISIAE-BASED MFCS

In general, yeast-based MFCs perform better than cyanobacteria but still have a lower power output than bacterial fuel cells [138]. It was shown that methylene blue-mediated *S. cerevisiae* MFC can give a power density of 1.5 W m^{-2}[141], which is less than the maximum of 6.86 W m^{-2} reported by Fan et al. [135] for a mixed-culture MFC. The corresponding volumetric density, based on the specified anodic chamber volume of 10 ml, is 90 W m^{-3}.

A recent MFC that employs *S. cerevisiae* as the electron donor in the anodic half-cell and *C. vulgaris* as the electron acceptor in the cathodic half-cell can reach a maximum power at 90 mV and a load of 5,000 Ω, giving a power density of 0.95 mW m^{-2} of electrode surface area [142]. This power density is still very low. Another study investigated the possibility of *S. cerevisiae* to transfer electrons to an extracellular electron acceptor through DET mode and found that the cells that adhered to the anode were able to sustain power generation in a mediator-less MFC. However, the power performance of this MFC was extremely low (0.003 W m^{-2}) [143].

2.4.5 GEOBACTER SULFURREDUCENS

G. sulfurreducens are comma-shaped Gram-negative, anaerobic bacteria capable of coupling oxidization of organic compounds to reduction of metals. This organism is one of the predominant metal-reducing bacteria in soil and hence plays an important ecological role in biotechnologically exploitable bioremediation. The activity of *Geobacter* species in sub-surface can be stimulated to remove organic and metal contaminants such as aromatic hydrocarbons and uranium from groundwater [144–146].

The genome sequence of *G. sulfurreducens* is available, and a system for genetic manipulation has been developed for this organism [147]. Since it was discovered in 1994 [148], this bacterium has been extensively studied for MFC applications. It has been reported that (1) *G. sulfurreducens* can completely oxidise electron donors by using only an electrode as the electron acceptor, (2) it can quantitatively transfer electrons to electrodes in the absence of electron mediators, and (3) this electron transfer is similar to those observed for electron transport to Fe(III) citrate [47].

2.4.5.1 ADVANTAGES OF G. SULFURREDUCENS

G. sulfurreducens is the most abundant species on anode surfaces in MFCs grown with more than one bacterial species [149–151]. It can form biofilms on the anodes, which make all the cells participate in electron transport to the anode and thus increase the current production [152]. *G. sulfurreducens* is an anaerobe but can withstand low levels of oxygen and may use oxygen as an electron acceptor to support growth under aerobic conditions [153].

This *Geobacter* species can produce large amounts of electrical energy since it possesses multiple mechanisms that involve either pili or c-type cytochromes to facilitate the electron transfer to electrode in MFCs (discussed before in the 'Microbial fuel cells' section) [149]. Also, with the electron transfer to electrodes, the *Geobacter* species can effectively oxidise acetate [47, 154]. A current density of 4.56 A m^{-2}, corresponding to power densities of 1.88 W m^{-2} and 43 W m^{-3}, measured for *G. sulfurreducens* is among the highest reported for a pure culture [155]. By reducing the anode compartment volume to a fraction of a millilitre, the volumetric density was in fact increased to 2.15 kW m^{-3}. While the lower value is more realistic for comparison to other studies, this does show that very high densities are achievable in principle with this organism. In addition, *G. sulfurreducens* converts acetate to current with coulombic efficiencies of over 90% [151, 155].

Previous studies have shown that when a high selective pressure for high rates of current production at high coulombic efficiencies is imposed on complex microbial communities, it is the organisms closely related to *G. sulfurreducens* that are routinely enriched on anodes of the MFCs [55, 154, 156–158]. Thus, *G. sulfurreducens* can also be used to study adaptation for enhanced power production.

2.4.5.2 METABOLISM OF GEOBACTER SPECIES

The metabolism of *G. sulfurreducens* was investigated by constraint-based modelling [159]. In contrast to *Escherichia coli*, which primarily produces energy and biosynthetic precursors through sugar fermentation, *Geobacter* completely oxidises acetate and other electron donors via the TCA cycle [160, 161], which makes it necessary to transfer electrons to terminal electron acceptors for regeneration of cytoplasmic and intramembrane electron acceptors and ATP synthesis. In *G. sulfurreducens*, this is accomplished by electron transfer to extracellular electron acceptors, i.e. Fe(III) oxides [162].

Since the rate of cytoplasmic proton consumption is lower than that of proton production during the reduction of extracellular electron acceptors such as Fe(III), the energy consumption with extracellular electron acceptors is lower compared to that associated with intracellular acceptors [159]. The use of extracellular electron shuttles makes the *Geobacter* species circumvent the metabolic cost of producing the electron shuttles and consequently more energetically competitive than shuttle-producing Fe(III) reducers in sub-surface environments [159].

In silico analysis suggested that the metabolic network of *G. sulfurreducens* contains pyruvate-ferredoxin oxidoreductase, which catalyzes synthesis of pyruvate from acetate and carbon dioxide in a single step, indicating that the synthesis of amino acids in *G. sulfurreducens* is more efficient than in *E. coli*[159].

2.4.5.3 LIMITS AND APPLICATIONS

MFCs powered by *G. sulfurreducens* are far away from being commercialized as a practical biofuel source [152], because up until now the current levels of these MFCs are around 14 mA which could be used to power very simple components [149] but still not big enough to drive complex mechanisms. However, the actual current densities that could be generated from MFCs based on *G. sulfurreducens* are still unclear and require further investigation [151].

2.4.5.4 COMPARISON OF GEOBACTER SP. AND SHEWANELLA SP

Geobacteraceae and *Shewanellaceae* are classic models in MFC research as their metabolism and versatility have been studied extensively [72, 163]. As mentioned before in the 'Microbial fuel cells' section, they are both capable of being used for DET mode in MFCs, because both *Shewanella sp.* and *Geobacter sp.* possess nanowires, electrically conductive bacterial appendages, to transport electrons from cells to solid electron acceptors such as graphite anodes in MFCs [34, 35, 162]. Despite those similarities, differences also exist when compared regarding the engineering design and performance of the MFCs.

Geobacter-based MFCs generate high coulombic efficiencies [164] but require strict anaerobic conditions which limit their applicability. In contrast, Shewanella-based MFCs can be operated with air-exposed cultures [27]. Unlike Geobacter sp. that requires direct contact to the electrode surface [72], Shewanella sp. can use additional mediators to facilitate electron transfer outside the cell membrane [27]. Importantly, besides utilizing nanowires to mediate the electron transport [165], they can synthesize their own redox mediators (i.e. flavins) for extracellular electron transfer under diverse environmental conditions [163, 166]. These two electron-mediated mechanisms determine the efficiency of the current generation in Shewanella-containing MFCs [167].

A maximum power density of 24 mW m^{-2} (in the presence of an additional mediator) was reported for Shewanella[168]. This value appears low in comparison to bacterial cells, but that is because it was referred to the true microscopic area of a porous electrode. When expressed, as customary, in terms of macroscopic MFC dimensions, the equivalent power densities are 3 W m^{-2} and 500 W m^{-3}. These values compare favourably with Geobacter. When dissolved oxygen was deliberately fed into the anode chamber, Shewanella-based MFC was still able to produce a power output of 6.5 mW m^{-2} and 13 mA m^{-2}. The MFC used lactate as the fuel source and relied on self-excreted mediators of Shewanella[169].

In fact, the previously described Shewanella-based system would not be directly applicable to powering electronics and is required to use aerobic water [170]. For instance, a complex pumping system is necessary to continuously recirculate the anolyte between the anode and the large anolyte reservoir, but this pumping system at the anode could consume more power than the Shewanella-based MFC produces. Since Shewanella sp. cannot use oxygen as the electron acceptor, ferricyanide needs to be added as catholyte. However, ferricyanide is a non-renewable and toxic electron acceptor and can thus not be deployed in the field in the long term. Moreover, the coulombic efficiencies were found to be low (<6%), when calculated based on the incomplete oxidation of lactate to acetate [170].

Conversely, *G. sulfurreducens* can effectively oxidise acetate with electron transfer to electrodes [47, 154] and convert acetate to current with coulombic efficiencies of more than 90% [151, 155]. *G. sulfurreducens* is an anaerobe that can withstand low levels of oxygen and may use oxygen as an electron acceptor [153]. It has recently been shown that with a new configuration, MFCs based on *G. sulfurreducens* can become 100% aerobic, allowing for floating and/or untethered applications. At the same time, the performance of the MFCs is similar to their anaerobic/aerobic counterparts [170]. It is expected that with this aerobic configuration, power could be produced in a *G. sulfurreducens* MFC suspended in aerobic seawater [170].

2.5 CONCLUSIONS

Electricity generation in MFCs is based on the metabolic activity of living microorganisms at the anode. The selection of microorganisms is based on many criteria, but the power output, electron transfer ability and biological functions such as photosynthesis and hydrogen production are particularly important. These important properties, in different combinations, are exemplified by the four representative microorganisms discussed above, and their referential facts for modelling are compared in Table 2. Studying these individually, and in combination, should reveal significant insights in the quest for higher power output MFCs.

Most MFC researchers have been active in engineering designs, i.e. how to create scalable and economical architectures and engineer more efficient hardware and how different microbes interact with the anodes/cathodes when transporting electrons [127]. Such research covers optimizing anodic conditions, housing constructs and component materials, learning more about microbial community ecology and isolating vigorous biocatalysts [9]. Biological aspects of MFCs have also received some attention, such as the anodic activity of different organisms. However, very little research has been done on the biochemical interface between the engineering design and the biological aspects (see Figure 1).

We conclude that future studies are required to work on that interface, i.e. how to enhance the anodic activity by means of adjusting the metabolic activity of biocatalysts, for example, utilizing metabolic network analysis. The genome-scale metabolic networks are quite new concepts and have only been produced in the last few years. The analysis of the metabolic network through modelling approaches, such as flux balance analysis, plays an important role in filling the gap between genotypes and phenotypes of microorganisms to provide a full picture of the biological system.

ENDNOTE

[a]Representative microorganisms chosen by this article and discussed in the 'Microorganisms for in silico study of MFC functioning' section.

REFERENCES

1. Potter MC: Electrical effects accompanying the decomposition of organic compounds. Proc. R. Soc. London Series B 1911,84(571):260–276.
2. Rittmann BE: Opportunities for renewable bioenergy using microorganisms. Biotechnol. Bioeng. 2008,100(2):203–212.
3. Pandyaswargo A, Onoda H, Nagata K: Energy recovery potential and life cycle impact assessment of municipal solid waste management technologies in Asian countries using ELP model. Int. J. Energy Environ. Eng. 2012,3(1):28.
4. Eddine B, Salah M: Solid waste as renewable source of energy: current and future possibility in Algeria. Int. J. Energy Environ. Eng. 2012,3(1):17.
5. Sharma V, Kundu PP: Biocatalysts in microbial fuel cells. Enzyme Microb. Technol. 2010,47(5):179–188.
6. Harnisch F, Schröder U: From MFC to MXC: chemical and biological cathodes and their potential for microbial bioelectrochemical systems. Chem. Soc. Rev. 2010,39(11):4433–4448.
7. Girguis PR, Nielsen ME, Figueroa I: Harnessing energy from marine productivity using bioelectrochemical systems. Curr. Opin. Biotechnol. 2010,21(3):252–258.
8. Li H, Opgenorth PH, Wernick DG, Rogers S, Wu T-Y, Higashide W, Malati P, Huo Y-X, Cho KM, Laio JC: Integrated electromicrobial conversion of CO2 to higher alcohols. Science 2012,335(6076):1596.
9. Schwartz K: Microbial fuel cells: design elements and application of a novel renewable energy source. MMG 445 Basic Biotechnology eJournal 2007,3(1):20–27.
10. Logan BE: Scaling up microbial fuel cells and other bioelectrochemical systems. Appl. Microbiol. Biotechnol. 2010,85(6):1665–1671.

11. Bullen RA, Arnot TC, Lakeman JB, Walsh FC: Biofuel cells and their development. Biosens. Bioelectron. 2006,21(11):2015–2045.

12. Rupprecht J: From systems biology to fuel– Chlamydomonas reinhardtii as a model for a systems biology approach to improve biohydrogen production. J. Biotechnol. 2009,142(1):10–20.

13. Mukhopadhyay A, Redding AM, Rutherford BJ, Keasling JD: Importance of systems biology in engineering microbes for biofuel production. Curr. Opin. Biotechnol. 2008,19(3):228–234.

14. Price ND, Reed JL, Palsson BO: Genome-scale models of microbial cells: evaluating the consequences of constraints. Nature Reviews Microbiology 2004,2(11):886–897.

15. Rocha I, Forster J, Nielsen J: Design and application of genome-scale reconstructed metabolic models. Methods Mol. Biol. 2008, 416:409–431.

16. Jaqaman K, Danuser G: Linking data to models: data regression. Nature Reviews Molecular Cell Biology 2006,7(11):813–819.

17. Terzer M, Maynard ND, Covert MW, Stelling J: Genome-scale metabolic networks. Wiley Interdiscip. Rev. Syst. Biol. Med. 2009,1(3):285–297.

18. Lee JM, Gianchandani EP, Eddy JA, Papin JA: Dynamic analysis of integrated signaling, metabolic, and regulatory networks. PLoS Comput. Biol. 2008,4(5):e1000086.

19. MathSciNetCovert MW, Xiao N, Chen TJ, Karr JR: Integrating metabolic, transcriptional regulatory and signal transduction models in *Escherichia coli* . Bioinformatics 2008,24(18):2044–2050.

20. Jamshidi N, Palsson B: Mass action stoichiometric simulation models: incorporating kinetics and regulation into stoichiometric models. Biophys. J. 2010,98(2):175–185.

21. Thiele I, Palsson BO: A protocol for generating a high-quality genome-scale metabolic reconstruction. Nat. Protoc. 2010,5(1):93–121.

22. Orth JD, Thiele I, Palsson BO: What is flux balance analysis? Nat Biotech 2010,28(3):245–248.

23. Fernie AR, Geigenberger P, Stitt M: Flux an important, but neglected, component of functional genomics. Curr. Opin. Plant Biol. 2005,8(2):174–182.

24. Saha R, Suresha S, Park W, Lee D-Y, Karimi IA: Strain improvement and mediator selection for microbial fuel cell by genome scale in silico model. In 17th European Symposium on Computer Aided Process Engineering – ESCAPE17. Edited by: Pleşu V, Agachi PS. Bucharest; 2007.

25. Rabaey K, Verstraete W: Microbial fuel cells: novel biotechnology for energy generation. Trends Biotechnol. 2005,23(6):291–298.

26. Du Z, Li H, Gu T: A state of the art review on microbial fuel cells: a promising technology for wastewater treatment and bioenergy. Biotechnol. Adv. 2007,25(5):464–482.

27. Schröder U: Anodic electron transfer mechanisms in microbial fuel cells and their energy efficiency. Phys. Chem. Chem. Phys. 2007,9(21):2619–2629.

28. Kim HJ, Hyun MS, Chang IS, Kim BH: A microbial fuel cell type lactate biosensor using a metal-reducing bacterium, *Shewanella* putrefaciens . J. Microbiol. Biotechnol. 1999,9(3):365–367.

29. Kim HJ, Park HS, Hyun MS, Chang IS, Kim M, Kim BH: A mediator-less microbial fuel cell using a metal reducing bacterium, *Shewanella* putrefaciens . Enzyme Microb. Technol. 2002,30(2):145–152.
30. Lovley DR: Microbial fuel cells: novel microbial physiologies and engineering approaches. Curr. Opin. Biotechnol. 2006,17(3):327–332.
31. Chaudhuri SK: Lovley. DR: Electricity generation by direct oxidation of glucose in mediatorless microbial fuel cells. Nat. Biotechnol. 2003,21(10):1229–1232.
32. Chang IS, Moon H, Bretschger O, Jang JK, Park HI, Nealson KH, Kim BH: Electrochemically active bacteria (EAB) and mediator-less microbial fuel cells. J. Microbiol. Biotechnol. 2006,16(2):163–177.
33. Schaetzle O, Barriere F, Baronian K: Bacteria and yeasts as catalysts in microbial fuel cells: electron transfer from micro-organisms to electrodes for green electricity. Energy Environ. Sci. 2008,1(6):607–620.
34. Gorby YA, Yanina S, McLean JS, Rosso KM, Moyles D, Dohnalkova A, Beveridge TJ, Chang IS, Kim BH, Kim KS, Culley DE, Reed SB, Romine MF, Saffarini DA, Hill EA, Shi L, Elias DA, Kennedy DW, Pinchuk G, Watanabe K, Ishii S, Logan B, Nealson KH, Fredrickson JK: Electrically conductive bacterial nanowires produced by *Shewanella* oneidensis strain MR-1 and other microorganisms. Proc. Natl. Acad. Sci. 2006,103(30):11358–11363.
35. Reguera G, Nevin KP, Nicoll JS, Covalla SF, Woodard TL, Lovely DR: Biofilm and nanowire production leads to increased current in *Geobacter* sulfurreducens fuel cells. Appl. Environ. Microbiol. 2006,72(11):7345–7348.
36. Li P, Ki J-P, Liu H: Analysis and optimization of current collecting systems in PEM fuel cells. Int. J. Energy Environ. Eng. 2012,3(1):2.
37. Cooney MJ, Roschi E, Marison IW, Comninellis C, von Stockar U: Physiologic studies with the sulfate-reducing bacterium Desulfovibrio desulfuricans : evaluation for use in a biofuel cell. Enzyme Microb. Technol. 1996,18(5):358–365.
38. Scholz F, Schröder U: Bacterial batteries. Nat. Biotechnol. 2003,21(9):3–4.
39. Rabaey K, Lissens G, Verstraete W: Microbial fuel cells: performances and perspectives. In Biofuels for Fuel Cells: Biomass Fermentation Towards Usage in Fuel Cells. Edited by: Lens P, Westermann P, Haberbauer M, Moreno A. London: IWA; 2005:377–399.
40. Rosenbaum M, He Z, Angenent LT: Light energy to bioelectricity: photosynthetic microbial fuel cells. Curr. Opin. Biotechnol. 2010,21(3):259–264.
41. Davis F, Higson SPJ: Biofuel cells–recent advances and applications. Biosens. Bioelectron. 2007,22(7):1224–1235.
42. Bennetto HP, Delany GM, Mason JR, Roller SD, Stirling JL, Thurston CF: The sucrose fuel cell: efficient biomass conversion using a microbial catalyst. Biotechnol. Lett. 1985,7(10):699–704.
43. Rabaey K, Boon N, Höfte M, Verstraete W: Microbial phenazine production enhances electron transfer in biofuel cells. Environ. Sci. Technol. 2005,39(9):3401–3408.
44. Park HS, Kim BH, Kim HS, Kim HJ, Kim GT, Kim M, Chang IS, Park YK, Chang HI: A novel electrochemically active and Fe(III)-reducing bacterium phylogenetically related to Clostridium butyricum isolated from a microbial fuel cell. Anaerobe 2001,7(6):297–306.

45. Kim BH, Park HS, Kim HJ, Kim GT, Chang IS, Lee J, Phung NT: Direct electrode reaction of Fe(III)-reducing bacterium, *Shewanella* putrefaciens . J. Microbiol. Biotechnol. 1999,9(2):127–131.

46. Logan BE, Regan JM: Electricity-producing bacterial communities in microbial fuel cells. Trends Microbiol. 2006,14(12):512–518.

47. Bond DR, Lovley DR: Electricity production by *Geobacter* sulfurreducens attached to electrodes. Appl. Environ. Microbiol. 2003,69(3):1548–1555.

48. Lanthier M, Gregory KB, Lovley DR: Growth with high planktonic biomass in *Shewanella* oneidensis fuel cells. FEMS Microbiol. Lett. 2008,278(1):29–35.

49. Logan BE, Regan JM: Microbial fuel cells, challenges and applications. Environ. Sci. Technol. 2006,40(17):5172–5180.

50. Fan Y, Hu H, Liu H: Enhanced Coulombic efficiency and power density of air-cathode microbial fuel cells with an improved cell configuration. J. Power. Sources 2007,171(2):348–354.

51. Zhang F, Cheng S, Pant D, Van Bogaert G, Logan BE: Power generation using an activated carbon and metal mesh cathode in a microbial fuel cell. Electrochem. Commun. 2009,11(11):2177–2179.

52. Zuo Y, Cheng S, Logan BE: Ion exchange membrane cathodes for scalable microbial fuel cells. Environ. Sci. Technol. 2008,42(18):6967–6972.

53. Chen G-W, Choi SJ, Lee TH, Lee GY, Cha JH, Kim CW: Application of biocathode in microbial fuel cells: cell performance and microbial community. Appl. Microbiol. Biotechnol. 2008,79(3):379–388.

54. Cheng S, Logan BE: Ammonia treatment of carbon cloth anodes to enhance power generation of microbial fuel cells. Electrochem. Commun. 2006, 9:492–496.

55. Liu JL, Lowy DA, Baumann RG, Tender LM: Influence of anode pretreatment on its microbial colonization. J. Appl. Microbiol. 2007,102(1):177–183.

56. Cao X, Huang X, Boon N, Liang P, Fan M: Electricity generation by an enriched phototrophic consortium in a microbial fuel cell. Electrochem. Commun. 2008,10(9):1392–1395.

57. Yang Y, Sun G, Xu M: Microbial fuel cells come of age. J. Chem. Technol. Biotechnol. 2011,86(5):625–632.

58. Lovley DR: Extracellular electron transfer: wires, capacitors, iron lungs, and more. Geobiology 2008,6(3):225–231.

59. Shi L, Squier TC, Zachara JM, Fredrickson JK: Respiration of metal (hydr)oxides by *Shewanella* and *Geobacter* : a key role for multihaem c-type cytochromes. Mol. Microbiol. 2007,65(1):12–20.

60. Oh ST, Kim JR, Premier GC, Lee TH, Kim C, Sloan WT: Sustainable wastewater treatment: how might microbial fuel cells contribute. Biotechnol. Adv. 2010,28(6):871–881.

61. Yuan Y, Ahmed J, Zhou L, Zhao B, Kim S: Carbon nanoparticles-assisted mediator-less microbial fuel cells using Proteus vulgaris . Biosens. Bioelectron. 2011,27(1):106–112.

62. He Z, Kan J, Mansfeld F, Angenent LT, Nealson KH: Self-sustained phototrophic microbial fuel cells based on the synergistic cooperation between photosynthetic microorganisms and heterotrophic bacteria. Environ. Sci. Technol. 2009,43(5):1648–1654.

63. Malik S, Drott E, Grisdela P, Lee J, Lee C, Lowy DA, Gray S, Tender LM: A self-assembling self-repairing microbial photoelectrochemical solar cell. Energy Environ. Sci. 2009,2(3):292–298.
64. Cho YK, Donohue TJ, Tejedor I, Anderson MA, McMahon KD, Noguera DR: Development of a solar-powered microbial fuel cell. J. Appl. Microbiol. 2008,104(3):640–650.
65. Rosenbaum M, Schröder U, Scholz F: In situ electrooxidation of photobiological hydrogen in a photobioelectrochemical fuel cell based on Rhodobacter sphaeroide. Environ. Sci. Technol. 2005,39(16):6328–6333.
66. Xing D, Zuo Y, Cheng S, Regan JM, Logan BE: Electricity generation by Rhodopseudomonas palustris DX-1. Environ. Sci. Technol. 2008,42(11):4146–4151.
67. Larsen K, Ibrom A, Beier C, Jonasson S, Michelsen A: Ecosystem respiration depends strongly on photosynthesis in a temperate heath. Biogeochemistry 2007,85(2):201–213.
68. Stal LJ, van Gemerden H, Krumbein WE: Structure and development of a benthic marine microbial mat. FEMS Microbiol. Lett. 1985,31(2):111–125.
69. Strik D, Terlouw H, Hamelers HV, Buisman CJ: Renewable sustainable biocatalyzed electricity production in a photosynthetic algal microbial fuel cell (PAMFC). Appl. Microbiol. Biotechnol. 2008,81(4):659–668.
70. Zou Y, Pisciotta J, Billmyre RB, Baskakov IV: Photosynthetic microbial fuel cells with positive light response. Biotechnol. Bioeng. 2009,104(5):939–946.
71. Yagishita T, Horigome T, Tanaka K: Effects of light, CO 2 and inhibitors on the current output of biofuel cells containing the photosynthetic organism Synechococcus sp. J. Chem. Technol. Biotechnol. 1993,56(4):393–399.
72. Lovley DR: Bug juice: harvesting electricity with microorganisms. Nature Reviews. Microbiology 2006,4(7):497–508.
73. Raghavulu SV, Goud RK, Sarma PN, Mohan SV: *Saccharomyces* cerevisiae as anodic biocatalyst for power generation in biofuel cell: influence of redox condition and substrate load. Bioresour. Technol. 2011,102(3):2751–2757.
74. Rosenbaum M, Schröder U: Photomicrobial solar and fuel cells. Electroanalysis 2010,22(7–8):844–855.
75. Dal'Molin CG, Quek LE, Palfreyman RW, Nielsen LK: AlgaGEM - a genome-scale metabolic reconstruction of algae based on the Chlamydomonas reinhardtii genome. BMC Genomics 2011,12(Suppl 4):S5.
76. Chang RL, Ghamsari L, Manichaikul A, Hom EF, Balaji S, Fu W, Shen Y, Hao T, Palsson BØ, Salehi-Ashtiani K, Papin JA: Metabolic network reconstruction of Chlamydomonas offers insight into light-driven algal metabolism. Mol. Syst. Biol. 2011, 7:518.
77. Yoshikawa K, Kojima Y, Nakajima T, Furusawa C, Hirasawa T, Shimizu H: Reconstruction and verification of a genome-scale metabolic model for Synechocystis sp. PCC6803. Appl. Microbiol. Biotechnol. 2011,92(2):347–358.
78. Nogales J, Gudmundsson S, Knight EM, Palsson BO, Thiele I: Detailing the optimality of photosynthesis in cyanobacteria through systems biology analysis. Proc. Natl. Acad. Sci. 2012,109(7):2678–2683.

79. Sun J, Sayyar B, Butler JE, Pharkya P, Fahland TR, Famili I, Schilling CH, Lovley DR, Mahadevan R: Genome-scale constraint-based modeling of *Geobacter* metallireducens . BMC Syst. Biol. 2009, 3:15.

80. Heavner BD, Smallbone K, Barker B, Mendes P, Walker LP: Yeast 5 - an expanded reconstruction of the *Saccharomyces* cerevisiae metabolic network. BMC Syst. Biol. 2012, 6:55.

81. Adams M: General growth and mating. http://nutmeg.easternct.edu/~adams/ChlamyTeach/growingchlamy.html (2012). Accessed 12 Feb 2013

82. Sheng J, Kim HW, Badalamenti JP, Zhou C, Sridharakrishnan S, Krajmalnik-Brown R, Rittmann BE, Vannela R: Effects of temperature shifts on growth rate and lipid characteristics of Synechocystis sp. PCC6803 in a bench-top photobioreactor. Bioresour. Technol. 2011,102(24):11218–11225.

83. Watson K: Temperature relations. In The Yeasts. Edited by: Rose AH, Harrison JS. London: Academic; 1987:41–71.

84. Trinh N, Park J, Kim B-W: Increased generation of electricity in a microbial fuel cell using *Geobacter* sulfurreducens . Korean J. Chem. Eng. 2009,26(3):748–753.

85. Yoon HS, Hackett JD, Ciniglia C, Pinto G, Bhattacharya D: A molecular timeline for the origin of photosynthetic eukaryotes. Mol. Biol. Evol. 2004,21(5):809–818.

86. Merchant SS, Prochnik SE, Vallon O, Harris EH, Karpowicz SJ, Witman GB, Terry A, Salamov A, Fritz-Laylin LK, Maréchal-Drouard L, Marshall WF, Qu LH, Nelson DR, Sanderfoot AA, Spalding MH, Kapitonov VV, Ren Q, Ferris P, Lindquist E, Shapiro H, Lucas SM, Grimwood J, Schmutz J, Cardol P, Cerutti H, Chanfreau G, Chen CL, Cognat V, Croft MT, Dent R: The Chlamydomonas genome reveals the evolution of key animal and plant functions. Science 2007,318(5848):245–250.

87. Harris EH: Chlamydomonas as a model organism. Annu. Rev. Plant Physiol. Plant Mol. Biol. 2001,52(1):363–406.

88. Manichaikul A, Ghamsari L, Hom EF, Lin C, Murray RR, Chang RL, Balaji S, Hao T, Shen Y, Chavali AK, Thiele I, Yang X, Fan C, Mello E, Hill DE, Vidal M, Salehi-Ashtiani K, Papin JA: Metabolic network analysis integrated with transcript verification for sequenced genomes. Nat. Methods 2009,6(8):589–592.

89. Keller LC, Romijn EP, Zamora I, Yates JR 3rd, Marshall WF: Proteomic analysis of isolated chlamydomonas centrioles reveals orthologs of ciliary-disease genes. Current Biol.: CB 2005,15(12):1090–1098.

90. Pazour GJ, Agrin N, Walker BL, Witman GB: Identification of predicted human outer dynein arm genes: candidates for primary ciliary dyskinesia genes. J. Med. Genet. 2006,43(1):62–73.

91. Vilchez C, Garbayo I, Markvicheva E, Galván F, León R: Studies on the suitability of alginate-entrapped Chlamydomonas reinhardtii cells for sustaining nitrate consumption processes. Bioresour. Technol. 2001,78(1):55–61.

92. Ghirardi ML, Posewitz MC, Maness PC, Dubini A, Yu J, Seibert M: Hydrogenases and hydrogen photoproduction in oxygenic photosynthetic organisms. Annu. Rev. Plant Biol. 2007, 58:71–91.

93. Kosourov SN, Seibert M: Hydrogen photoproduction by nutrient-deprived Chlamydomonas reinhardtii cells immobilized within thin alginate films under aerobic and anaerobic conditions. Biotechnol. Bioeng. 2009,102(1):50–58.

94. The Chlamydomonas Center: Chlamydomonas connection. http://www.chlamy.org (2013). Accessed 12 Feb 2013

95. Chen P, Min M, Chen Y, Wang L, Li Y, Chen Q, Wang C, Wan Y, Wang X, Cheng Y, Deng S, Hennessy K, Lin X, Liu Y, Wang Y, Martinez B, Ruan R: Review of biological and engineering aspects of algae to fuels approach. Int. J. Agric. Biol. Eng. 2009,2(4):1–30.

96. Pienkos PT, Darzins A: The promise and challenges of microalgal-derived biofuels. Biofpr. 2009, 3:431–440.

97. Campbell MN: Biodiesel: algae as a renewable source for liquid fuel. Guelph Eng. J. 2008,1(1916–1107):207.

98. Velasquez-Orta SB, Curtis TP, Logan BE: Energy from algae using microbial fuel cells. Biotechnol. Bioeng. 2009,103(6):1068–1076.

99. Rosenbaum M, Schröder U, Scholz F: Utilizing the green alga Chlamydomonas reinhardtii for microbial electricity generation: a living solar cell. Appl. Microbiol. Biotechnol. 2005,68(6):753–756.

100. Strik DPBTB, Hamelers HVM, Buisman CJN: Solar energy powered microbial fuel cell with a reversible bioelectrode. Environ. Sci. Technol. 2009,44(1):532–537.

101. Harnisch F, Schröder U: Selectivity versus mobility: separation of anode and cathode in microbial bioelectrochemical systems. ChemSusChem 2009,2(10):921–926.

102. Melis A, Happe T: Hydrogen production. Green algae as a source of energy. Plant Physiol. 2001,127(3):740–748.

103. Rupprecht J, Hankamer B, Mussgnug JH, Ananyev G, Dismukes C, Kruse O: Perspectives and advances of biological H 2 production in microorganisms. Appl. Microbiol. Biotechnol. 2006,72(3):442–449.

104. Hankamer B, Lehr F, Rupprecht J, Mussgnug JH, Posten C, Kruse O: Photosynthetic biomass and H2 production by green algae: from bioengineering to bioreactor scale-up. Physiol. Plant. 2007,131(1):10–21.

105. Gaffron H, Rubin J: Fermentative and photochemical production of hydrogen in algae. J. Gen. Physiol. 1942,26(2):219–240.

106. Kruse O, Rupprecht J, Mussgnug JH, Dismukes GC, Hankamer B: Photosynthesis: a blueprint for solar energy capture and biohydrogen production technologies. Photochem. Photobiol. Sci. 2005,4(12):957–970.

107. Melis A, Zhang L, Forestier M, Ghirardi ML, Seibert M: Sustained photobiological hydrogen gas production upon reversible inactivation of oxygen evolution in the green alga Chlamydomonas reinhardtii . Plant Physiol. 2000,122(1):127–136.

108. Ghirardi ML, Zhang L, Lee JW, Flynn T, Seibert M, Greenbaum E, Melis A: Microalgae: a green source of renewable H2. Trends Biotechnol. 2000,18(12):506–511.

109. Esquível MG, Amaro HM, Pinto TS, Fevereiro PS, Malcata FX: Efficient H2 production via Chlamydomonas reinhardtii . Trends Biotechnol. 2011,29(12):595–600.

110. Mus F, Dubini A, Seibert M, Posewitz MC, Grossman AR: Anaerobic acclimation in Chlamydomonas reinhardtii . J. Biol. Chem. 2007,282(35):25475–25486.

111. Dubini A, Mus F, Seibert M, Grossmman AR, Posewitz MC: Flexibility in anaerobic metabolism as revealed in a, mutant of Chlamydomonas reinhardtii lacking hydrogenase activity. J. Biol. Chem. 2009,284(11):7201–7213.

112. Posewitz MC, Dubini A, Meuser JE, Seibert M, Ghirardi ML: Hydrogenases, hydrogen production, and anoxia. In The Chlamydomonas Sourcebook. 2nd edition. Edited by: Elizabeth HH, Stern DB, Witman GB. London: Academic; 2009:217–255.

113. Timmins M, Thomas-Hall SR, Darling A, Zhang E, Hankamer B, Marx UC, Schenk PM: Phylogenetic and molecular analysis of hydrogen-producing green algae. J. Exp. Bot. 2009,60(6):1691–1702.

114. Doebbe A, Keck M, La Russa M, Mussgnug JH, Hankamer B, Tekçe E, Niehaus K, Kruse O: The interplay of proton, electron, and metabolite supply for photosynthetic H2 production in Chlamydomonas reinhardtii . J. Biol. Chem. 2010,285(39):30247–30260.

115. Grossman AR, Catalanotti C, Yang W, Dubini A, Magneschi L, Subramanian V, Posewitz MC, Seibert M: Multiple facets of anoxic metabolism and hydrogen production in the unicellular green alga Chlamydomonas reinhardtii . New Phytol. 2011,190(2):279–288.

116. Markov SA, Eivazova ER, Greenwood J: Photostimulation of H2 production in the green alga Chlamydomonas reinhardtii upon photoinhibition of its O2-evolving system. Int. J. Hydrogen Energy 2006,31(10):1314–1317.

117. Seibert M, King PW, Posewitz MC, Melis A, Ghirardi ML: Photosynthetic water-splitting for hydrogen production. In Bioenergy. Edited by: Wall JD, Harwood CS, Demain A. Washington D.C: ASM; 2008:273–291.

118. Zorina A, Mironov K, Stepanchenko N, Sinetova M, Koroban N, Zinchenko V, Kupriyanova E, Allakhverdiev S, Los D: Regulation systems for stress responses in cyanobacteria. Russian J. Plant Physiol. 2011,58(5):749–767.

119. Waterbury JB, Watson SW, Guillard RL, Brand LE: Widespread occurrence of a unicellular, marine, planktonic, cyanobacterium. Nature 1979,277(5694):293–294.

120. Carr NG, Whitton BA: The Biology of Cyanobacteria. Berkeley: University of California Press; 1982.

121. Martin W, Rujan T, Richly E, Hansen A, Cornelsen S, Lins T, Leister D, Stoebe B, Hasegawa M, Penny D: Evolutionary analysis of Arabidopsis , cyanobacterial, and chloroplast genomes reveals plastid phylogeny and thousands of cyanobacterial genes in the nucleus. Proc. Natl. Acad. Sci. U.S.A. 2002,99(19):12246–12251.

122. Dutta D, De D, Chaudhuri S, Bhattacharya SK: Hydrogen production by Cyanobacteria. Microb. Cell Fact. 2005, 4:36.

123. Kaneko T, Sato S, Kotani H, Tanaka A, Asamizu E, Nakamura Y, Miyajima N, Hirosawa M, Sugiura M, Sasamoto S, Kimura T, Hosouchi T, Matsuno A, Muraki A, Nakazaki N, Naruo K, Okumura S, Shimpo S, Takeuchi C, Wada T, Watanabe A, Yamada M, Yasuda M, Tabata S: Sequence analysis of the genome of the unicellular cyanobacterium Synechocystis sp. strain PCC6803. II. Sequence determination of the entire genome and assignment of potential protein-coding regions. DNA Res. 1996,3(3):109–136.

124. Nakamura Y, Kaneko T, Hirosawa M, Miyajima N, Tabata S: CyanoBase, a www database containing the complete nucleotide sequence of the genome of Synechocystis sp. strain PCC6803. Nucleic Acids Res. 1998,26(1):63–67.

125. Atsumi S, Higashide W, Liao JC: Direct photosynthetic recycling of carbon dioxide to isobutyraldehyde. Nat. Biotechnol. 2009,27(12):1177–1180.

126. Johnson CH, Stewart PL, Egli M: The cyanobacterial circadian system: from bio-physics to bioevolution. Annu. Rev. Biophys. 2011, 40:143–167.
127. Lovley DR: The microbe electric: conversion of organic matter to electricity. Curr. Opin. Biotechnol. 2008,19(6):564–571.
128. Pisciotta JM, Zou Y, Baskakov IV: Role of the photosynthetic electron trans-fer chain in electrogenic activity of cyanobacteria. Appl. Microbiol. Biotechnol. 2011,91(2):377–385.
129. Pisciotta JM, Zou Y, Baskakov IV: Light-dependent electrogenic activity of cyano-bacteria. PLoS One 2010,5(5):e10821.
130. Madiraju KS, Lyew D, Kok R, Raghavan V: Carbon neutral electricity production by Synechocystis sp. PCC6803 in a microbial fuel cell. Bioresour. Technol. 2012, 110:214–218.
131. McCormick AJ, Bombelli P, Scott AM, Philips AJ, Smith AG, Fisher AC, Howe CJ: Photosynthetic biofilms in pure culture harness solar energy in a mediatorless bio-photovoltaic cell (BPV) system. Energy Environ. Sci. 2011,4(11):4699–4709.
132. Dequin S, Casaregola S: The genomes of fermentative Saccharomyces . Comptes Rendus Biologies 2011,334(8–9):687–693.
133. Kurtzman CP: Phylogenetic circumscription of Saccharomyces , Kluyveromyces and other members of the Saccharomycetaceae, and the proposal of the new genera Lachancea , Nakaseomyces , Naumovia . Vanderwaltozyma and Zygotorulaspora. FEMS Yeast Res. 2003,4(3):233–245.
134. Vaughan-Martini A, Martini A: Facts, myths and legends on the prime industrial microorganism. J. Ind. Microbiol. 1995,14(6):514–522.
135. Fan Y, Sharbrough E, Liu H: Quantification of the internal resistance distribution of microbial fuel cells. Environ. Sci. Technol. 2008,42(21):8101–8107.
136. Peralta-Yahya PP, Keasling JD: Advanced biofuel production in microbes. Biotech-nol. J. 2010,5(2):147–162.
137. Wilkinson S: "Gastrobots" - benefits and challenges of microbial fuel cells in food powered robot applications. Auton. Robots 2000,9(2):99–111.
138. Haslett ND, Rawson FJ, Barrière F, Kunze G, Pasco N, Gooneratne R, Baronian KH, Haslett ND, Rawson FJ, Barrière F, Kunze G, Pasco N, Gooneratne R, Baronian KH: Characterisation of yeast microbial fuel cell with the yeast Arxula adeninivorans as the biocatalyst. Biosens. Bioelectron. 2011,26(9):3742–3747.
139. Gunawardena A, Fernando S, To F: Performance of a yeast-mediated biological fuel cell. Int. J. Mol. Sci. 2008,9(10):1893–1907.
140. Feldmann H: Yeast Molecular Biology: A Short Compendium on Basic Features and Novel Aspects. Munich: Adolf-Butenandt-Institut, University of Munich; 2005.
141. Ganguli R, Dunn BS: Kinetics of anode reactions for a yeast-catalysed microbial fuel cell. Fuel Cells 2009,9(1):44–52.
142. Powell EE, Evitts RW, Hill GA, Bolster JC: A microbial fuel cell with a photo-synthetic microalgae cathodic half cell coupled to a yeast anodic half cell. Energy Sources Part a-Recovery Utilization Environ. Effects 2011,33(5):440–448.
143. Sayed ET, Tsujiguchi T, Nakagawa N: Catalytic activity of baker's yeast in a media-torless microbial fuel cell. Bioelectrochemistry 2012, 86:97–101.
144. Anderson RT, Vrionis HA, Ortiz-Bernad I, Resch CT, Long PE, Dayvault R, Karp K, Marutzky S, Metzler DR, Peacock A, White DC, Lowe M, Lovley DR: Stimulating

the in situ activity of *Geobacter* species to remove uranium from the groundwater of a uranium-contaminated aquifer. Appl. Environ. Microbiol. 2003,69(10):5884–5891.

145. Holmes DE, Finneran KT, O'Neil RA, Lovley DR: Enrichment of members of the family *Geobacter*aceae associated with stimulation of dissimilatory metal reduction in uranium-contaminated aquifer sediments. Appl. Environ. Microbiol. 2002,68(5):2300–2306.

146. Lovley DR, Anderson RT: Influence of dissimilatory metal reduction on fate of organic and metal contaminants in the subsurface. Hydrogeol. J. 2000, 8:77–88.

147. Coppi MV, Leang C, Sandler SJ, Lovley DR: Development of a genetic system for *Geobacter* sulfurreducens . Appl. Environ. Microbiol. 2001,67(7):3180–3187.

148. Caccavo F Jr, Lonergan DJ, Lovley DR, Stolz JF, McInerney MJ: *Geobacter* sulfurreducens sp. nov., a hydrogen- and acetate-oxidizing dissimilatory metal-reducing microorganism. Appl. Environ. Microbiol. 1994,60(10):3752–3759.

149. Nevin KP, Kim BC, Glaven RH, Johnson JP, Woodard TL, Methé BA, Didonato RJ, Covalla SF, Franks AE, Liu A, Lovley DR: Anode biofilm transcriptomics reveals outer surface components essential for high density current production in *Geobacter* sulfurreducens fuel cells. PLoS One 2009,4(5):e5628.

150. Chae KJ, Choi MJ, Lee JW, Kim KY, Kim IS: Effect of different substrates on the performance, bacterial diversity, and bacterial viability in microbial fuel cells. Bioresour. Technol. 2009,100(14):3518–3525.

151. Yi H, Nevin KP, Kim BC, Franks AE, Klimes A, Tender LM, Lovley DR: Selection of a variant of *Geobacter* sulfurreducens with enhanced capacity for current production in microbial fuel cells. Biosens. Bioelectron. 2009,24(12):3498–3503.

152. Salgado CA: Microbial fuel cells powered by *Geobacter* sulfurreducens . MMG 445 Basic Biotechnology eJournal 2009, 5:96–101.

153. Lin WC, Coppi MV, Lovley DR: *Geobacter* sulfurreducens can grow with oxygen as a terminal electron acceptor. Appl. Environ. Microbiol. 2004,70(4):2525–2528.

154. Bond DR, Holmes DE, Tender LM, Lovley DR: Electrode-reducing microorganisms that harvest energy from marine sediments. Science 2002,295(5554):483–485.

155. Nevin KP, Richter H, Covalla SF, Johnson JP, Woodard TL, Orloff AL, Jia H, Zhang M, Lovley DR: Power output and columbic efficiencies from biofilms of *Geobacter* sulfurreducens comparable to mixed community microbial fuel cells. Environ. Microbiol. 2008,10(10):2505–2514.

156. Gregory KB, Bond DR, Lovley DR: Graphite electrodes as electron donors for anaerobic respiration. Environ. Microbiol. 2004,6(6):596–604.

157. Holmes DE, Bond DR, O'Neil RA, Reimers CE, Tender LR, Lovley DR: Microbial communities associated with electrodes harvesting electricity from a variety of aquatic sediments. Microb. Ecol. 2004,48(2):178–190.

158. Tender LM, Reimers CE, Stecher HA, Holmes DE, Bond DR, Lowy DA, Pilobello K, Fertig SJ, Lovley DR: Harnessing microbially generated power on the seafloor. Nat. Biotechnol. 2002,20(8):821–825.

159. Mahadevan R, Bond DR, Butler JE, Esteve-Nuñez A, Coppi MV, Palsson BO, Schilling CH, Lovley DR: Characterization of metabolism in the Fe(III)-reducing organism *Geobacter* sulfurreducens by constraint-based modeling. Appl. Environ. Microbiol. 2006,72(2):1558–1568.

160. Champine JE, Underhill B, Johnston JM, Lilly WW, Goodwin S: Electron transfer in the dissimilatory iron-reducing bacterium *Geobacter* metallireducens . Anaerobe 2000,6(3):187–196.

161. Galushko AS, Schink B: Oxidation of acetate through reactions of the citric acid cycle by *Geobacter* sulfurreducens in pure culture and in syntrophic coculture. Arch. Microbiol. 2000,174(5):314–321.

162. Reguera G, McCarthy KD, Mehta T, Nicoll JS, Tuominen MT, Lovley DR: Extracellular electron transfer via microbial nanowires. Nature 2005,435(7045):1098–1101.

163. Fredrickson JK, Romine MF, Beliaev AS, Auchtung JM, Driscoll ME, Gardner TS, Nealson KH, Osterman AL, Pinchuk G, Reed JL, Rodionov DA, Rodrigues JLM, Saffarini DA, Serres MH, Spormann AF, Zhulin IB, Tiedje JM: Towards environmental systems biology of *Shewanella* . Nature Reviews Microbiology 2008,6(8):592–603.

164. Call DF, Wagner RC, Logan BE: Hydrogen production by *Geobacter* species and a mixed consortium in a microbial electrolysis cell. Appl. Environ. Microbiol. 2009,75(24):7579–7587.

165. El-Naggar MY, Gorby YA, Xia W, Nealson KH: The molecular density of states in bacterial nanowires. Biophys. J. 2008,95(1):L10-L12.

166. Marsili E, Baron DB, Shikhare ID, Coursolle D, Gralnick JA, Bond DR: *Shewanella* secretes flavins that mediate extracellular electron transfer. Proc. Natl. Acad. Sci. 2008,105(10):3968–3973.

167. Newton GJ, Mori S, Nakamura R, Hashimoto K, Watanabe K: Analyses of current-generating mechanisms of *Shewanella* loihica PV-4 and *Shewanella* oneidensis MR-1 in microbial fuel cells. Appl. Environ. Microbiol. 2009,75(24):7674–7681.

168. Ringeisen BR, Henderson E, Wu PK, Pietron J, Ray R, Little B, Biffinger JC, Jones-Meehan JM: High power density from a miniature microbial fuel cell using *Shewanella* oneidensis DSP10. Environ. Sci. Technol. 2006,40(8):2629–2634.

169. Ringeisen BR, Ray R, Little B: A miniature microbial fuel cell operating with an aerobic anode chamber. J. Power. Sources 2007,165(2):591–597.

170. Nevin KP, Zhang P, Franks AE, Woodard TL, Lovley DR: Anaerobes unleashed: aerobic fuel cells of *Geobacter* sulfurreducens . J. Power. Sources 2011,196(18):7514–7518.

CHAPTER 3

FROM TINY MICROALGAE TO HUGE BIOREFINERIES

LUISA GOUVEIA

3.1 INTRODUCTION

Biofuel and bioproduct production from microalgae have several advantages when compared to the 1st and 2nd biofuel generation having: high areal productivity, minimal competition with conventional agriculture, environmental benefits by recycling nutrients (N and P) from waste waters and mitigating carbon dioxide from air emissions. In addition, all components of microalgae can be separated and transformed into different valuable products. The high metabolic versatility of microalgae and cyanobacteria metabolisms, offer interesting applications in several fields such as nutrition (human and animal), nutraceuticals, therapeutic products, fertilizers, plastics, isoprene, biofuels and environment (such as water stream bioremediation and carbon dioxide mitigation).

The high content of antioxidants and pigments (carotenoids such as fucoxanthin, lutein, betacarotene and/or astaxanthin and phycobilliproteins) and the presence of long-chain Polyunsaturated Fatty Acids (PUFAs) and proteins (essential amino acids methionine, threonine and tryptophan), makes microalgae an excellent source of nutritional compounds. Coextraction of other high-value products (PUFAs, such as Eicosapentaenoic

Acid (EPA), Docosahexaenoic Acid (DHA), and Arachidonic Acid (AA)) will also be evaluated since these compounds may enhance the nutritional or nutraceutical value of the microalgal oil.

Microalgae have also been screened for new pharmaceutical compounds with biological activity, such as antibiotics, antiviral, anticancer, enzyme inhibitory agents and other therapeutic applications. They have been reported to potentially prevent or reduce the impact of several lifestyle-related diseases [1-3] with antimicrobial (antibacterial, antifungal, antiprotozoal) and antiviral (including anti-HIV) functions and they also have cytotoxic, antibiotic, and anti-tumour properties as well as having biomodulatory effects such as immunosuppressive and anti-inflammatory roles [4,5]. Chlorella has also been used against infant malnutrition and neurosis [6], as well as being a food additive. Furthermore, algae are believed to have a positive effect on the reduction of cardio-circulatory and coronary diseases, atherosclerosis, gastric ulcers, wounds, constipation, anaemia, hypertension, and diabetes [6,7].

The microalgae compounds, such as carotenoids have also been associated and claimed to reduce the risk of: (1) certain cancers [8-11], (2) cardiovascular diseases [12,13], (3) macular degeneration and cataract formation [14,15] and possibly may have an effect on the immune system and may influence chronic diseases [16,17].

Besides nutritional, nutraceutical and therapeutic compounds, microalgae can also synthesize polysaccharides that can be used as an emulsion stabilizer or as biofloculants and polyhydroxyalkanoate, which are linear polyesters used in the production of bioplastics. Microalgae biomass has been demonstrated to improve the physical and thermal properties of plastic by replacing up to 25% of polymers, which increases the biodegradability of the final bioplastic. Microalgae can also produce isoprene, which is a key intermediate compound for the production of synthetic rubber and adhesives, including car and truck tires. It is also an important polymer building block for the chemical industry, such as for a wide variety of elastomers used in surgical gloves, rubber bands, golf balls, and shoes [18].

Furthermore, the aminoacids produced by microalgae can be used as biofertilizers and therefore assist higher plant growth. Amino acid-based fertilization supplies plants with the necessary elements to develop their structures by adding nutrients through the natural processes of nitrogen fixation, solubilizing phosphorus, and stimulating plant growth through the synthesis of growth-promoting substances [19-21]. Bio-fertilizers provide eco-friendly organic agroinput and are more cost-effective than chemical fertilizers.

Finally, regarding biofuels, they can be obtained from the microalgae biomass leftovers after the extraction of added-value compounds. According to the composition of the "waste" biomass, it can be used for the production of liquid biofuels (bioethanol, biodiesel, biobutano and bio-oil) [22,23] or gaseous biofuels (biomethane, biohydrogen, syngas etc.) [24-26]. The technology used to produce biofuels efficiently is not yet established, thus different biological and thermochemical processes still need to be studied and improved.

Unfortunately, the economic viability of algae-based biofuels is still unfeasible. However, the high metabolic versatility of microalgae and cyanobacteria metabolisms allow the production of the several mentioned non-fuel products, which have a very high value and could play a major role in turning economic and energy balances more favorable. This versatility and huge potential of tiny microalgae could support a microalgae-based biorefinery and microalgae-based bioeconomy opening up vast opportunities in the global algae business.

The microalgae could play an important response to the worldwide biofuel demand, together with the production of high value-added products and assisting some other environmental issues such as water stream bioremediation and carbon dioxide mitigation.

Only the co-production of high added value products and environmental benefits could eventually off-set the high production costs of mass microalgae cultivation and support a microalgae-based bioeconomy. In fact, a microalgae-based biorefinery should integrate several processes and related industries, such as food, feed, energy, pharmaceutical, cosmetic, and chemical. Such an approach, in addition to the biomass, will take advantage of the various products synthesized by the microalgae. This adds value to the whole process which has a minimal environmental impact by recycling the nutrients and water, and by mitigating the CO_2 from the flue gases (Figure 1).

This review highlights the potential of the tiny autotrophic microalgae for the production of several products in an experimental (lab scale) Biorefinery. The production contains biofuel(s) and other high value-added compounds which could be used for different applications and markets.

3.2 FROM (TINY) MICROALGAE TO (HUGE) BIOREFINERIES

The main bottleneck of the biorefinery approach is to separate the different fractions without damaging one or more of the product fractions. There is a need for mild, inexpensive and low energy consumption separation techniques to overcome these bottlenecks [27,28]. They should also be applicable for a variety of end products which have a sufficient quality but are also available in large quantities [29,30].

Some of the biorefinery techniques appropriate for metabolite separation and extraction are ionic liquids or surfactants [28,31]. These techniques are relatively new and should therefore be studied thoroughly before commercial use will be possible.

TABLE 1: Nannochloropsis sp. composition.

Composition	(%)
Crude fat	41
Total sugars	17
Total minerals	13
Others	29

3.2.1 NANNOCHLOROPSIS SP. *BIOREFINERY*

Nobre et al. [31] used *Nannochloropsis sp.* microalga and developed a Biorefinery with the extraction of carotenoids and fatty acids (mainly EPA) for food and the feed industry as well as lipids for biodiesel production. The biomass composition is present in Table 1.

The fractionated recovery of the different compounds was done by Supercritical Extraction using CO_2 and ethanol as an entrainer. From the biomass leftovers and using *Enterobacter aerogeneses* through dark fermentation, $bioH_2$ was also produced (Figure 2), yielding a maximum of 60.6 mL H_2/g alga [31].

The energy consumption and CO_2 emissions emitted during the whole process (microalgae cultivation, harvesting, dewatering, milling, extraction and leftover biomass fermentation), as well as the economic factors were evaluated [25]. The authors showed five pathways and two biorefineries which were analysed (Figure 3):

1. Path # 1: Oil extraction by soxhlet (oil SE);
2. Path #2: Oil and pigment extraction and fractionation through Supercritical Fluid Extraction (oil and pigment SFE);
3. Path #3: Hydrogen production through dark fermentation of the leftover biomass after soxhlet extraction ($bioH_2$ via SE);
4. Path #4: Hydrogen production by dark fermentation from the leftover biomass after Supercritical Fluid extraction ($bioH_2$ via SFE);
5. Path #5: Hydrogen production from the whole biomass through dark fermentation ($bioH_2$ using the whole biomass).

Where path #1 and path #3 are the Biorefinery 1, path #2 and path #4 are the Biorefinery 2 and path #5 is the direct bioH2 production.

The analysis of pathways #1, #2 and #5 considers a system boundary from the *Nannochloropsis sp.* microalgal culture to the final product output (oil, pigments, or $bioH_2$, respectively). For pathways #3 and #4, the $bioH_2$ production from the leftover biomass from SE and SFE respectively was evaluated.

The authors concluded that the oil production pathway by SE shows the lowest energy consumption, 176-244 MJ/MJprod, and CO_2 emissions, 13-15 kg CO_2/MJprod.

However, economically the most favourable biorefinery was the one producing oil, pigments and H_2 via Supercritical Fluid Extraction (SFE).

From the net energy balance and the CO_2 emission analysis, Biorefinery 1 (biodiesel SE + $bioH_2$) presented the better results. Biorefinery 2 (biodiesel SFE + $bioH_2$) showed results in the same range of those in Biorefinery 1. However, the use of SFE produced high-value pigments in addition to the fact that it is a clean technology which does not use toxic organic solvents.

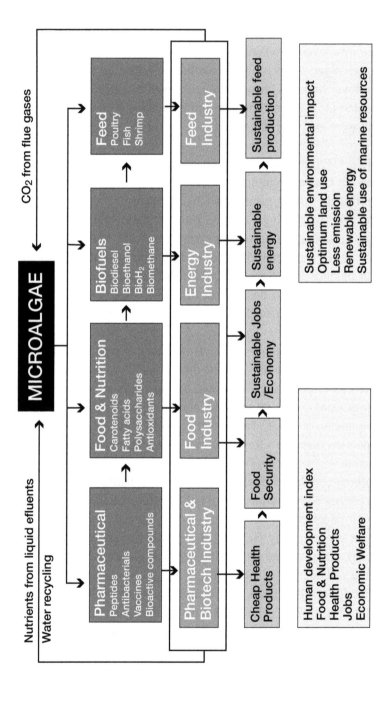

FIGURE 1: Example of a microalgae based biorefinery and how it integrates several related industries (adapted from Subhadra [46]).

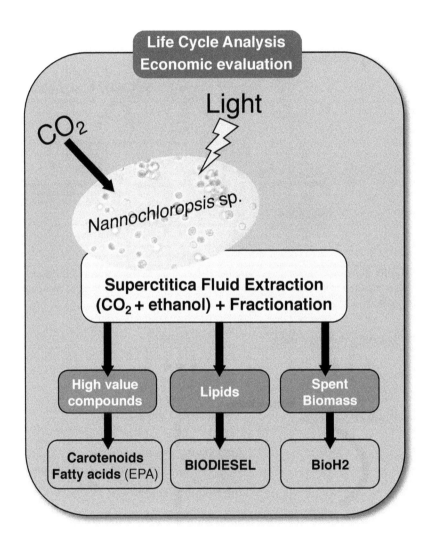

FIGURE 2: Nannochloropsis sp. biorefinery.

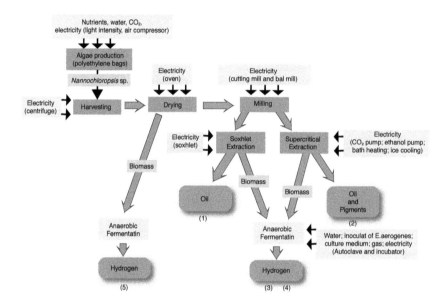

FIGURE 3: *Nannochloropsis sp.* biorefinery (including all steps, material and energy, and different pathw ays) to the production of oil, pigments and bioHydrogen) (adapted from Ferreira, et al. [50]).

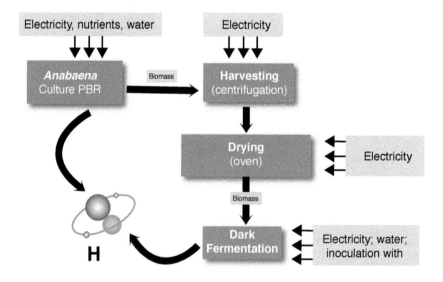

FIGURE 4: *Anabaena* sp. biorefinery (including both H_2 production: autotrophically and by dark fermentation (through Enterobacter aerogenes) (adapted from Ferreira, et al. [32]).

Therefore, Biorefinery 2 was the best in terms of energy/CO_2/ and it being the most economically advantageous solution.

3.2.2 ANABAENA SP. *BIOREFINERY*

The experimental biohydrogen production by photoautrotophic cyanobacterium *Anabaena sp.* was studied by Marques et al. [24]. Hydrogen production from the *Anabaena* biomass leftovers was also achieved by fermentation through the *Enterobacter aerogenes* bacteria and was reported by Ferreira et al. [32] (Figure 4).

Different culture conditions and gas atmospheres were tested in order to maximize the autotrophic bioH_2 yield versus the energy consumption and CO_2 emissions. The authors stated that the best conditions included an Ar+CO_2+20% N_2 gas atmosphere and medium light intensity (384 W) [32]. The yielded H_2 could be increased using the biomass leftovers through a fermentative process; however this would mean higher energy consumption as well as an increase in CO_2 emissions.

3.2.3 CHLORELLA VULGARIS *BIOREFINERIES*

Quite a few reported works describe biorefineries from *Chlorella vulgaris* and these are stated below:

1. Cv1: An integrating process for lipid recovery from the biomass of *Chlorella vulgaris* and methane production from the remaining biomass (after lipid extraction) was worked on by Collet et al. [33]. The authors demonstrated that, in terms of Life Cycle Assessment (LCA), the methane from algae (algal methane) is the worst case, compared to algal biodiesel and diesel, in terms of abiotic depletion, ionizing radiation, human toxicity, and possible global warming. These negative results are mainly due to a strong demand for electricity. For the land use category, algal biodiesel also had a lesser impact compared to algal methane. However, algal methane is a much better option in terms of acidification and eutrophication.

2. Cv2: Another work concerning the simultaneous production of bio-diesel and methane in a biorefinery concept was done by Ehimen et al. [34]. The authors obtained biodiesel from a direct transesterification on the *Chlorella* microalgal biomass, and from the biomass residues they obtained methane through anaerobic digestion. For a temperature of 40°C and a C/N mass ratio of 8.53, a maximum methane concentration of 69% (v/v) with a specific yield of 0.308 m^3 CH_4/kg VS was obtained. However, in this work the biodiesel yield was not reported.

3. Cv3: In another work, the *Chlorella vulgaris* biorefinery approach was studied by Gouveia, et al. [35] and it included a Photosynthetic Algal Microbial Fuel Cell (PAMFC), where the microalga *Chlorella vulgaris* are present in the cathode compartment (Figure 5). The study demonstrated the simultaneous production of bioelectricity and added-value pigments, with possible wastewater treatment. The authors proved that the light intensity increases the PAMFC power and augments the carotenogenesis process in the cathode compartment. The maximum power produced was 62.7 mW/m^2 with a light intensity of 96 $\mu E/(m^2.s)$.

4. Cv4: A bioethanol-biodiesel-microbial fuel cell was reported by Powel and Hill [36] and basically consisted in an integration of photosynthetic *Chlorella vulgaris* (in the cathode) that captured CO_2 emitted by yeast (in the anode) fermenters, creating a microbial fuel cell. The study demonstrated the possibility of electrical power generation and oil for biodiesel, in a bioethanol production facility. The remaining biomass after oil extraction could also be used in animal feed supplement [36].

3.2.4 CHLORELLA PROTOTHECOIDS *BIOREFINERY*

The biorefinery stated by Campenni et al. [37] used *Chlorella protothecoides* as a source of lipids and carotenoids, it was grown autotrophically and with nitrogen deprivation and the addition of a 20 g/l NaCl solution (Figure 6).

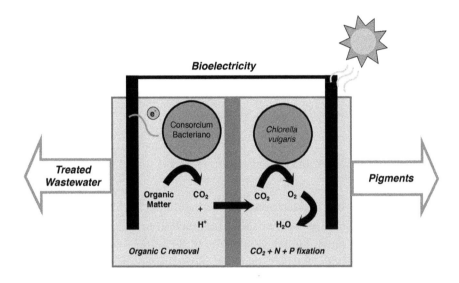

FIGURE 5: *Chlorella vulgaris* biorefinery (Photosynthetic Algal Microbial Fuel Cell) (adapted from da Silva TL, et al. [30]).

The total carotenoid content was 0.8% (w/w) (canthaxanthin (23.3%), echinenone (14.7%), free astaxanthin (7.1%) and lutein/ zeaxanthin (4.1%)) which can be used for food applications. Furthermore, the total lipid content reached 43.4% (w/w), with a fatty acid composition of C18:1 (33.6%), C16:0 (23.3%), C18:2 (11.5%), and C18:3 (less than 12%), which is needed to fulfil the biodiesel EN 14214 quality specifications [38] and can be used for the biofuel (biodiesel) industry.

The leftover biomass is still available for hydrogen or bioethanol production in a biorefinery approach, as the residue still contains sugar taking advantage of all the *C. protothecoids* gross composition.

3.2.5 CHLORELLA REINHARDTII *BIOREFINERY*

The production of biohydrogen and the consequent biogas (methane) production by anaerobic fermentation of the residue of *Chlorella reinhardtii* biomass were achieved by Mussgnug et al. [39]

The authors reported that using the biomass, after the hydrogen production cycle instead of using the fresh biomass, would increase the biogas production by 123%. The authors attributed these results to the storage compounds, such as starch and lipids with a high fermentative potential which is the key in the microalgae-based integrated process and could be used for more value-added applications.

3.2.6 DUNALIELLA SALINA BIOREFINERY

Sialve et al. [40] attested the production of methane from the leftover biomass of *Dunaliella salina* after the oil extraction to make biodiesel. The authors found a much higher yield (around 50%) for a shorter hydraulic retention time (HRT, 18 days), than the corresponding values reported by Collet et al. [33] using the *Chlorella vulgaris* biomass.

3.2.7 DUNALIELLA TERTIOLECTA BIOREFINERY

The chemoenzymatic saccharification and bioethanol fermentation of the residual biomass of *Dunaliella tertiolecta* after lipid extraction (for biodiesel production purposes) were investigated by Kim et al. [41]. The bioethanol was produced from the enzymatic hydrolysates without pretreatment by *S. cerevisiae,* resulting in yields of 0.14 g ethanol/g residual biomass and 0.44 g ethanol/g glucose produced from the residual biomass.

According to these authors, the residual biomass generated during microalgal biodiesel production, could be used for bioethanol production in order to improve the economic feasibility of a microalgaebased integrated process.

3.2.8 ARTHROSPIRA (SPIRULINA) BIOREFINERY

Olguin [42] highlighted that the biorefinery strategy offers new opportunities for a cost-effective and competitive production of biofuels along with nonfuel compounds. The author studied an integrated system where the production of biogas, biodiesel, hydrogen and other valuable products (e.g. PUFAs, phycocyanin, and fish feed) could be possible.

3.2.9 SPIROGYRA SP. BIOREFINERY

Pacheco et al. [43] pointed a biorefinery from *Spirogyra sp.*, a sugar-rich microalga, for $bioH_2$ production as well as pigments (Figure 7). The economic and life cycle analysis of the whole process, allowed the authors to conclude that it is crucial to increase the sugar content of the microalgae to increase the $bioH_2$ yield. Furthermore, it is important to reduce the centrifugation needs and use alternative methods for pigment extraction other than using acetone solvents. The electrocoagulation and solar drying were used for harvesting and dewatering, respectively, and were able to reduce energy requirements by 90%. Overall, centrifugation of the microalgal biomass and heating of the fermentation vessel are still major energy consumers and CO_2 contributors to this process. Pigment production is necessary to improve the economic benefits of the biorefinery, but it is mandatory to reduce its extraction energy requirements that are demanding 62% of the overall energy.

Mostafa et al. [44] evaluated the growth and lipid, glycerol, and carotenoid content of nine microalgae species (green and blue green microalgae) grown in domestic wastewater obtained from the Zenein Wastewater Treatment Plant in the Giza governorate in Egypt (Figure 8). The authors cultivated the different species under different conditions, such as without treatment after sterilization, with nutrients and sterilization, and with nutrients without sterilization, at $25 \pm 1°C$, under continuous shaking (150 rpm) and illumination (2,000 lx), for 15 days. The highest biodiesel production from algal biomass cultivated in wastewater was obtained by Nostoc humifusum (11.80%) when cultivated in wastewater without treatment and the lowest (3.8%) was recorded by *Oscillatoria sp.* when cultivated on the sterilized domestic wastewater. The authors concluded that cultivating microalgae on domestic wastewater, combines nutrient removal and algal lipid production which has a high potential in terms of biodiesel feedstock. This methodology is suitable and non-expensive compared to the conventional cultivation methods for sustainable biodiesel and glycerol.

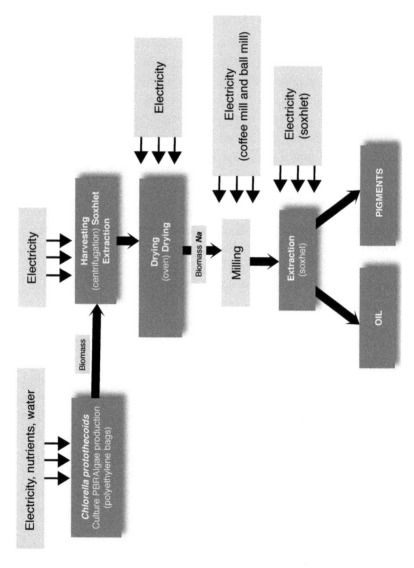

FIGURE 6: Chlorella protothecoids biorefinery (adapted from Campenni', et al. [37]).

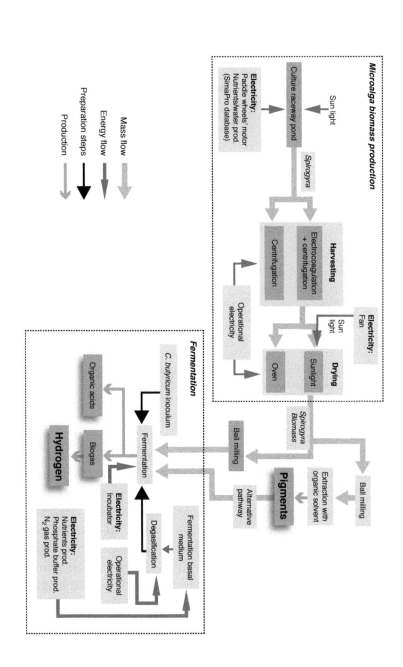

FIGURE 7: Spirogyra sp. biorefinery (adapted from Pacheco, et al. [43]).

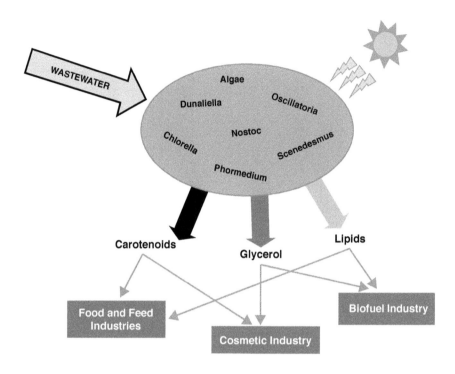

FIGURE 8: Biorefinery from several microalgae using w astew ater w ith lipid, carotenoid and glycerol production (adapted from Mostafa, et al. [44]).

According to Subhadra and Edwards [45] (Figure 9), an integrated Renewable Energy Park (IREP) approach can be envisaged by combining different renewable energy industries, in resource-specific regions, for synergetic electricity and liquid biofuel production, with zero net carbon emissions. Choosing the appropriate location, an IREP design, combining a wind power plant with solar panels and algal growth facilities to harness additional solar energy, could greatly optimize land. Biorefineries configured within these IREPs can produce about 50 million gallons of biofuel per year, providing many other value-added co-products and having almost no environmental impact [46] (Figure 9).

Clarens et al. [47] suggested that the results from algae-to-energy systems can be either net energy positive or negative depending on the specific combination of cultivation and conversion processes used addressed the shortcoming "well-to-wheel", including the conversion of each biomass into transportation energy sources. The algal conversion pathway resulted in a combination of biodiesel and bioelectricity production for transportation, evaluated by Vehicle Kilometers Traveled (VKT) per hectare. In this study, it was assumed that bioelectricity and biodiesel are used in commercially available Battery Electric Vehicles (BEVs) and Internal Combustion Vehicles (ICVs), respectively. The authors depicted four pathways:

1. A: Methane-derived bioelectricity from the bulk algae biomass by anaerobic digestion
2. B: Biodiesel from the algae lipids and methane-derived bioelectricity from the residual biomass by anaerobic digestion
3. C: Biodiesel from the algae lipids and bioelectricity from the residual biomass by direct combustion
4. D: Bioelectricity from the bulk algae biomass by direct combustion

The four pathways follow various nutrient sources (e.g., virgin commercial CO_2, CO_2 from a coal-fired power plant, compressed CO_2 from flue gas, commercial fertilizers, and wastewater supplementation).

The authors found that algae-to-energy systems depend on the combination of cultivation and conversion processes used. They concluded that the conversion pathways involving direct combustion for the production of bioelectricity generally outperformed systems involving anaerobic digestion and biodiesel production. They ranked the four pathways as D>A>C>B in terms of energy return on investment.

The authors found an algae bioelectricity (D) generation of 1,402,689 MJ/km and algae biodiesel + bioelectricity (C) generation of 1,110 MJ/km. These algae-to-energy systems generate 4 and 15 times as VKT per hectare as switch grass or canola, respectively [47].

Subhadra and Edwards [48] analyzed the water footprint of two simulated algal biorefineries for the production of biodiesel, algal meal, and omega-3 fatty acids. The authors highlighted the advantages of multiproducts to attain a high operational profit with a clear return on investment. The energy return of algal biodiesel for different scenarios ranged between 0.016–0.042 MJ.

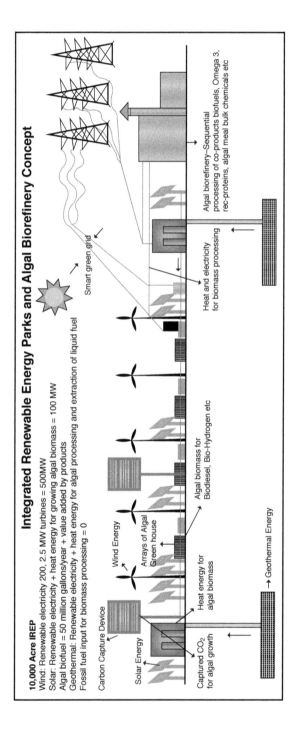

FIGURE 9: Integrated renewable energy and an Algal Biorefinery concept: a framework for the production of biofuel and high-value products with zero net carbon emissions (from Subhadra and Edwards [45]).

Park, et al. [49,50] also studied algae which are grown as a byproduct of High-Rate Algal Ponds (HRAPs) operated for wastewater treatment. In addition to significantly better economics, algal biofuel production from wastewater treatment HRAPs has a much smaller environmental footprint compared to commercial algal production HRAPs which consume freshwater and fertilizers.

3.3 CONCLUSION

Biomass, as a renewable source, is attracting worldwide attention to satisfy the so called bioeconomy demand. Microalgae could be the appropriate feedstock as they did not compete with food and feed production, in terms of either land or water. Furthermore, microalgae remove/recycle nutrients from wastewater and flue-gases providing additional environmental benefits.

Due to their efficient sunlight utilization, microalgae are projected as living-cell factories with simple growth requirements. Their potential for energy and value-added products production is widely recognized. Nevertheless, to be economically sustainable the tiny microalgae should supply a huge biorefinery. Technical advances combined with the several advantages such as CO_2 capture, wastewater bioremediation and the extraction of value added-products will greatly increase algal bioproduct profitability.

The versatility and the huge potential of the tiny microalgae could support a microalgae biorefinery and microalgae-based bioeconomy, opening up a huge increase of opportunities in the global algae business.

REFERENCES

1. Shibata S, Natori Y, Nishihara T, Tomisaka K, Matsumoto K, et al. (2003) Antioxidant and anti-cataract effects of Chlorellaon rats with streptozotocin-induced diabetes. J Nutr Sci Vitaminol 49: 334-339.
2. Shibata S, Sansawa H (2006) Preventive effects of heterotrophically cultured Chlorella regularis on lifestyle-associated diseases. Annu Rep Yakult Central Inst Microbiol Res 26: 63-72.
3. Shibata S, Hayakawa K, Egashira Y, Sanada H (2007) Hypocholesterolemic mechanism of Chlorella: Chlorella and its indigestible fraction enhance hepatic cholesterol

catabolism through up-regulation of cholesterol 7alpha-hydroxylase in rats. Biosci Biotechnol Biochem 71: 916-925.

4. Burja AM, Banaigs B, Abou-Mansour EB, Wright PC (2001) Marine cyanobacteria - a prolific source of natural products. Tetrahedron 57: 9347-9377.

5. Singh S, Kate BN, Banerjee UC (2005) Bioactive compounds from cyanobacteria and microalgae: an overview. Crit Rev Biotechnol 25: 73-95.

6. Yamaguchi K (1996) Recent advances in microalgal bioscience in Japan, with special reference to utilization of biomass and metabolites: A review. J Appl Phycol 8: 487-502.

7. Nuño K, Villarruel-López A, Puebla-Pérez AM, Romero-Velarde E, Puebla-Mora AG, et al. (2013) Effects of the marine microalgae Isochrysis galbanaand Nannochloropsis oculata in diabetic rats. J Funct Foods 5: 106-115.

8. Gerster H (1993) Anticarcinogenic Effect of Common Carotenoids. Int J Vitam Nutr Res 63: 93-121.

9. Willett WC (1994) Micronutrients and Cancer Risk. Am J Clin Nutr 59: 1162-1165.

10. Lupulescu A (1994) The role of vitamin-A, vitamin-beta-carotene, vitamin-E and vitamin-C in cancer cell biology. Int J Vitam Nutr Res 64: 3-14.

11. Tanaka T, Shnimizu M, Moriwaki H (2012) Cancer Chemoprevention by Carotenoids. Molecules 17: 3202-3242.

12. Kohlmeier L, Hastings SB (1995) Epidemiologic Evidence of a Role of Carotenoids in Cardiovascular-Disease Prevention. Am J Clin Nutr 62: 1370S-1376S.

13. Giordano P, Scicchitano P, Locorotondo M, Mandurino C, Ricci G, et al. (2012) Carotenoids and Cardiovascular Risk. Curr Pharm Des 18: 5577-5589.

14. Snodderly DM (1995) Evidence for Protection against Age-Related Macular Degeneration by Carotenoids and Antioxidant Vitamins. Am J Clin Nutr 62: 1448S-1461S.

15. Weikel KA, Chiu CJ, Taylor A (2012) Nutritional modulation of age-related macular degeneration. Mol Aspects Med 33: 318-375.

16. Meydani SN, Wu DY, Santos MS, Hayek MG (1995) Antioxidants and Immune-Response in Aged Persons - Overview of Present Evidence. Am J Clin Nutr 62: 1462S-1476S.

17. Park JS, Chyun JH, Kim YK, Line LL, Chew BP (2010) Astaxanthin decreased oxidative stress and inflammation and enhanced immune response in humans. Nutr Metab (Lond) 7: 18.

18. Matos CT, Gouveia L, Morais AR, Reis A, Bogel-Lukasik R (2013) Green metrics evaluation of isoprene production by microalgae and bacteria. Green Chem 15: 2854-2864.

19. Painter TJ (1993) Carbohydrate polymers in desert reclamation. The potential of microalgal biofertilizers. Carbohyd Polym 20: 77-86.

20. Dey K (2011) Production of Biofertilizer (Anabaena and Nostoc) using CO2. Presentation on Roll: DURJ BT No.2011/2. Regn.No: 660.

21. Sahu D, Priyadarshani I, Rath B (2012) Cyanobacteria - as potential biofertilizer. CIBTech Journal Microbiol 1: 20-26.

22. Gouveia L, Oliveira C (2009) Microalgae as a raw material for biofuels production. J Ind Microbiol Biotechnol 36: 269-274.

23. Miranda JR, Passarinho PC, Gouveia L (2012) Bioethanol production from Scenedesmus obliquussugars: the influence of photobioreactors and culture conditions on biomass production. Appl Microbiol Biotechnol 96: 555-564.

24. Marques AE, Barbosa TA, Jotta J, Tamagnini P, Gouveia L (2011) Biohydrogen production by *Anabaena*sp. PCC 7120 wild-type and mutants under different conditions: Light, Nickel and CO2. Biomass and Bioenergy 35: 4426-4434.

25. Ferreira AF, Ortigueira J, Alves L, Gouveia L, Moura P, et al. (2013a) Energy requirement and CO2 emissions of bioH2 production from microalgal biomass. Biomass & Bioenergy 49: 249-259.

26. Batista AP, Moura P, Marques PASS, Ortigueira J, Alves L, et al. (2014) Scenedesmus obliquusas a feedstock for bio-hydrogen production by Enterobacter aerogenesand Clostridium butyricumby dark fermentation. Fuel 117: 537-543.

27. Wijffels RH, Barbosa MJ, Eppink MHM (2010) Microalgae for bulk chemicals and biofuels. Biofuels Bioprod Bioref 4: 287-295.

28. Vanthoor-Koopmans M, Wijffels RH, Barbosa MJ, Eppink MHM (2013) Biorefinery of microalgae for food and fuel. Bioresour Technol 135: 142-149.

29. Brennan L, Owende P (2010) Biofuels from microalgae - A review of technologies for production, processing, and extractions of biofuels and co-products. Ren Sustain Energy Rev 14: 557-577.

30. da Silva TL, Gouveia L, Reis A (2014) Integrated microbial processes for biofuels and high added value products: the way to improve the cost effectiveness of biofuel production. Appl Microbiol Biotechnol. doi:10.1007/s00253-013-5389-5.

31. Nobre BP, Villalobos F, Barragán BE, Oliveira AC, Batista AP, et al. (2013) A biorefinery from Nannochloropsissp. microalga - Extraction of oils and pigments. Production of biohydrogen from the leftover biomass. Bioresour Technol 135: 128-136.

32. Ferreira AF, Marques AC, Batista AP, Marques PASS, Gouveia L, et al. (2012) Biological hydrogen production by *Anabaena*sp. - yield, energy and CO2 analysis including fermentative biomass recovery. Int J Hydrogen Energ 37: 179-190.

33. Collet P, Hélias A, Lardon L, Ras M, Goy RA, et al. (2011) Lifecycle assessment of microalgae culture coupled to biogas production. Bioresour Technol 102: 207-214.

34. Ehimen EA, Sun ZF, Carrington CG, Birch EJ, Eaton-Rye JJ (2011) Anaerobic digestion of microalgae residues resulting from the biodiesel production process. Appl Energ 88: 3454-3463.

35. Gouveia L, Neves C, Sebastião D, Nobre BP, Matos CT (2014) Effect of light on the production of bioelectricity and pigments by a Photosynthetic Alga Microbial Fuel Cell. Bioresour Technol 154: 171-177.

36. Powel EE, Hill GA (2009) Economic assessment of an integrated bioethanolbiodiesel-microbial fuel cell facility utilizing yeast and photosynthetic algae. Chem Eng Res Design 87: 1340-1348.

37. Campenni' L, Nobre BP, Santos CA, Oliveira AC, Aires-Barros AR, et al. (2013) Carotenoids and lipids production of autotrophic microalga *Chlorella protothecoides* under nutritional, salinity and luminosity stress conditions. Appl Microbiol Biotechnol 97: 1383-1393.

38. EN 14214 (2008) Automotive fuels—fatty acid methyl esters (FAME) for diesel engines—requirements and test methods.

39. Mussgnug JH, Klassen V, Schlüter A, Kruse O (2010) Microalgae as substrates for fermentative biogas production in a combined biorefinery concept. J Biotechnol 150: 51-56.

40. Sialve B, Bernet N, Bernard O (2009) Anaerobic digestion of microalgae as a necessary step to make microalgal biodiesel sustainable. Biotechnol Adv 27: 409-416.

41. Kim AL, Lee OK, Seong DH, Lee GG, Jung YT, et al. (2013) Chemoenzymatic saccharification and bioethanol fermentation of lipid-extracted residual biomass of the microalga Dunaliella tertiolecta. Bioresour Technol 132: 197-201.

42. Olguín EJ (2012) Dual purpose microalgae-bacteria-based systems that treat wastewater and produce biodiesel and chemical products within a biorefinery. Biotechnol Adv 30: 1031-1046.

43. Pacheco R, Ferreira AF, Pinto T, Nobre BP, Loureiro D, et al. (2014) Life Cycle Assessment of a *Spirogyra sp.* biorefinery for the production of pigments, hydrogen and leftovers energy valorisation. Applied Energy.

44. Mostafa SSM, Shalaby EA, Mahmoud GI (2012) Cultivating microalgae in domestic wastewater for biodiesel production. Nat Sci Biol 4: 56-65.

45. Subhadra B, Edwards M (2010) Algal biofuel production using integrated renewable energy park approach in United States. Energ Policy 38: 4897-4902.

46. Subhadra BG (2010) Sustainability of algal biofuel production using integrated renewable energy park (IREP) and algal biorefinery approach. Energy Policy 38: 5892-5901.

47. Clarens AF, Nassau H, Resurreccion EP, White MA Colosi LM (2011) Environmental Impacts of Algae-Derived Biodiesel and Bioelectricity for Transportation. Environ Sci Technol 45: 7554-7560.

48. Subhadra B, Edwards B (2011) Coproduct market analysis and water footprint of simulated commercial algal biorefineries. Appl Energ 88: 3515-3523.

49. Park JBK, Craggs RJ, Shilton AN (2011) Wastewater treatment high rate algal ponds for biofuel production. Bioresour Technol 102: 35-42.

50. Ferreira AF, Ribeiro L, Batista AP, Marques PASS, Nobre BP, et al. (2013b) A Biorefinery from Nannochloropsissp. microalga - Energy and CO_2 emission and economic analyses. Bioresour Technol 138: 235-244

CHAPTER 4

YEAST BIOTECHNOLOGY: TEACHING THE OLD DOG NEW TRICKS

DIETHARD MATTANOVICH, MICHAEL SAUER, AND BRIGITTE GASSER

4.1 BACKGROUND

The use of yeast for food processing and fermentation of alcoholic beverages is traditionally marked as the primary inventive step of biotechnology, dating back several millennia. The discovery of microbial metabolic activities from the 19th century onward initiated the target-oriented development of yeast bioprocesses which were the prototype of modern biotechnological processes. A major hallmark was the development of the "Zulaufverfahren" for efficient baker's yeast production [1]. Today this process is employed under the term "fed-batch" to avoid overflow metabolism in the majority of industrial bioproductions. Efficient fermentative processes developed for ethanol production served as a model for the production of different metabolic products from yeasts, filamentous fungi and bacteria.

Bacteria and filamentous fungi have taken over the lead role in the development of bioprocesses around mid of the 20th century [2]. However since then novel developments of recombinant protein production, metabolic engineering, and systems and synthetic biology, paired by the demand

Yeast Biotechnology: Teaching the Old Dog New Tricks. © Mattanovich D, Sauer M, and Gasser B. Microbial Cell Factories 13,34 (2014), doi:10.1186/1475-2859-13-34. Licensed under Creative Commons Attribution 2.0 Generic License, http://creativecommons.org/licenses/by/2.0/.

for many products which can be synthesized by yeasts enable a plethora of new applications of yeasts in biotechnology. We see three major fields of application for yeasts in modern biotechnology: production of metabolites, production of recombinant proteins, and in vivo biotransformations.

Traditionally "yeast" denotes *Saccharomyces cerevisiae* and its close relatives, used for alcoholic fermentation and baking. However today about 1500 yeast species have been identified (a variable number due to current reclassifications). Biotechnologists have summarized all non-*S. cerevisiae* yeasts which they use as "non-conventional" yeasts. What unifies them is a lower degree of fermentative overflow metabolism [3] and a rather short history of genetic and biological characterization. The lifestyle of *S. cerevisiae* is characterized by flourishing in extremely high sugar concentrations—disposing most of it as the fermentative by-product ethanol. Most natural habitats however do not provide such extreme substrate conditions so that most non-conventional yeasts provide alternative metabolic routes for substrate utilization and product formation, and different regulatory patterns. A few species of major interest are *Pichia pastoris* (syn. *Komagataella pastoris*), *Hansenula polymorpha* (syn. *Ogataea parapolymorpha*), *Yarrowia lipolytica*, *Pichia stipitis* (syn. *Scheffersomyces stipitis*), or *Kluyveromyces marxianus*.

The classical carbon substrates for yeast processes are glucose or sucrose, derived mainly from corn starch and cane sugar. Extrapolating the successful expansion of industrial biotechnology, and most importantly considering the food requirements of mankind lets us envisage a shortage of these classical substrates, driving research towards the utilization of alternative carbon sources. Lignocellulose hydrolysate constitutes such an abundant carbon source, requiring yeasts that can utilize xylose and arabinose (the major constituents of hemicellulose). These are either natural pentose assimilating yeasts like e.g. *Pichia stipitis*[4] or *Hansenula polymorpha*[5], or *S. cerevisiae* strains with engineered pentose utilization pathways [6]. Alternatively, glycerol as an abundant by-product of biodiesel production is explored as a substrate for yeast processes. While also *S. cerevisiae* can utilize glycerol, the uptake and assimilation is much higher in other yeasts like *Pachysolen tannophilus*[7], *Y. lipolytica*[8,9], or *P. pastoris*[10].

Different substrates and products of yeast biotechnology are summarized in Figure 1. In the following, the main current applications of yeasts are discussed.

4.2 YEASTS AS PLATFORMS FOR METABOLITE PRODUCTION

Due to the foreseeable limitation of mineral oil resources the interest in biotechnological production of chemicals by microbial metabolic activities is ever increasing. While the first wave of metabolite production processes used natural producers of desired molecules (e.g. production of astaxanthin by *Phaffia rhodozyma* (syn. *Xanthophyllomyces dendrorhous*) or production of riboflavin by *Pichia guilliermondii*), current concepts aim at engineering a few platform strains for the production of many chemicals. Several physiological features predestine yeasts as such a platform [11]: high substrate uptake rates, potentially high metabolic rates, robustness against stressful process conditions. Additionally, there is also a revival in engineering of natural production hosts for improved productivity as the increasing availability of yeast genome sequences enables better understanding and relief of rate-limiting steps based on the results of Omics data and metabolic modelling.

Ethanol made by yeast is by far the largest biotech product [12]. The development of more efficient and robust strains using different substrates has been a major driving force for the development of the yeast platform. However, ethanol as a biofuel suffers from a low substrate yield and a rather low energy content, so that the production of other alcohols like butanol or isobutanol is attempted today [13-16].

Several short chain organic acids are valued precursor chemicals. Production of the free acid requires low pH of the culture broth, so that the acid tolerance of yeasts is a valuable feature. Low pH lactic acid production has been achieved in *S. cerevisiae*[17], *P. stipitis*[18], *Candida boidinii*[19] and *Candida sorenensis*[20] and is reaching industrial scale. Succinic acid production with engineered *S. cerevisiae* is employed in a process announced to reach 30,000 t/y scale in 2015 [21].

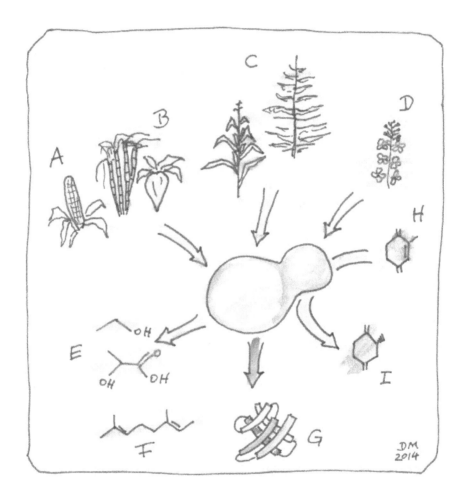

FIGURE 1: Substrates and products of yeast bioprocesses. Main carbon sources employed in yeast bioprocesses are derived from (A) corn starch, (B) cane or beet sugar, (C) lignocellulose (corn stover, straw, wood etc.) and (D) crude glycerol from biodiesel production. Different native and engineered yeast strains convert the substrate to products of (E) primary or (F) secondary metabolism, or (G) recombinant proteins. Whole cell biocatalysis is a special case where (H) a complex substrate is biochemically transformed to (I) a product by the metabolic activity of yeast cells. Chemical structures are illustrative images only.

Efficient pathways for production of phenolic substances such as flavonoids [22] and stilbenoids [23] have been developed but still need further increase of productivity and yield.

Isoprenoids are a universal class of molecules all based on the same building blocks. This universality enables to design novel pathways in yeasts, using the native core structures with specific conversions carried out by heterologous pathways. Thereby, often recombinant genes from different species are combined to obtain the intended variety. Isoprenoids encompass more than 40,000 plant secondary metabolites, a number of them with pharmaceutical activity. Recently yeast based production of the antimalaria agent artemisinin reached commercial production [24]. Isoprenoids produced with recombinant yeast have also been proposed as biobased jet fuel [25].

Polyketides are complex biomolecules mainly of bacterial or fungal origin. Recombinant expression of polyketide synthases in yeasts enables the study of their complex function [26] and the development of heterologous production strains [27,28]. The heterologous production of synthetic penicillins in yeasts has been suggested as well [29].

4.3 RECOMBINANT PROTEIN PRODUCTION IN YEASTS

S. cerevisiae has been the first yeast employed for production of heterologous proteins [30]. In the early 1980s this was the only yeast species with significant molecular genetic characterization which explains its wide commercial use in the following years for production of human insulin and hepatitis B surface antigen. It has turned out however that other yeasts are more efficient in the production of many recombinant proteins [31-33]. A current literature survey indicates that most work on recombinant protein production in yeasts is performed with *P. pastoris* and *H. polymorpha,* followed by *S. cerevisiae* and *Y. lipolytica*. In 2009, about 20% of the biopharmaceutical products approved in U.S.A. and Europe were produced in *S. cerevisiae*[34]. Other yeast platforms play an important role in clinical studies and have begun to enter the biopharma market in the recent years.

Secretion of recombinant proteins to the culture supernatant constitutes a major bottleneck of yeast production hosts [35], favouring some
non-conventional yeasts over *S. cerevisiae*[32,33]. A genomic comparison
of the secretory pathway of 8 yeast species indicates that *S. cerevisiae*
and its close relatives have lost some functions of secretory protein quality control [36]. Engineering of folding and secretion related genes is a
valuable strategy to enhance the secretory capacity of yeasts [35,37-39],
however the comparison to mammalian cells like Chinese hamster ovary
cells shows that there is still a lot of room for improvement [40].

Systems biology has strongly contributed to our current understanding
of limitations of protein production [41]. Genome scale transcriptomics
and proteomics revealed physiological reactions to protein overproduction
[42]. Overexpression has a severe impact on primary metabolism reflect-
ing a higher demand for energy and reducing equivalents [43,44] and free
amino acids [45,46]. Metabolic engineering may further channel the flux
towards required precursers (own unpublished data), and may also contribute to enhanced protein secretion by providing sufficient cofactors, e.g.
heme [47].

4.4 WHOLE CELL BIOCATALYSIS

To differentiate from microbial metabolite production, whole cell biocatalysis may be defined as the conversion of organic compounds by enzymatic activities of life cells. The main advantages compared to classical
biocatalysis are the cheap production of the required enzymatic setup, and/
or the use of the cellular metabolism for cofactor regeneration. Recombinant yeast whole cell biocatalysts have been developed for the conversion
of cephalosporins [48] and steroids [49], and the asymmetric reduction of
α-keto esters [50]. Whole cell biocatalysts usually exert their activity intra-
cellularly. However also secreted enzymes may act in in vivo biotransformations. E.g. D-tagatose has been produced from D-lactose by secretory
production of bacterial β-D-galactosidase and L-arabinose isomerase [51].
Yeast surface display of lipase enabled the whole cell based production of
phospholipids and fatty acid methyl esters [52].

4.5 CONCLUSIONS

Recent research has generated exciting new developments of products and bioprocesses using yeasts. Common patterns of successful yeast based processes are the efficient use of substrate, and a closed energy and redox balance where metabolic engineering may serve to meet the extra demand of product formation. Engineering the protein secretory pathway solves specific problems in overproduction of recombinant proteins.

To provide a forum for scientific discourse, Microbial Cell Factories has initiated a thematic series on Yeast Biotechnology [53]. This virtual series will continue to compile the most relevant papers in yeast research published in Microbial Cell Factories, to serve research in this field for the benefit of mankind.

REFERENCES

1.   Internatl. Yeast Co. Ltd: Verfahren zur Herstellung von Hefe nach dem Zulaufver-fahren. 1933. German Patent DE583760
2.   Porro D, Gasser B, Fossati T, Maurer M, Branduardi P, Sauer M, Mattanovich D: Production of recombinant proteins and metabolites in yeasts: when are these systems better than bacterial production systems? Appl Microbiol Biotechnol 2011, 89:939-948.
3.   Rozpędowska E, Hellborg L, Ishchuk OP, Orhan F, Galafassi S, Merico A, Wool-fit M, Compagno C, Piskur J: Parallel evolution of the make-accumulate-consume strategy in Saccharomyces and Dekkera yeasts. Nat Commun 2011, 2:302.
4.   Toivola A, Yarrow D, van den Bosch E, van Dijken JP, Scheffers WA: Alcoholic fermentation of d-Xylose by yeasts. Appl Environ Microbiol 1984, 47:1221-1223.
5.   Ryabova OB, Chmil OM, Sibirny AA: Xylose and cellobiose fermentation to ethanol by the thermotolerant methylotrophic yeast Hansenula polymorpha. FEMS Yeast Res 2003, 4:157-164.
6.   Walfridsson M, Hallborn J, Penttilä M, Keränen S, Hahn-Hägerdal B: Xylose-metabolizing Saccharomyces cerevisiae strains overexpressing the TKL1 and TAL1 genes encoding the pentose phosphate pathway enzymes transketolase and transaldolase. Appl Environ Microbiol 1995, 61:4184-4190.
7.   Liu X, Mortensen UH, Workman M: Expression and functional studies of genes involved in transport and metabolism of glycerol in Pachysolen tannophilus. Microb Cell Fact 2013, 12:27. L
8.   Workman M, Holt P, Thykaer J: Comparing cellular performance of Yarrowia lipolytica during growth on glucose and glycerol in submerged cultivations. AMB Express 2013, 3:58. L

9. Celińska E, Grajek W: A novel multigene expression construct for modification of glycerol metabolism in Yarrowia lipolytica. Microb Cell Fact 2013, 12:102. L

10. Mattanovich D, Graf A, Stadlmann J, Dragosits M, Redl A, Maurer M, Kleinheinz M, Sauer M, Altmann F, Gasser B: Genome, secretome and glucose transport highlight unique features of the protein production host Pichia pastoris. Microb Cell Fact 2009, 8:29. L

11. Nielsen J, Larsson C, van Maris A, Pronk J: Metabolic engineering of yeast for production of fuels and chemicals. Curr Opin Biotechnol 2013, 24:398-404.

12. Mussatto SI, Dragone G, Guimarães PM, Silva JP, Carneiro LM, Roberto IC, Vicente A, Domingues L, Teixeira JA: Technological trends, global market, and challenges of bio-ethanol production. Biotechnol Adv 2010, 28:817-830.

13. Steen EJ, Chan R, Prasad N, Myers S, Petzold CJ, Redding A, Ouellet M, Keasling JD: Metabolic engineering of Saccharomyces cerevisiae for the production of n-butanol. Microb Cell Fact 2008, 7:36. L

14. Lan EI, Liao JC: Microbial synthesis of n-butanol, isobutanol, and other higher alcohols from diverse resources. Bioresour Technol 2013, 135:339-349.

15. Branduardi P, Longo V, Berterame NM, Rossi G, Porro D: A novel pathway to produce butanol and isobutanol in Saccharomyces cerevisiae. Biotechnol Biofuels 2013, 6:68. L

16. Matsuda F, Ishii J, Kondo T, Ida K, Tezuka H, Kondo A: Increased isobutanol production in Saccharomyces cerevisiae by eliminating competing pathways and resolving cofactor imbalance. Microb Cell Fact 2013, 12:119. L

17. Porro D, Bianchi MM, Brambilla L, Menghini R, Bolzani D, Carrera V, Lievense J, Liu CL, Ranzi BM, Frontali L, Alberghina L: Replacement of a metabolic pathway for large-scale production of lactic acid from engineered yeasts. Appl Environ Microbiol 1999, 65:4211-4215.

18. Ilmén M, Koivuranta K, Ruohonen L, Suominen P, Penttilä M: Efficient production of L-lactic acid from xylose by Pichia stipitis. Appl Environ Microbiol 2007, 73:117-123.

19. Osawa F, Fujii T, Nishida T, Tada N, Ohnishi T, Kobayashi O, Komeda T, Yoshida S: Efficient production of L-lactic acid by Crabtree-negative yeast Candida boidinii. Yeast 2009, 26:485-496.

20. Ilmén M, Koivuranta K, Ruohonen L, Rajgarhia V, Suominen P, Penttilä M: Production of L-lactic acid by the yeast Candida sonorensis expressing heterologous bacterial and fungal lactate dehydrogenases. Microb Cell Fact 2013, 12:53. L

21. Bio-Based News http://bio-based.eu/news/commercial-production-worlds-largest-scale-bio-succinic-acid-plant-2015 webcite

22. Koopman F, Beekwilder J, Crimi B, van Houwelingen A, Hall RD, Bosch D, van Maris AJ, Pronk JT, Daran JM: De novo production of the flavonoid naringenin in engineered Saccharomyces cerevisiae. Microb Cell Fact 2012, 11:155. L

23. Wang Y, Halls C, Zhang J, Matsuno M, Zhang Y, Yu O: Stepwise increase of resveratrol biosynthesis in yeast Saccharomyces cerevisiae by metabolic engineering. Metab Eng 2011, 13:455-463.

24. Paddon CJ, Westfall PJ, Pitera DJ, Benjamin K, Fisher K, McPhee D, Leavell MD, Tai A, Main A, Eng D, Polichuk DR, Teoh KH, Reed DW, Treynor T, Lenihan J, Fleck M, Bajad S, Dang G, Dengrove D, Diola D, Dorin G, Ellens KW, Fickes S,

Galazzo J, Gaucher SP, Geistlinger T, Henry R, Hepp M, Horning T, Iqbal T, Jiang H, Kizer L, Lieu B, Melis D, Moss N, Regentin R, Secrest S, Tsuruta H, Vazquez R, Westblade LF, Xu L, Yu M, Zhang Y, Zhao L, Lievense J, Covello PS, Keasling JD, Reiling KK, Renninger NS, Newman JD: High-level semi-synthetic production of the potent antimalarial artemisinin. Nature 2013, 496:528-532.

25. Peralta-Yahya PP, Ouellet M, Chan R, Mukhopadhyay A, Keasling JD, Lee TS: Identification and microbial production of a terpene-based advanced biofuel. Nat Commun 2011, 2:483.

26. Ishiuchi K, Nakazawa T, Ookuma T, Sugimoto S, Sato M, Tsunematsu Y, Ishikawa N, Noguchi H, Hotta K, Moriya H, Watanabe K: Establishing a new methodology for genome mining and biosynthesis of polyketides and peptides through yeast molecular genetics. Chembiochem 2012, 13:846-854.

27. Gao L, Cai M, Shen W, Xiao S, Zhou X, Zhang Y: Engineered fungal polyketide biosynthesis in Pichia pastoris: a potential excellent host for polyketide production. Microb Cell Fact 2013, 12:77. L

28. Rugbjerg P, Naesby M, Mortensen UH, Frandsen RJ: Reconstruction of the biosynthetic pathway for the core fungal polyketide scaffold rubrofusarin in *Saccharomyces cerevisiae*. Microb Cell Fact 2013, 12:31. L

29. Gidijala L, Kiel JA, Douma RD, Seifar RM, van Gulik WM, Bovenberg RA, Veenhuis M, van der Klei IJ: An engineered yeast efficiently secreting penicillin. PLoS One 2009, 4:e8317.

30. Hitzeman RA, Hagie FE, Levine HL, Goeddel DV, Ammerer G, Hall BD: Expression of a human gene for interferon in yeast. Nature 1981, 293:717-722.

31. Mattanovich D, Branduardi P, Dato L, Gasser B, Sauer M, Porro D: Recombinant protein production in yeasts. Methods Mol Biol 2012, 824:329-358.

32. Dragosits M, Frascotti G, Bernard-Granger L, Vázquez F, Giuliani M, Baumann K, Rodríguez-Carmona E, Tokkanen J, Parrilli E, Wiebe MG, Kunert R, Maurer M, Gasser B, Sauer M, Branduardi P, Pakula T, Saloheimo M, Penttilä M, Ferrer P, Luisa Tutino M, Villaverde A, Porro D, Mattanovich D: Influence of growth temperature on the production of antibody Fab fragments in different microbes: a host comparative analysis. Biotechnol Prog. 2011, 27:38-46.

33. Mack M, Wannemacher M, Hobl B, Pietschmann P, Hock B: Comparison of two expression platforms in respect to protein yield and quality: Pichia pastoris versus Pichia angusta. Protein Expr Purif 2009, 66:165-171.

34. Ferrer-Miralles N, Domingo-Espín J, Corchero JL, Vázquez E, Villaverde A: Microbial factories for recombinant pharmaceuticals. Microb Cell Fact 2009, 8:17. L

35. Delic M, Göngrich R, Mattanovich D, Gasser B: Engineering of protein folding and secretion - strategies to overcome bottlenecks for efficient production of recombinant proteins. Antioxid Redox Signal 2014. [Epub ahead of print] doi:10.1089/ars.2014.5844

36. Delic M, Valli M, Graf AB, Pfeffer M, Mattanovich D, Gasser B: The secretory pathway: exploring yeast diversity. FEMS Microbiol Rev 2013, 37:872-914.

37. Idiris A, Tohda H, Kumagai H, Takegawa K: Engineering of protein secretion in yeast: strategies and impact on protein production. Appl Microbiol Biotechnol 2010, 86:403-17.

38. Damasceno LM, Huang CJ, Batt CA: Protein secretion in Pichia pastoris and advances in protein production. Appl Microbiol Biotechnol 2012, 93:31-9.

39. Hou J, Tyo KE, Liu Z, Petranovic D, Nielsen J: Metabolic engineering of recombinant protein secretion by *Saccharomyces cerevisiae*. FEMS Yeast Res 2012, 12:491-510.

40. Maccani A, Landes N, Stadlmayr G, Maresch D, Leitner C, Maurer M, Gasser B, Ernst W, Kunert R, Mattanovich D: Pichia pastoris secretes recombinant proteins less efficiently than Chinese hamster ovary cells but allows higher space-time yields for less complex proteins. Biotechnol J 2014. [Epub ahead of print] doi: 10.1002/biot.201300305

41. Graf A, Dragosits M, Gasser B, Mattanovich D: Yeast systems biotechnology for the production of heterologous proteins. FEMS Yeast Res 2009, 9:335-48.

42. Vanz AL, Lünsdorf H, Adnan A, Nimtz M, Gurramkonda C, Khanna N, Rinas U: Physiological response of Pichia pastoris GS115 to methanol-induced high level production of the Hepatitis B surface antigen: catabolic adaptation, stress responses, and autophagic processes. Microb Cell Fact 2012, 11:103. L

43. Jordà J, Jouhten P, Cámara E, Maaheimo H, Albiol J, Ferrer P: Metabolic flux profiling of recombinant protein secreting Pichia pastoris growing on glucose:methanol mixtures. Microb Cell Fact 2012, 11:57. L

44. Klein T, Lange S, Wilhelm N, Bureik M, Yang TH, Heinzle E, Schneider K: Overcoming the metabolic burden of protein secretion in Schizosaccharomyces pombe - A quantitative approach using (13)C-based metabolic flux analysis. Metab Eng 2014, 21:34-45.

45. Carnicer M, Ten Pierick A, van Dam J, Heijnen JJ, Albiol J, van Gulik W, Ferrer P: Quantitative metabolomics analysis of amino acid metabolism in recombinant Pichia pastoris under different oxygen availability conditions. Microb Cell Fact 2012, 11:83.

46. Heyland J, Fu J, Blank LM, Schmid A: Carbon metabolism limits recombinant protein production in Pichia pastoris. Biotechnol Bioeng 2011, 108:1942-53.

47. Liu L, Martínez JL, Liu Z, Petranovic D, Nielsen J: Balanced globin protein expression and heme biosynthesis improve production of human hemoglobin in *Saccharomyces cerevisiae*. Metab Eng 2014, 21:9-16.

48. Abad S, Nahalka J, Bergler G, Arnold SA, Speight R, Fotheringham I, Nidetzky B, Glieder A: Stepwise engineering of a Pichia pastoris D-amino acid oxidase whole cell catalyst. Microb Cell Fact 2010, 9:24. L

49. Braun A, Geier M, Bühler B, Schmid A, Mauersberger S, Glieder A: Steroid biotransformations in biphasic systems with Yarrowia lipolytica expressing human liver cytochrome P450 genes. Microb Cell Fact 2012, 11:106. L

50. Kratzer R, Egger S, Nidetzky B: Integration of enzyme, strain and reaction engineering to overcome limitations of baker's yeast in the asymmetric reduction of alpha-keto esters. Biotechnol Bioeng 2008, 101:1094-101.

51. Wanarska M, Kur J: A method for the production of D-tagatose using a recombinant Pichia pastoris strain secreting β-D-galactosidase from Arthrobacter chlorophenolicus and a recombinant L-arabinose isomerase from Arthrobacter sp. 22c. Microb Cell Fact 2012, 11:113. L

52. Hama S, Yoshida A, Nakashima K, Noda H, Fukuda H, Kondo A: Surfactant-modified yeast whole-cell biocatalyst displaying lipase on cell surface for enzymatic production of structured lipids in organic media. Appl Microbiol Biotechnol 2010, 87:537-43.

53. Yeast biotechnology http://www.microbialcellfactories.com/series/Yeast Biotechnology.

CHAPTER 5

MICROBIAL COMMUNITY STRUCTURES DIFFERENTIATED IN A SINGLE-CHAMBER AIR-CATHODE MICROBIAL FUEL CELL FUELED WITH RICE STRAW HYDROLYSATE

ZEJIE WANG, TAEKWON LEE, BONGSU LIM, CHANSOO CHOI, AND JOONHONG PARK

5.1 BACKGROUND

Microbial fuel cells (MFCs) are devices to produce electric energy from organic matters and treat wastewaters in both anode and cathode chambers [1,2]. Pure organic compound, real wastewater, and biomass have been successfully used as fuel for power generation in MFCs [3]. Rice straw is one of the most abundant biomasses, mainly composed of cellulose, hemicellulose, and some lignin [4]. The hemicellulose can be easily degraded to its constituent sugars through acidic and/or enzymatic hydrolysis; the produced sugars can be further used as substrate to produce organic acids or bioethanol [5,6]. Therefore, the sugars produced from the rice straw hydrolysate might be used as a useful fuel for power generation from MFCs.

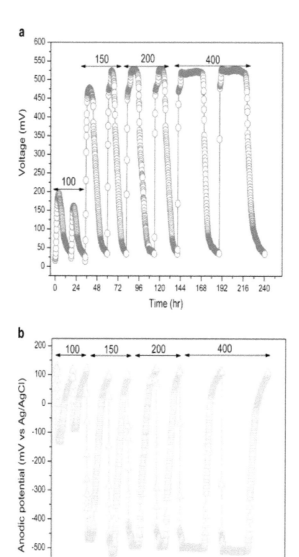

FIGURE 1: Voltage response of the microbial fuel cell (MFC) to chemical oxygen demand (COD) under batch-mode operation. (a) Voltage output and (b) anodic potential as a function of COD concentration at a fixed external resistance of 500 Ω. Figures on the plot represent the COD concentrations (mg/L).

In MFCs, microbes play crucial roles in energy output and organic contaminants removal [7]. The ability of microbes to transfer electrons in the anode can significantly affect the performance of MFCs. Anodic microbial communities were reported to be significantly related with the types of substrates fed into the anode chamber [8]. For example, *Acetobacterium* species (sp.), *Geobacter sp.*, and *Arcobacter sp.* were detected in the anodic biofilm fed with formate [9]; however, *Enterobacter sp.* was the dominant bacterial species in the MFC with glucose as substrate [10]. For air-cathode MFCs, biofilm was commonly formed on the water-facing side of the cathode. It was discovered that the formation of biofilm on the Pt-loaded air-cathode could decrease the power output due to the increased cathodic resistance and limited proton transfer rate [11]; however, recent research demonstrated that the biofilm formation on a bare air-cathode could enhance the electric power output from air-cathode MFCs [12]. The different research conclusions might be caused by different air-cathode configurations. Moreover, the cathodic biofilm in a Pt-loaded air-cathode was observed to be capable of removing nitrogen, with enhanced removal efficiency due to the pre-accumulation of nitrifying biofilm [13]. The aforementioned results indicate that the cathodic biofilm deserves further research.

Therefore, the purpose of the present study was to evaluate the availability of diluted acid-treated rice straw hydrolystate as fuel for an air-cathode MFC. In addition, microbial analysis at high resolution level using 454 pyrosequencing was carried out to evaluate the effect of the rice straw hydrolystate and niches on the microbial diversity and community.

5.2 RESULTS AND DISCUSSION

5.2.1 PERFORMANCE OF THE MFC

After addition of the rice straw hydrolysate as an anodic solution, cell voltage was immediately increased with no lag time. Stable voltage increased from 177.6 ± 17.3 mV for chemical oxygen demand (COD) of 100 mg/L to 524.7 ± 3.2 mV for COD of 400 mg/L, in response to the decrease in anodic potential from -110.5 ± 21.6 mV to -508.7 ± 6.9 mV (Figure 1a and b). The results indicated that organic matters produced from the hydrolysate could be easily utilized by anodic microorganism and release electrons, decreasing the anodic potential and consequently increasing the cell voltage [14]. The stable anodic potential was appropriately −300 mV (versus standard hydrogen electrode), similar to that of −340 mV observed by Wang et al. [15]. For relatively low COD concentrations (100 and 150 mg/L), the cell voltage showed a large difference between two separate cycles; while the COD concentration increased to 200 mg/L, the voltage was well reproduced, suggesting that the saturated COD for voltage output was 200 mg/L. The saturated COD concentration in the present study was far lower than that of the 1,000 mg/L observed when wheat straw hydrolysate was used as fuel in a dual-chamber MFC [16]. A lower saturated COD concentration indicated higher capacity of power production from the fuel based on the same quantity. Furthermore, the discharge time increased from 17.1 ± 1.2 hours to 49.8 ± 2.4 hours relying on the increased COD concentration.

Coulombic efficiency (CE) indicates the ratio of total electrons recovered as electric current from organic matter. In this study, the CE was calculated in the range from 8.5% to 17.9%, and the COD removal efficiency ranged from $49.4 \pm 4.0\%$ to $72.0 \pm 1.7\%$ (Figure 2). The present CE was lower than that of wheat straw- [16] or corn stover biomass-fuelled [17] MFCs, which were 15.5% to 37.1% and 19.3% to 25.6%, respectively; whereas it was higher than that previously reported using real wastewater as fuels, which for example, is less than 1% for fermented wastewater [18], and a maximum CE of 8% for starch processing wastewater [19]. For air-cathode MFCs, oxygen diffused to the anode chamber can aerobically consume substrates other than anaerobically generated electrons, which would be the reason for the low CE of air-cathode MFCs [20].

The maximum power density (P_{max}) was determined as 137.6 ± 15.5 mW/m^2 for COD of 400 mg/L, at a current density of 0.28 A/m^2 (Figure 3a). It was further promoted to 293.3 ± 7.9 mW/m^2 at a current density of 0.90 A/m^2 when the solution conductivity was adjusted from 5.6 mS/cm to 17 mS/cm through addition of NaCl. Consequently, the internal resistance (R_{int}) of the MFC was decreased from $714.4 \pm 19.1\Omega$ to $229.9 \pm 14.5\Omega$. Moreover, conductivity adjustment enhanced the anodic performance, relieving the mass transfer limitation while limiting the cathode performance (Figure 3b).

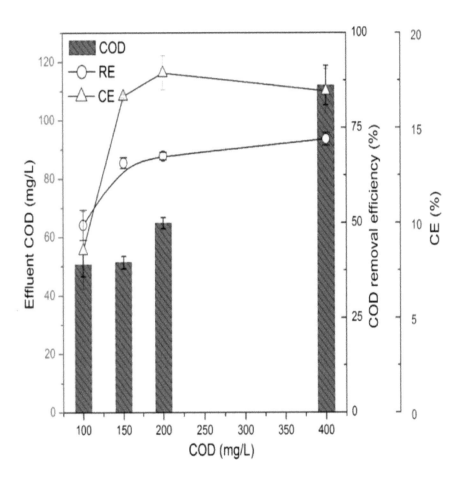

FIGURE 2: Effluent chemical oxygen demand (COD) concentration and COD removal efficiency (RE) and coulombic efficiency (CE). The initial COD concentration was 100, 150, 200, and 400 mg/L with an external resistance of 500 Ω.

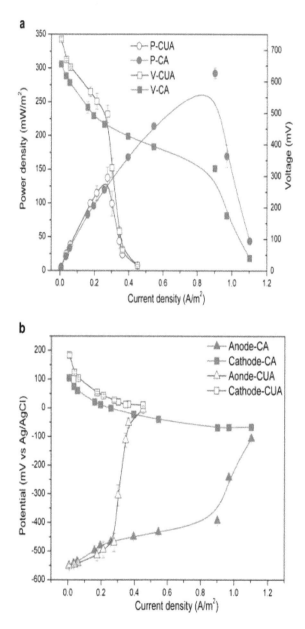

FIGURE 3: Power density and electrode potentials as a function of current density. (a) Polarization curves, and (b) electrode potentials for conductivity unadjusted (CUA) and adjusted (CA) solutions with hydrolysate of a COD concentration of 400 mg/L. The conductivity was adjusted from 5.6 mS/cm to 17 mS/cm with addition of NaCl.

The P_{max} (293.33 ± 7.89 mW/m^2) was larger than that of wheat straw hydrolysate (123 mW/m^2). Based on the maximum power of 1.08 mW (293.33 mW/m^2) and the stable discharge time of 24 hours, 2.592×10^{-5} kWh of electric power can be extracted from a COD concentration of 400 mg/L. The worldwide generation rate of rice straw was about 731 million dry tons in 2007. Therefore, a total amount of 1.51×10^{18} mg/L COD can be produced, from which 9.78×10^{10} kWh of electric power can be extracted. In 2007, the worldwide per capita consumption of electric power was 2,752 kWh (http://www.chinaero.com.cn). Therefore, the electric power extracted from rice straw can meet the annual demand of 35.56 million people for electric power.

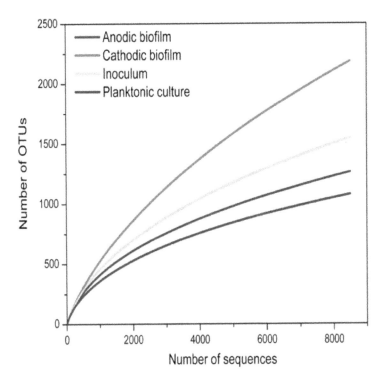

FIGURE 4: Rarefaction curves based on pyrosequencing of bacterial communities. Samples were collected from anodic and cathodic samples, and planktonic culture. Inoculum was also analyzed to observe the change in microbial structure after microbial fuel cell operation. The operational taxonomic units (OTUs) were defined at 0.03 distances.

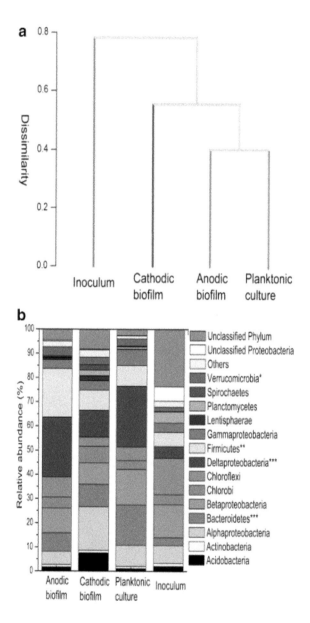

FIGURE 5: Microbial community analysis of the collected microbial samples from anodic and cathodic biofilm, planktonic culture, and inoculum. (a) Jaccard clustering results of bacterial communities defined at 0.03 distances, and (b) microbial community composition at phylum level. Phyla accounting for less than 1% of the total composition in all four libraries were classified as others. *P <0.05, **P <0.01, ***P <0.001.

Some factors, including reactor design, distance between electrodes, solution conductivity, and electrode material, can affect the performance of MFCs. In the present study, a graphite brush anode with a large specific surface area for attachment of exoelectrogens was adopted, which was previously reported as very beneficial for power output [21]; whereas on the other hand, it is difficult to control the distance between electrodes with the cylindrical type of brush anode, because electrode distance has been verified as an important factor capable of affecting the R_{int} of MFCs [22]. The R_{int} of MFCs can be separated into that of the anode, electrolyte, and cathode. The conductivity of the electrolyte can affect the ion transport within the electrolyte. Conductivity adjustment increased the mass transfer rate of organic matters to the anodic biofilm, providing sufficient electron donors and thus, improved the performance of the anodic exoelectrogens. On the other hand, the addition of NaCl to the electrolyte reduced the performance of the cathode. It has been reported that Cl^- can cause deterioration in the performance of Pt towards the oxygen reduction reaction due to the strong interaction between adsorbed Cl^- and Pt, which could suppress the adsorption of O_2 and the formation of Pt site-pairs acquired for breaking the O-O band [23]. Moreover, increasing the solution conductivity could promote the ionic transfer rate and thus decrease the solution resistance [24]. Therefore, the overall effects of NaCl on the anode and electrolyte exceeded that of the cathode, reducing the overall R_{int} of the MFC and consequently promoting the P_{max} output.

5.2.2 MICROBIAL RICHNESS AND DIVERSITY

Four 16S rRNA gene libraries were constructed for the 454-pyrosequencing. The qualified sequences for the microbial samples were 8,615 to 14,308 (Figure 4). The sequences were clustered to represent operational taxonomic units (OTUs) with 3% nucleotide dissimilarity. To reduce sequencing efforts to compare the number of OTUs exactly, 8,500 sequences were randomly selected to calculate the number of OTUs in each sample. The total observed OTUs were 1,242, 2,200, 1,549 and 1,085 for anodic and cathodic biofilm, inoculums, and planktonic culture, respectively (Additional file 1: Table S1). The results illustrated that MFC operation increased the microbial richness of the cathodic biofilm, whereas the richness of the anodic biofilm and planktonic culture were reduced. The Shannon index provides information on species richness and diversity [25]. The cathodic biofilm showed the largest Shannon index (6.5), followed by the inoculum (6.2), anodic biofilm (5.9), and the planktonic culture (5.7). The Shannon index further confirmed that the operation reduced the microbial diversity of the anodic biofilm, which might be due to the selection of exoelectrogens caused by the generation of electricity [26]. However, the microbial diversity of the cathodic biofilm was increased compared to the inoculum. In an air-cathode MFC, ambient air transfers through the diffusion layer into the anolyte; however, the formation of cathodic biofilm decreases the transfer rate of air into the anolyte, preventing mass transfer of organic matter, and the transport of OH⁻ out of the biofilm [27]. Therefore, the formation of cathodic biofilm resulted in respiratory environments with different levels of oxygen, organic matter concentration, and pH. This could be the reason why there was an increase in microbial diversity in the cathodic biofilm. In an air-diffusion biocathode MFC, however, the microbial diversity in cathodic biofilm was smaller than that in planktonic culture [28], possibly due to the different MFC configuration and carbon source.

5.2.3 MICROBIAL COMMUNITY ANALYSIS

A taxonomic supervised dendrogram was prepared to examine the overall variation in the microbial community of the samples. As demonstrated in Figure 5a, the anodic biofilm and planktonic culture were relatively well-clustered, and the cathodic biofilm was clustered closer to the inoculum in comparison to the anodic and planktonic culture. The results indicated that the electrode reactions and niches could influence the microbial community structure which was significantly differentiated from that of the inoculum.

Figure 5b shows the relative abundance of the four microbial communities at phylum level. *Proteobacteria* including α-, β-, δ- and γ- was the predominant phylum accounting for 44.2% , 41.9%, 55.2% and 29.8% of the total abundance in the anodic and cathodic biofilm, planktonic culture and inoculum, respectively. Among the *Proteobacteria*, the δ- class showed a great increase from 5.0% in the inoculum to 11.1% to 25.0% in the MFC samples; the α- class increased from 7.5% in the inoculum to 8.8% in the planktonic culture, and 17.9% in the cathodic biofilm, whereas it decreased to 5.6% in the anodic biofilm. Moreover, *Bacteroidetes* increased from 3.3% in the inoculum to 7.7% in the anodic biofilm, 9.2% in the cathodic biofilm, and 16.6% in the planktonic culture; *Firmicutes* were enriched from 5.8% in the inoculum to 20.1%, 8.3%, and 8.6% in the anodic biofilm, cathodic biofilm and planktonic culture, respectively. *Bacteroidetes* and *Firmicutes* made up the subdominant members accounting for 27.7% in the anodic biofilm, 17.5% in the cathodic biofilm, and 25.2% in the planktonic culture, respectively. *Chloroflexi* was reduced after MFC operation from 15.0% in the inoculum to 8.4%, 3.8%, and 5.7% in the anodic biofilm, cathodic biofilm, and the planktonic culture, respectively. Furthermore, unclassified phylum was significantly reduced to 2.6% to 8.3% in the MFC samples from 23.7% in the inoculum. As previously reported, the species, such as *Shewanella putrefacies* IR-1 [29], *Geobacter sulfurreducens*[30], and *Ochrobacrum anthropi* YZ-1 [31] belonging to *Proteobacteria* were the most important exoelectrogens in the anodic biofilm. Moreover, a few isolated exoelectrogens belonging to *Firmicutes* such as *Clostridium butyricum* EG3 [32], *Desulfitobacterium hafniense* strain DCB2 [33], *Thermincola sp.* strain JR [34], and *Thermincola ferriacetica*[35], were known as a source of exoelectrogens. The present results differed from that of the observed exoelectrogens of a two-chambered MFC using wheat straw biomass for which the predominant culture was *Bacteroidetes* with 40% of sequences [16]. And *Bacteroidetes* and γ-proteobacteria were the most abundant phylum in the anodic biofilm of an air-cathode dual-chamber MFC fed with glucose and glutamate [36]. In MFCs, the microbial community was greatly influenced by the substrates, operation time, and architecture of the cell [16,37,38]. Rice straw hydrolyte generally consists of glucose, xylose, arabinose, acetic acid, and small amount of furfural and 5-hydroxymethyl-furfural [39]. The component of rice straw hydrolyte and single-chamber design in the present study is proposed the factor resulting in different microbial community to the previous studies.

   Proteobacteria (41.9%, α-17.9%, δ-11.1%, β-9.0%, and γ-3.8%), *Bacteroidetes* (9.2%), *Firmicutes* (8.3%), *Acidobacteria* (7.6%), and *Chlorobi* (6.9%) made up the dominant groups of the cathodic biofilm in this study.

In a two-chamber air-diffusion biocathode MFC, the dominant groups were observed as *Proteobacteria* (39.9%, α- 31.7%, γ- 3.8%, β- 2.5%, and δ- 1.1%), *Planctomycetes* (29.9%) and *Bacteroidetes* (13.3%) [28]. The cathodic biofilm in an air-cathode single-chamber MFC can utilize organic carbon sources whereas the cathodic biofilm in a two-chamber air-diffusion MFC was fed with an inorganic carbon source, such as $NaHCO_3$. This should be the reason for the different dominant groups in the two cathodic biofilms.

A total of 484 genera were obtained using the Ribosomal Database Project (RDP) classifier, of which 114 genera were commonly shared by all samples (Figure 6a). They accounted for 85.9%, 79.8%, 82.1% and 79.5% of the anodic and cathodic biofilm, planktonic culture and inoculum, respectively (Figure 6b). There were 182 genera that appeared in only one sample, accounting for 0.4%, 3.1%, 0.1% and 5.3% of the classified sequences in the anodic and cathodic biofilm, planktonic culture and inoculum, respectively. The results suggested that the differentiation of microbial community structure in the samples was caused by a minor portion of the genus.

5.2.4 EXOELECTROGENS

Exoelectrogens play a key role in the generation of electric power through transferring electrons produced from organic matters to the surface of anode materials. Exoelectrogens observed in the present study with greater than 1% abundance are summarized in Table 1. In anodic biofilm, five genera of known exoelectrogens accounted for 23.5% in total communities, including *Desulfobulbus* (11.0%), *Geobacter* (5.3%), *Desulfovibrio* (3.6%), *Pseudononas* (2.3%), and *Comamonas* (1.4%). The genera of exoelectrogens in planktonic culture included *Desulfoblbus* (19.0%), *Desulfovibrio* (4.5%), *Pseudomonas* (2.9%), and *Comamonas* (2.2%), accounting for 28.5% of the total genus. For the cathodic biofilm, however, exoelectrogen species was reduced to three genera, including *Desulfoblbus* (7.8%), *Comamonas* (1.9%), and *Desulfovibrio* (1.3%), with a total abundance of 11.0%. In the inoculum, only one genus with an abundance greater than 1% (*Clostridium*, 1.6%) was discovered, demonstrating that the electricity production can influence the richness of exoelectrogens in MFCs. On the other hand, *Clostridium* was significantly reduced after MFC operation, indicating that the genus of *Clostridium* lost the competition of electron generation compared to other exoelectrogens. The genus of *Desulfobulbus* was the common predominant exoelectrogen in anodic and cathodic biofilm, and planktonic culture, demonstrating that this genus was well adopted in the operating conditions and might attribute to the power generation as a main exoelectrogen in the electrodes. *Pseudomonas* was shared in the anodic biofilm and planktonic culture, but not in the cathodic biofilm, indicating that *Pseudomonas* was not suitable to utilize external electron transferred from the anode. *Geobacter*, an important genus attributing to the power generation in an MFC [40,41], was only dominant in the anodic biofilm. Among the *Geobacter* species, *G. sulfurreducens* and *G. metallireducens,* which were revealed to transfer electrons through nanowire to electron acceptors [42], accounted for 2.9% and 0.2% of the total abundance, respectively. These results suggested the attribution of *Geobacter* species to current generation in the present study. In MFCs, an opposite electron flow of the anode and cathode reaction occurred, and the physical and chemical environment in the anodic and cathodic biofilm was also different. This might be the reason for the different predominant exoelectrogens observed in the anodic and cathodic biofilm.

TABLE 1: Exoelectrogenic genus observed in different MFC samples and the inoculum (>1%)

Genus	Anodic biofilm, %	Planktonic culture, %	Cathodic biofilm, %	Inoculum, %
Clostridium	-	-	-	1.6
Comamonas	1.4	2.2	1.9	-
Desulfobulbus	11.0	19.0	7.8	-
Desulfovibrio	3.6	4.5	1.3	-
Geobacter	5.3	-	-	-
Pseudomonas	2.3	2.9	-	-

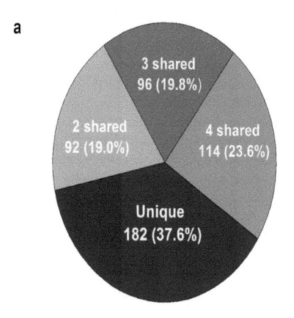

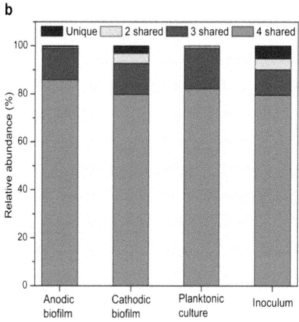

FIGURE 6: Distribution of detected genus in the samples. (a) Overlapping microbial communities of anodic and cathodic biofilm, planktonic culture and inoculum at genus level, and (b) ratio of shared genus to the total genus of different samples.

5.2.5 SULFATE REDUCING BACTERIA

The diluted-acid treatment of rice straw produced sulfate as a main component of the rice straw hydrolysate. The sulfate-reducing bacteria (SRB) could serve a main role in sulfate reduction in this study. Twenty genera of putative sulfate and sulfur-reducing bacteria were detected, and belonged to δ-proteobacteria and Firmicutes. The dominant members were *Desulfobulbus, Desulfovibrio, Desulfomicobium, Desulforhabdus* and *Geobacter* (Additional file 1: Table S2), of which *Desulfobulbus, Desulfovibrio* and *Geobacter* are well-known to play an important role in transferring electrons to the anode. The abundance of SRB was remarkably increased in the three MFC samples compared with the inoculum community. The SRB from the anodic biofilm and planktonic culture were approximately three times more abundant than in the cathodic biofilm. A relatively low abundance of SRB in the cathodic biofilm can be explained by the diffused oxygen to the cathodic biofilm which might have competed with sulfate as an electron acceptor. Interestingly, a considerably higher number of SRB such as *Desulfobulbus* and *Desulfovibrio* were observed in the planktonic culture than in the anodic biofilm. The results showed opposite trends in other dominant groups such as *Desulfomicobium, Desulforhabdus,* and *Geobacter,* representing greater abundance in the anodic biofilm than in the planktonic culture. Although SRB was reported to play an important role in transferring electrons on the anode, some groups of SRB that could not form biofilm on the electrode may also reduce sulfate in the planktonic niche [42]. It has already been reported that SRB demonstrate functional dynamics, including electron transfer, sulfate reduction, and converting organic matters, such as acetic and butyric acids to alcohols and acetone via direct electron transfer [43-45]. In the present study, therefore, high-abundance groups (*Desulfobulbus* and *Desulfovibrio*) may perform both electron transfer on the electrode and sulfate reduction in the electrolyte, whereas low-abundance groups may focus primarily on transferring electrons in the electrode.

5.3 CONCLUSIONS

The present study confirmed that rice straw hydrolysate is flexible as fuel for MFCs to generate electric power. The microbial community was differentiated after the MFC operation. Furthermore, the differentiation in the microbial community structure resulted from a small portion of the genus. The microbial community of the anodic biofilm had a similar microbial structure to the planktonic culture, but it was different to the cathodic biofilm. The known exoelectrogens were differently enriched depending on the niches caused by the different electrode reaction and respiratory environment. Sulfate-reducing bacteria were greatly abundant due to sulfate production by the dilute-acid treatment of the rice straw. They might play different roles in different niches.

5.4 MATERIALS AND METHODS

5.4.1 RICE STRAW HYDROLYSATE PREPARATION

The rice straw collected from Daejeon, Korea was rinsed with tap water and then distilled water, and further dried in an oven at 50°C. The dried rice straw was milled with a juice extractor and was then hydrolyzed using the diluted-acid method (1:10, w/v) [46]. The hydrolysate was further treated by an over-liming process as previously reported to reduce toxic inhibitors [47]. The pH of the resultant hydrolysate was adjusted to approximately 7.0 with concentrated H_2SO_4, and the residue was separated by centrifugation (3,000 rpm, 3 minutes). The COD concentration of the resultant hydrolysate was 20,689 mg/L.

5.4.2 MFC FABRICATION

The MFC reactor was made of Plexi-glass (6 cm × 6 cm × 6.5 cm), with 220 mL of working volume. The carbon brush anode (length 2 cm, diameter 3 cm) was prepared by twisting carbon fiber (PANEX® 35, Zoltek) with stainless steel wire. The cathode was a commercially available ELAT® gas diffusion electrode (Lot #LT 120E-W: 090205) with Pt catalyst (20%). The catalyst side of the cathode was coated with Nafion® polymer dispersion (5%, Aldrich) and dried in air, leading to a Nafion loading rate of 0.5 mg/cm^2. The effective area of the cathode was 36 cm^2. To determine the anodic potential, an Ag/AgCl (3.3 M KCl) was introduced to the electrolyte.

5.4.3 CULTURE INOCULATION AND MFC OPERATION

To start up the MFC, 10 mL anaerobic sludge collected from the anaerobic digester of a wastewater treatment plant in Orgchen, Korea was mixed with 210 mL medium solution and pumped into the MFC. The medium solution contained 0.1 g/L KCl, 0.5 g/L NH_4Cl, 0.1 g/L $MgCl_2$, 0.1 g/L $CaCl_2$, 0.3 g/L KH_2PO_4, 0.5 g/L $NaHCO_3$, and 1.36 g/L $CH_3COONa \cdot 3H_2O$ as an electron donor. Each time the voltage at a fixed external resistance of 500 Ω dropped below 35 mV, sodium acetate was added to the solution until there was a repeatable voltage output. After the cell was successfully started up, the solution was switched to rice straw hydrolysate which was diluted to different COD concentrations (100, 150, 200, and 400 mg/L) with distilled water and then buffered with 1.05 g/L NH_4Cl, 1.5 g/L KH_2PO_4, and 2.2 g/L K_2HPO_4. All the experiments were performed by batch mode and the solution was stirred with a magnetic stirring bar. Temperature was controlled at $30 \pm 1°C$ in an incubator.

5.4.4 ANALYSIS AND CALCULATION

COD was analyzed following the standard method of Korea after filtering through a glass filter paper to remove bacteria. Voltage (V) was monitored using a LabView program every ten minutes. Power density was normalized to the cathode surface area (A) as follows:

$$P = V^2 / R \cdot A,$$

and current density (j) was calculated as follows:

$$j = V / R \cdot A,$$

where R is the external resistance. In order to determine the P_{max}, the external resistance was varied between 30 k and 10 Ω. To observe the effect of conductivity on the performance of the MFC, the conductivity of the electrolyte was adjusted from 5.6 mS/cm to 17 mS/cm using NaCl. The CE was calculated as:

$$CE = C_i / C_o,$$

where C_i is the coulomb collected from the passed current and C_o is the coulomb generated from the consumed organic matters. R_{int} was determined as the slope of the i-V curve according to:

$$V = U_{cell} - iR_{int},$$

where U_{cell} is the electromotive force of the cell [15].

5.4.5 MICROBIAL ANALYSIS

5.4.5.1 DNA EXTRACTION, PCR, AND FLX TITANIUM PYROSEQUENCING

Samples were collected from inoculum sludge, anodic and cathodic biofilm, and planktonic culture for microbial analysis. DNA was extracted using PowerSoil™ DNA Isolation Kit (MoBio, Carlsbad, CA, USA) following the manufacturer's instructions. The following universal 16S rRNA primers were used for the PCR reactions: F563 (AYTGGGYD-

TAAAGNG) and BSR926 (CCGTCAATTYYTTTRAGTTT). Barcode sequences (AGCATCTG, AGCATGAG, AGCTCAGC and AGCTCATG for anodic biofilm, cathodic biofilm, planktonic culture, and inoculum, respectively) were attached between the 454 adaptor sequence and the forward primers. Each PCR reaction was carried out with two of the 25-μl reaction mixtures containing 60 ng of DNA, 10 μM of each primer (Macrogen, Seoul, Korea), and AccuPrime™ Taq DNA Polymerase High Fidelity (Invitrogen, Madison, WI, USA) in order to obtain the following final concentrations: 1.25 U of Taq polymerase, 50 mM of $MgSO_4$, and 10× of the PCR buffer. A C1000TM thermal cycler (Bio-rad, Hercules, CA, USA) was used for the PCR as follows: (i) an initial denaturation step at 94°C for 3 minutes, (ii) 30 cycles of annealing and extending (each cycle occurred at 95°C for 60 s followed by 55°C for 45 s and an extension step at 72°C for 60 s), and (iii) the final extension at 72°C for 5 minutes. After this PCR amplification, the amplicons were purified by one-time gel electrophoresis/isolation and two-time purifications using a QIAquick Gel extraction kit (Qiagen, Valencia, CA, USA) and QIAquick PCR purification kit (Qiagen). In order to recover a sufficient amount of purified amplicons from the purification steps, two 25-μl reaction mixtures were combined into one prior to amplicon purification. All four amplicons were pooled and amplicon pyrosequencing was performed by Macrogen Inc. (Seoul, South Korea) using a 454/Roche GS-FLX titanium instrument (Roche, Nutley, NJ, USA).

5.4.5.2 MICROBIAL COMMUNITY AND CLASSIFICATION

Sequences were analyzed following the modified protocol using Mothur. The range of flow was modified from 450 to 720 to obtain highly accurate sequences. Chimera sequences were removed by the Uchime algorithm with self-references. Filtered sequences were aligned to Silva Gold aligned sequences, and clustered with the furthest algorithm at 0.03 distances. Sequences were classified using the RDP training set (Version 9) with a threshold of 50%. Classified sequences were analyzed into phylotype at phylum and genus level. To reduce sequencing efforts from samples, the smallest sequence numbers were selected to measure the alpha diversity, such as observed OTUs, the Shannon index, and rarefaction curves. A taxonomy-supervised dendrogram was constructed to compare microbial communities from samples using relative abundances. After calculating the relative abundance of genus including unclassified sequences at genus level, a distance matrix (vegdist) was produced and clustered through the average algorithm (hclust) using Vegan package from R.

REFERENCES

1. Zhao F, Rahunen N, Varcoe JR, Chandra A, Avignone-Rossa C, Thumser AE, Slade RCT: Activated carbon cloth as anode for sulfate removal in a microbial fuel cell. Environ Sci Technol 2008, 42:4971-4976.
2. Wang Z, Lim B, Choi C: Removal of Hg2+ as an electron acceptor coupled with power generation using a microbial fuel cell. Bioresour Technol 2011, 102:6304-6307.
3. Pant D, Van Bogaert G, Diels L, Vanbroekhoven K: A review of the substrates used in microbial fuel cells (MFCs) for sustainable energy production. Bioresour Technol 2010, 101:1533-1543.
4. He Y, Pang Y, Liu Y, Li X, Wang K: Physicochemical characterization of rice straw pretreated with sodium hydroxide in the solid state for enhancing biogas production. Energy Fuel 2008, 22:2775-2781.
5. Elmekawy A, Diels L, De Wever H, Pant D: Valorization of cereal based biorefinery byproducts: reality and expectations. Environ Sci Technol 2013, 47:9014-9027.
6. Roberto IC, Mussatto SI, Rodrigues RCLB: Dilute-acid hydrolysis for optimization of xylose recovery from rice straw in a semi-pilot reactor. Industrial Crops and Products 2003, 17:171-176.
7. Pant D, Singh A, Van Bogaert G, Irving Olsen S, Singh Nigam P, Diels L, Vanbroekhoven K: Bioelectrochemical systems (BES) for sustainable energy production and product recovery from organic wastes and industrial wastewaters. RSC Advances 2012, 2:1248.
8. Chae KJ, Choi MJ, Lee JW, Kim KY, Kim IS: Bioresour Technol. Effect of different substrates on the performance, bacterial diversity, and bacterial viability 2009, 100:3518-3525.
9. Ha PT, Tae B, Chang IS: Performance and bacterial consortium of microbial fuel cell fed with formate. Energy Fuel 2008, 22:164-168.
10. Luo Y, Zhang R, Liu G, Li J, Li M, Zhang C: Electricity generation from indole and microbial community analysis in the microbial fuel cell. J Hazard Mater 2010, 176:759-764.
11. Zhang XY, Cheng SA, Wang X, Huang X, Logan BE: Separator characteristics for increasing performance of microbial fuel cells. Environ Sci Technol 2009, 43:8456-8461.
12. Cristiani P, Carvalho ML, Guerrini E, Daghio M, Santoro C, Li B: Cathodic and anodic biofilms in single chamber microbial fuel cells. Bioelectrochem 2013, 92:6-13.

13. Yan H, Saito T, Regan JM: Nitrogen removal in a single-chamber microbial fuel cell with nitrifying biofilm enriched at the air cathode. Water Res 2012, 46:2215-2224.

14. Sevda S, Dominguez-Benetton X, Vanbroekhoven K, De Wever H, Sreekrishnan TR, Pant D: High strength wastewater treatment accompanied by power generation using air cathode microbial fuel cell. Appl Energy 2013, 105:194-206.

15. Wang Z, Lim B, Lu H, Fan J, Choi C: Cathodic reduction of Cu2+ and electric power generation using a microbial fuel cell. B Korean Chem Soc 2010, 31:2025-2030.

16. Zhang Y, Min B, Huang L, Angelidaki I: Generation of electricity and analysis of microbial communities in wheat straw biomass-powered microbial fuel cells. Appl Environ Microbiol 2009, 75:3389-3395.

17. Zuo Y, Maness P, Logan BE: Electricity production from steam-exploded corn stover biomass. Energy Fuel 2006, 20:1716-1721.

18. Nam JY, Kim HW, Lim KH, Shin HS: Effects of organic loading rates on the continuous electricity generation from fermented wastewater using a single-chamber microbial fuel cell. Bioresour Technol 2010, 101:S33-S37.

19. Lu N, Zhou S, Zhuang L, Zhang J, Ni J: Electricity generation from starch processing wastewater using microbial fuel cell technology. Biochem Eng J 2009, 43:246-251.

20. Liu H, Logan BE: Electricity generation using an air-cathode single chamber microbial fuel cell in the presence and absence of a proton exchange membrane. Environ Sci Technol 2004, 38:4040-4046.

21. Logan B, Cheng S, Valerie W, Estadt G: Graphite fiber brush anodes for increased power production in air-cathode microbial fuel cells. Environ Sci Technol 2007, 41:3341-3346.

22. Liu H, Cheng SA, Logan BE: Power generation in fed-batch microbial fuel cells as a function of ionic strength, temperature, and reactor configuration. Environ Sci Technol 2005, 39:5488-5493.

23. Stamenkovic V, Markovic NM, Ross JPN: Structure-relationships in electrocatalysis oxygenreduction and hydrogen oxidation reactions on Pt(111) and Pt(100) in solutions containing chloride ions. J Electroanal Chem 2001, 500:44-51.

24. Elmekawy A, Hegab HM, Dominguez-Benetton X, Pant D: Internal resistance of microfluidic microbial fuel cell: challenges and potential opportunities. Bioresour Technol 2013, 142:672-682.

25. Shannon CE, Weaver W: The Mathematical Theory of Communication. Urbana: University of Illinois Press; 1949.

26. Jong BC, Kim BH, Chang IS, Liew PWY, Choo YF, Kang GS: Enrichment, performance, and microbial diversity of a thermophilic mediatorless microbial fuel cell. Environ Sci Technol 2006, 40:6449-6454.

27. Yuan Y, Zhou S, Tang J: In situ investigation of cathode and local biofilm microenvironments reveals important roles of OH- and oxygen transport in microbial fuel cells. Environ Sci Technol 2013, 47:4911-4917.

28. Wang Z, Zheng Y, Xiao Y, Wu S, Wu Y, Yang Z, Zhao F: Analysis of oxygen reduction and microbial community of air-diffusion biocathode in microbial fuel cells. Bioresour Technol 2013, 144:74-79.

29. Kim B, Kim H, Hyun M, Park D: Direct electrode reaction of Fe(III)-reducing bacterium, Shewanella putrefaciens. J Microbiol Biotechnol 1999, 9:127-131.

30. Bond DR, Holmes DE, Tender LM, Lovley DR: Electrode-reducing microorganisms that harvest energy from marine sediments. Science 2002, 295:483-485.

31. Zuo Y, Xing D, Regan JM, Logan BE: Isolation of the exoelectrogenic bacterium ochrobactrum anthropi YZ-1 by using a U-tube microbial fuel cell. Appl Environ Microbiol 2008, 74:3130-3137.

32. Park HS, Kim BH, Kim HS, Kim HJ, Kim GT, Kim M, Chang IS, Park YK, Chang HI: A novel electrochemically active and Fe(III)-reducing bacterium phylogenetically related to Clostridium butyricum isolated from a microbial fuel cell. Anaerobe 2001, 7:297-306.

33. Milliken CE, May HD: Sustained generation of electricity by the spore-forming, Gram-positive, Desulfitobacterium hafniense strain DCB2. Appl Microbiol Biotechnol 2006, 73:1180-1189.

34. Wrighton KC, Agbo P, Warnecke F, Weber KA, Brodie EL, DeSantis TZ, Hugenholtz P, Andersen GL, Coates JD: A novel ecological role of the Firmicutes identified in thermophilic microbial fuel cells. Isme J 2008, 2:1146-1156.

35. Marshall CW, May HD: Electrochemical evidence of direct electrode reduction by a thermophilic Gram-positive bacterium. Thermincola ferriacetica. Energy Environ Sci 2009, 2:699-705.

36. Kim GT, Webster G, Wimpenny JWT, Kim BH, Kim HJ, Weightman AJ: Bacterial community structure, compartmentalization and activity in a microbial fuel cell. J Appl Microbiol 2006, 101:698-710.

37. Zhang Y, Min B, Huang L, Angelidaki I: Electricity generation and microbial community response to substrate changes. Bioresour Technol 2011, 102:1166-1173.

38. Beecroft NJ, Zhao F, Varcoe JR, Slade RCT, Thumser AE, Avignone-Rossa C: Dynamic changes in the microbial community composition in microbial fuel cells fed with sucrose. Appl Microbiol Biotechnol 2011, 93:423-437.

39. Huang C, Zong M, Wu H, Liu Q: Microbial oil production from rice straw hydrolysate by Trichosporon fermentans. Bioresour Technol 2009, 100:4535-4538.

40. Bond DR, Lovley DR: Electricity production by Geobacter sulfurreducens attached to electrodes. Appl Environ Microbiol 2003, 69:1548-1555.

41. Nercessian O, Parot S, Délia M, Bergel A, Achouak W: Harvesting electricity with Geobacter bremensis isolated from compost. PLoS ONE 2012, 7:e34216.

42. Malvankar NS, Lovley DR: Microbial nanowires: a new paradigm for biological electron transfer and bioelectronics. ChemSusChem 2012, 5:1039-1046.

43. Santegoeds CM, Ferdelman TG, Muyzer G, Beer DD: Structural and functional dynamics of sulfate-reducing populations in bacterial biofilms. Appl Environ Microbiol 1998, 64:3731-3739.

44. Cordas C, Guerra L, Xavier C, Moura J: Electroactive biofilms of sulphate reducing bacteria. Electrochim Acta 2008, 54:29-34.

45. Sharma M, Aryal N, Sarma PM, Vanbroekhoven K, Lal B, Benetton XD, Pant D: Bioelectrocatalyzed reduction of acetic and butyric acids via direct electron transfer using a mixed culture of sulfate-reducers drives electrosynthesis of alcohols and acetone. Chem Commun 2013, 49:6495-6497.

46. Yoswathana N, Phuriphipat P, Treyawutthiwat P, Eshtiaghi MN: Bioethanol production from rice straw. Energy Res J 2010, 1:26-31.

47. Martinez A, Rodriguez ME, York SW, PRESTON JF, Ingram LO: Effects of Ca(OH)2 treatments ("overliming") on the composition and toxicity of bagasse hemicellulose hydrolysates. Biotechnol Bioeng 2000, 69:526-536.

There are several supplemental files that are not available in this version of the article. To view this additional information, please use the citation on the first page of this chapter.

CHAPTER 6

METATAXONOMIC PROFILING AND PREDICTION OF FUNCTIONAL BEHAVIOR OF WHEAT STRAW DEGRADING MICROBIAL CONSORTIA

DIEGO JAVIER JIMÉNEZ, FRANCISCO DINI-ANDREOTE, AND JAN DIRK VAN ELSAS

6.1 BACKGROUND

Efficient bioconversion of lignocellulosic substrates depends critically on the functioning of multispecies microbial consortia rather than single strains [1]. In such consortia, secretion of the enzymes involved in biodegradation, as affected by the interactions between the microbial players (bacteria-fungi), is of crucial importance [2,3]. Wheat straw, as the source of lignocellulose, can potentially serve to provide building blocks for production of plastics or energy in biofuels [4]. The conversion of lignocellulosic polymers into monomers that can be further processed involves the synergistic action of a range of secreted enzymes, that is, peroxidases, xylanases and endo/exoglucanases [5,6]. In spite of the fact that intricate

Metataxonomic Profiling and Prediction of Functional Behaviour of Wheat Straw Degrading Microbial Consortia. © Jiménez DJ, Dini-Andreote F, and van Elsas JD.; licensee BioMed Central Ltd. Biotechnology for Biofuels 7,92 (2014), doi:10.1186/1754-6834-7-92. Licensed under Creative Commons Attribution 4.0 International License, http://creativecommons.org/licenses/by/4.0/.

knowledge on the decomposition process is lacking, many bacteria are known to be capable of producing such enzymes. In particular, members of the *Gammaproteobacteria, Firmicutes* and *Bacteroidetes* have been implicated in lignocellulose biodegradation [7,8]. Moreover, fungi like *Trichosporon* and *Coniochaeta* are considered as potential sources of hydrolytic enzymes, in particular those involved in the bioconversion of (toxic) furanic compounds and in the production of unique secondary metabolites [9,10]. In addition, recent evidence suggests that, from the biotechnological perspective, *Penicillium, Acremonium* and *Trichoderma* species represent fungi that are applicable in the production of commercial lignocellulases [11].

The current literature indicates several strategies by which effective microbial consortia can be obtained [12]. In addition, the construction of target microbial communities can be aided using stable isotope probing (SIP) [13]. However, SIP suffers from drawbacks related to cross-feeding phenomena and/or the possible detection of bacterial or fungal predators of labeled cells, that is, those representing "microbial cheaters" [14]. Thus, a valid strategy to obtain efficient microbial consortia that degrade lignocellulosic matter is ex situ dilution to stimulation, using (partially unlocked) plant material as the unique energy and carbon source [15,16]. Due to selective processes, this last approach results in a stimulus of (biodegradation) function within the emerging consortia during succession [12]. The enrichment cultures produced can then provide a robust platform for biotechnological applications [17-19].

Unfortunately, cultivation-based analyses of complex microbial consortia are restrictive, as key organisms may be omitted. Thus, DNA-based high-throughput sequencing techniques have been recently applied to lignocellulolytic consortia [20,21]. The studies performed so far have, however, only addressed the role of bacteria, to the exclusion of fungal players. Fungi, either in the mycelial or yeast form, can have dominant roles in lignocellulose decomposition in plant litter and soil [22,23]. In lignocellulosic enrichment cultures, the bacterial and fungal diversities may be driven by the microbial source, available substrates, pH, redox potential, temperature and possible toxic compounds [24-26]. Thus, such consortia need to be assessed over time in relation to conditions and metabolic fluxes among key members, which is important for further "consortium engineering" [2].

The classical bacterial 16S rRNA gene and fungal ITS1 based markers are useful to describe community composition but do not provide information on the genes that are involved in lignocellulose deconstruction. Recently, Langille et al. (2013) [27] suggested a way to overcome such a limitation. They developed the software PICRUSt (Phylogenetic Investigation of Communities by Reconstruction of Unobserved States) to predict the occurrence of functions in microbial communities solely on the basis of bacterial 16S rRNA gene sequences. Although such an approach is theoretically fraught with uncertainties, realistic predictions of function in low-complexity environments were given. Thus, PICRUSt has been used to analyze the human intestinal mucosal surface microbiome, and the results correlated fairly well with the extant metabolome, suggesting a relationship between inferred function and metabolites found [28]. However, the method needs extreme caution in the interpretation of its outcomes, given the known impact of horizontal gene transfer (HGT) across the genomes of the members of most microbial communities. In addition, the quality of these functional predictions is largely dependent on the availability of annotated reference genomes.

In a previous study [29], we reported the construction of two novel bacterial-fungal consortia involved in the bioconversion of lignocellulose next to furanic compounds. We described their characteristics based on bacterial cell counts, quantitative PCR (qPCR), denaturing gradient gel electrophoresis (DGGE) analyses and isolation of some key consortium members. In addition, we designed a novel iodide oxidation method to detect 5-hydroxymethylfurfural oxidoreductase activity. In the current study, we expanded our previous work by focusing on the metataxonomic evaluation (based on bacterial 16S rRNA gene and fungal ITS1 pyrosequencing data) of two lignocellulolytic microbial consortia enriched on untreated versus pretreated wheat straw. We here analyze the successional microbial diversity and community composition of the two consortia and apply PICRUSt, thus predicting genes for functions involved in lignocellulose metabolism. Moreover, we evaluated the joint expression of some of these genes in the secretome, by quantification of specific (hemi)cellulolytic enzymatic activities. The two consortia constitute starting points for biotechnological applications in the light of their possible capacities in the conversion of lignin, (hemi)cellulose, furanic compounds and cello-oligosaccharides.

6.2 RESULTS

6.2.1 ANALYSIS OF THE COMMUNITY STRUCTURES AND DIVERSITIES OF TWO MICROBIAL CONSORTIA

Overall, 18,200 trimmed-rarefied sequences of bacterial 16S rRNA from the forest soil inoculum (SS) as well as the RWS (untreated wheat straw) and TWS (heat-treated wheat straw) consortia (n=13) were analyzed. The rarefied sequencing data (1,400 sequences per sample) were binned into 338, 109 (±6.5) and 102 (±0.3) abundant operational taxonomic units (OTUs), for SS, RWS and TWS (both at transfer 1 - T1), respectively (Figure 1A and B; Additional file 1). At T10 (transfer 10, approximately 70 days after setting up the first microcosm), we observed the presence of 100 (±9.5) and 47 (±2.7) OTUs for RWS and TWS, respectively. Based on the bacterial 16S rRNA gene sequences, the Faith's phylogenetic diversity (PD) and Chao richness estimator (CRE) values in SS were 20.66 and 340.90, respectively. In contrast, these values (at T10) in RWS were 4.15 (±0.68) and 151.87 (±3.09). For RWS, in particular the CRE values for 16S rRNA gene decreased slightly from T1 to T10. However, the PD and Shannon-Wiener index (SWI) values did not show large changes, for example with PD values of 3.91 (±0.23) and 4.15 (±0.68) for T1 to T10, respectively (Figure 1A). For TWS, the bacterial consortia also showed progressively decreasing CRE and PD values, with the higher ones in the SS and T1 (161,67±0.49 and 3.54±0.26 for CRE and PD, respectively) (Figure 1B).For fungal communities across all samples, 6,600 trimmed-rarefied sequences (n=12) were analyzed. One sample (T10 in RWS) was omitted due to low-quality reads. At the sequencing depth of 550 sequences per sample, 91, 54 (±16) and 61 (±1.2) different OTUs were identified in SS, T1/RWS and T1/TWS, respectively (Figure 1C and D). At T10, 36 and 50 (±7.4) OTUs were identified for RWS and TWS, respectively. In SS, the fungal consortia showed values of 43.02, 97.95 and 5.41 for PD, CRE and SWI, respectively. For RWS, the PD values were 33.52 (±8.71) and 26.20 (at T1 and T10, respectively) (Figure 1C). As expected, we observed a decrease of the PD values from SS (43.02) to T10 (22.81±2.00). The CRE values also showed a decrease, from 97.95 in SS to 89.33 (±13.02) at T1 and 74.93 (±6.28) at T10 (Figure 1D).

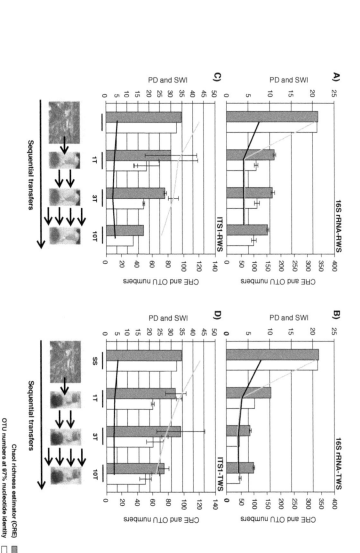

FIGURE 1: Diversity indices in the soil inoculum (SS) and in enriched cultures (RWS and TWS) along the sequential batches. Diversity indices and richness estimator measured using (A, B) rarefied bacterial 16S rRNA and (C, D) ITS1 region sequences. Bars refer to standard errors (n = 2). For SS and 10-RWS (ITS1) only one sample was analyzed. The arrows represent the number of parallel sequential transfers between 1T (transfer 1), 3T (transfer 3) and 10T (transfer 10) (for more detail see Methods).

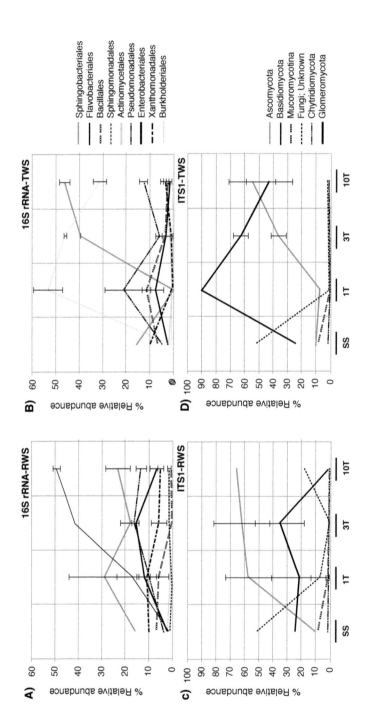

FIGURE 2: Relative abundance (bacteria and fungi) in the soil inoculum (SS) and in enriched cultures (RWS and TWS) along the sequential batches. Relative abundance (%) of the most abundant (A, B) bacterial orders and (C, D) fungal phylum members based on 1,400 (16S rRNA) and 550 (ITS1) sequences.

6.2.2 RELATIVE ABUNDANCE OF BACTERIAL AND FUNGAL TYPES IN THE MICROBIAL CONSORTIA

The source (SS) bacterial community was predominantly composed of Acidobacteria (19.28%), Gammaproteobacteria (18.07%), Bacteroidetes (17.5%) and Betaproteobacteria (10.21%) (Additional file 2A). For RWS at T10, increases of the relative abundances (RA) of Gammaproteobacteria and Bacteroidetes (67.64%±2.28 and 29.78%±2.28, respectively) were noted. For TWS, the RA of Gammaproteobacteria was high at T1 (80.5%±2.14), declining at T10 to 47.42% (±2.5) (Additional file 2A). Successive subcultivation led to the enrichment of OTUs mainly affiliated with nine bacterial orders (Figure 2 and Additional file 2B). For RWS, we observed an increase of the RA of Enterobacteriales from 0.71% in SS to 49.39% (±1.46) in T10. Pseudomonadales and Flavobacteriales showed a similar behaviour, with an increase of the RA from SS (3.71% and 1.92%, respectively) to T3 (15.96%±2.03 and 15.57%±1.57, respectively), with a small subsequent reduction in T10 (Figure 2A). For TWS, the RA of Enterobacteriales increased to 53.21% (±6.14) in T1, decreasing along the transfers to 31.32% (±2.67) in T10 (Figure 2B). In contrast, the Sphingobacteriales RA showed an opposite pattern, starting at 15.5% in SS and decreasing to 0.03% (±0.03%) at T1, with an increase to 46.25% (±2.17) at T10. Members of Bacillales and Flavobacteriales showed a similar tendency in TWS, with higher values at T1 and lower ones at T10. Pseudomonadales was the third most abundant order, with values of 12.60% (±1.60) and 20.96% (±7.96) at T1 and T3, respectively (Figure 2B).

The most abundant fungi in SS belonged to unclassified types (possibly uncultured Ascomycota) (52.18%), followed by Basidiomycota (24%), Ascomycota (9.63%), Mucoromycotina (8.18%) and Chytridiomycota (1.09%). Totals of 24, 172 and 97 sequences in SS, RWS and TWS, respectively, showed no affiliation against the UNITE and/or GenBank databases, and were removed from the analyses (Additional file 3A). For RWS, the RA of Ascomycota and Basidiomycota increased compared to that in SS. However, the replicate patterns were internally not very consistent (Figure 2C). On the other hand, TWS showed high consistency between replicates in each transfer (Additional file 3A). The RA of Basidiomycota was 90.45% (±0.09) in T1, decreasing along the transfers to

43.09% (±16.18) in T10. In contrast, the RA of Ascomycota increased from 7.54% (±0.27) in T1 to 54.09% (±16.18) in T10 (Figure 2D).

6.2.3 RELATIVE ABUNDANCE AT GENUS LEVEL AND STRUCTURE OF THE MICROBIAL CONSORTIA

To assess the RA of each genus in the RWS and TWS microbial consortia, we removed the least abundant OTUs (those containing less than 10 sequences in total). We thus used 92.64% (±1.50) to 98.50% (±0.57), and 81.63% (±0) to 95.65 (±1.09) of the 1,400 16S rRNA gene and 550 ITS1 sequences, respectively. On this basis, 24 bacterial and 13 fungal genera were detected across all samples in both enrichment strategies (omitting the source SS) (Figure 3). The bacterial communities at T1 for RWS and TWS showed eight abundant genera, defined as having an RA > 5%; these were *Stenotrophomonas, Sphingobacterium, Acinetobacter, Flavobacterium, Pseudomonas, Serratia, Klebsiella* and *Paenibacillus.*

After the third transfer (T3), when communities had stabilized, the bacterial genera in RWS with the highest RA were *Klebsiella* (about 35%), *Acinetobacter* (about 12%), *Flavobacterium* (about 9%), *Sphingobacterium* (about 9%), *Pedobacter* (about 8%) and *Enterobacter* (about 3%). This was followed by *Stenotrophomonas, Citrobacter, Sphingobium* and *Chitinophaga* (1 to 2%). For TWS, the most abundant bacterial genera at T3 and T10 were *Sphingobacterium* (about 42%), *Klebsiella* (about 38%) and *Pseudomonas* (about 8%), whereas the least abundant OTUs (1 to 2%) were affiliated with *Stenotrophomonas, Flavobacterium, Achromobacter* and *Paenibacillus* species. Concerning the fungi, for RWS the genera with highest RA at T1 and T3 were *Acremonium* (about 42%), *Malassezia* (about 25%) and *Coniochaeta* (~10%), whereas at T10 we observed highest RA for *Penicillium* (about 63%). For TWS, *Trichosporon* (66.45% ± 5.90) and *Malassezia* (16.81% ± 5.18) were most abundant at T1. After this stage, we observed an increase of the RA of *Coniochaeta* (about 33%), *Penicillium* (about 5%) and *Acremonium* (about 4%). In addition, the RA of *Trichosporon* (about 39%) was also high at T10 (Figure 3).

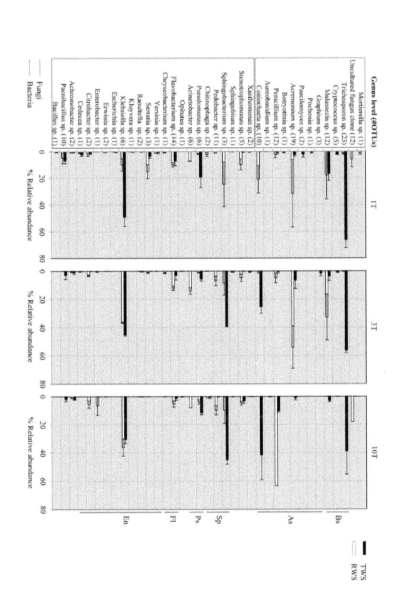

FIGURE 3: Relative abundances of the most abundant genera in the sequential batches enriched cultures (RWS and TWS). Abbreviations: Basidiomycota (Bs), Ascomycota (As), Sphingobacteriales (Sp), Pseudomonadales (Ps), Flavobacteriales (Fl), Enterobacteriales (En).

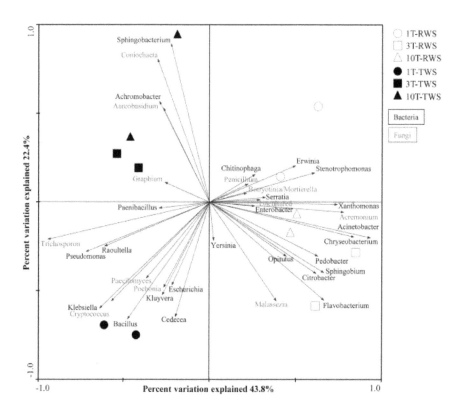

FIGURE 4: Principal components analysis (PCA) of the most abundant genera in the enriched cultures (RWS and TWS).

A principal components analysis (PCA) of the data showed that *Flavobacterium, Acinetobacter, Pedobacter, Citrobacter, Chryseobacterium, Opitutus, Sphingomonas, Acremonium* and *Malassezia* were preferentially selected at T10 for RWS, while *Sphingobacterium, Achromobacter, Coniochaeta* and *Aureobasidium* were increased at T10 for TWS (Figure 4).

6.2.4 BACTERIAL OTUS RELATED TO (HEMI)CELLULOLYTIC STRAINS

A total of 10 out of the 15 abundant bacterial OTUs detected by direct molecular assessment was recovered as isolates (Figure 5). Bacterial isolates were recovered by dilution plating on R2A agar and presumptively identified using 16S rRNA gene sequencing [29]. Among these, non-(hemi)cellulolytic strain 10w8 isolated from TWS matched *Pseudomonas putida*_OTU418 (99% identity). In RWS, two OTUs representing the genus *Acinetobacter* were found to be abundant. Sequence-wise, the isolated strains 8w3 and 8w5, which were found to have endoglucanase and xylanase activities, matched one OTU, affiliated with *Acinetobacter johnsonii* (OTU1927), whereas strain 10w16, which was devoid of any (hemi)cellulolytic activity (based on its activity on carboxymethylcellulose (CMC) and xylan from birchwood), matched *Acinetobacter calcoaceticus*_OTU636. The sequence of OTU1062 (*Klebsiella variicola*) represented the most abundant OTU in RWS (approximately 37%), and the RWS- and TWS-derived strains 10w11, 10w26, 10 t14 and 1 t2 matched this sequence type. Interestingly, the strains retrieved from RWS (10w11 and 10w26) showed (hemi)cellulolytic activity, whereas those from TWS (10 t14 and 1 t2) did not (Figure 5). In accordance with the phylogenetic tree and low identity (90%), we not consider that *Flavobacterium hercynium*_OTU838 represents the (hemi)cellulolytic bacterial strain 3w2. Moreover, strains 3 t5 and 3 t6, which likely represented the highly abundant *Sphingobacterium faecium*_OTU387, did not show CMC-ase and xylanase activities on agar plates.

6.2.5 PREDICTING FUNCTIONS INVOLVED IN LIGNOCELLULOSE DECONSTRUCTION

Totals of 25 and 17 genes related with plant biomass deconstruction were predicted to be consistently enriched from the SS to the RWS and TWS consortia at T10, respectively. Conversely, 34 and 18 predicted genes were enriched from the sequential batches (T1 to T10) in RWS and TWS, respectively (Table 1). Interestingly, predicted genes that codify for glycolate oxidase (EC:1.1.3.15), alpha-L-fucosidase (EC:3.2.1.51), alpha-N-arabinofuranosidase (EC:3.2.1.55), endo-1.4-beta-xylanase (EC:3.2.1.8), alpha-L-rhamnosidase (EC:3.2.1.40) and maltooligosyltrehalose trehalohydrolase (EC:3.2.1.93) decreased in number along the sequential transfers in RWS; however, they increased in TWS (Table 1).

Six predicted genes potentially related to lignin bioconversion were enriched in RWS at T10, while one (catalase - EC:1.11.1.6) was enriched in TWS at T10, compared with the SS source (Table 1). With respect to (hemi)cellulose bioconversion, the number of predicted genes that encode alpha-mannosidase (EC:3.2.1.24) and levanase (EC:3.2.1.65) increased along the sequential transfers (T1 to T10) in both cultures and also were higher than those in the SS source. The predicted genes that encode glucosidase (alpha and beta) enzymes also increased along the sequential transfer in both strategies. Moreover, many predicted genes that were already abundant in SS, such as those that encode beta-galactosidases (EC:3.2.1.23) and beta-glucosidases (EC:3.2.1.21), showed an extra increase by the sequential transfer in both cultures. Also, xylan 1.4-beta-xylosidase (EC:3.2.1.37) and endoglucanase (EC:3.2.1.4) related predicted genes were either enriched or depleted along the sequential transfers in RWS and TWS, respectively (Table 1). In addition, a beta-mannosidase (EC:3.2.1.25) related gene was found in both enrichment cultures. Regarding the accuracy of metagenome predictions, the Nearest Sequenced Taxon Index (NSTI), which quantifies the uncertainty of the prediction (lower values mean a better prediction), decreased from 0.18 in SS to 0.03 (RWS) and 0.09 (TWS) at T10 (Figure 6A). The NSTI metric represents the sum of phylogenetic distances for each organism in the OTU table to its nearest relative with a sequenced reference genome, measured in terms of substitutions per site in the 16S rRNA gene and weighted by the frequency of that organism in the OTU table [27].

6.2.6 QUANTIFICATION OF SPECIFIC ENZYMATIC ACTIVITIES RELATED TO (HEMI)CELLULOSE BIOCONVERSION

As shown by direct enzymatic assays, beta-xylosidases, beta-galactosidases, beta-mannosidases, cellobiohydrolases and beta-glucosidases were active in the secretome of both consortia. Interestingly, we observed high beta-mannosidase (0.35 nM 4-methylumbelliferyl (MUF)/min) and beta-galactosidase (1.3 nM MUF/min) activities in TWS but low ones in RWS. In addition, cellobiohydrolases showed higher activities in RWS (0.11 nM MUF/min) compared to TWS (0.03 nM MUF/min). Beta-galactosidases, beta-glucosidases and beta-xylosidases also showed higher activity values in both secretomes compared with the other two enzymes (Figure 6B). The bacterial abundance at T10 was evaluated by cell counts, and we observed increases from the inoculum level, around 5, to about 8.4 log bacterial cells/mL in both consortia (data not shown).

6.3 DISCUSSION

The use of microbial consortia in lignocellulose transformation is likely to reduce impairments in bioprocesses using lignocellulosic matter, such as incompletely synergistic enzymes, pH regulation, the presence of toxic compounds, end-product inhibition and tolerance to environmental fluc-tuations [2,18]. Previously, several different types of plant biomass, such as sugarcane bagasse, poplar wood chips and switchgrass [15,30,31], have been used in the selection of biodegradative microbial consortia. In our work, wheat straw, either torrefied (pretreated with heat at 240°C) or not, was used. However, torrefaction, which is proposed as a valuable step in waste plant biomass valorization [32], introduces furanic compounds and/ or cello-oligosaccharides in the medium [24]. Other studies suggested that lignocellulose-degrading soil communities are best addressed using SIP analysis based on 13C-substrates [33], whereas lignin-degradative micro-bial communities can be enriched directly in soil [34].

Lignocellulolytic microbes are important members of forest soil com-munities and aid in the degradation of litter [35,36]. Our dilution-to-stimu-lation approach from a forest soil source community resulted in a stimulus of biodegradative function within the emerging bacterial-fungal consortia. Moreover, qPCR showed that the bacterial 16S rRNA gene copy numbers were higher than the fungal ITS1 copy numbers (about 2 log units) in both substrates [29]. Notably, the microbial diversity became markedly reduced in both consortia compared to the source SS (Figure 1), suggesting the selection of particular taxa (including lignocellulose and possibly cello-oligosaccharide eaters) at the detriment of others. On the other hand, the apparently low microbial richness in SS could be related to the deletion of all singletons in our bioinformatic analysis. The bacterial CRE values decreased along the sequential batches, indicating enrichment of OTUs that grew consistently well in the enrichment. Interestingly, the TWS consortia showed low bacterial diversities compared to the RWS ones, possible due to the presence of toxic compounds such as furanic aldehydes. With respect to fungal diversity and richness, our data showed values that were similar between RWS and TWS, indicating that substrate type did not strongly affect fungal diversity. However, Faith's PD measure suggested the selection of particular lignocellulolytic fungi in both consortia. Moreover, both UniFrac unweighted distances and PCA (Figure 4; Additional file 4) showed that the consortia were highly influenced by substrate type (variation explained: 43.8%), which is similar to other reported data [21,37]. In addition, the structures of both consortia became less different after T3. In our previous analysis based on PCR-DGGE, stability of the dominant microorganisms after six transfers was reported [29].

The enrichment process increased the abundances of the bacterial orders Enterobacteriales, Pseudomonadales, Flavobacteriales and Sphingobacteriales (Figure 2A and B). In 2012 Eichorst and Kuske [38] performed SIP using ^{13}C-maize cellulose to evaluate the biodegradative microbial communities in different soils. They identified Burkholderiales, Caulobacterales, Rhizobiales, Sphingobacteriales and Xanthomonadales as the active bacterial groups. Enterobacteriales and Pseudomonadales were also enriched in insect herbivore microbiomes and lignin-enriched cultures, respectively [39,40]. Interestingly, in our RWS consortia the abundance of Enterobacteriales increased after T1, but the number of OTUs decreased, suggesting the selection of specific strains, for example, *K. variicola*, within this group. A similar pattern was observed for Sphingobacteriales in TWS, although the number of OTUs did not change after T3 (Additional file 5).

In soils, Ascomycota may be highly abundant in early stages of litter decomposition, whereas particular basidiomycetous yeasts increase in the later stages, possibly due to their capacity to degrade more recalcitrant compounds [41]. Members of the orders Atheliales, Agaricales, Helotiales, Chaetothyriales and Russulales were found to be abundant in (coniferous) forest soils [35]. Here, we found a high abundance of Malasseziales and Hypocreales. Abundant fungal orders present in SS were even more enriched in RWS. In contrast, only low-abundance orders were enriched in TWS, such as Coniochaetales and Tremellales (Additional file 3B). Štursová et al. (2012) [22], using SIP analysis in soil and litter plant samples with ^{13}C-cellulose, identified fungi affiliated with Dothideales, Leotiomycetes, Helotiales, Tremellales and Chaetothyriales. Thus, diverse fungi can be involved in lignocellulose bioconversion. We posit here that this diversity is dictated by the environmental source, substrate and methodology used for recovery and characterization.

Bacterial strains affiliated with an abundant *K. variicola*_OTU1062 showed (hemi)cellulolytic activity when isolated from RWS, but not from TWS, suggesting either HGT of mobile genes or differential gene expression between the isolates obtained from the two substrate types. Okeke and Lu (2011) [42] proposed that the capacity of *Klebsiella* types to degrade lignocellulose can be attributed to the acquisition of plasmids encoding (hemi)cellulolytic enzymes from the environment. However, Suen et al. (2010) [43] reported a chromosomal location in *K. variicola* At-22 of genes involved in plant biomass degradation, that is, beta-1,4-glucanase, alpha-xylosidases and alpha-mannosidases. Interestingly, the degradation of lignocellulose by an insect herbivore microbiome has been attributed to association between *Leucoagaricus gongylophorus* (Basidiomycota) with *Klebsiella* species [39]. Another possible explanation for such findings arises from the regulatory mechanism of (hemi)cellulolytic genes, which is ultimately mediated by environmental conditions, in this case the torrefied substrate.

Members of *Citrobacter, Enterobacter, Acinetobacter, Pseudomonas, Flavobacterium* and *Stenotrophomonas* have the capacity to degrade plant lignin, (hemi)cellulose and/or CMC [44-48,40]. The presence of *Sphingobacterium* and *Pedobacter* in both microbial consortia may suggest the production of beta-glucosidases, indicating that these organisms are acting as "cheaters" that remove the cello-oligosaccharides produced by polymer degraders [49]. However, such organisms might also be involved in the production of aryl alcohol oxidases or endoxylanases [50,51]. We recently confirmed the production of beta-mannosidases, beta-galactosidases and beta-glucosidases by characterization of *S. faecium* (similar to OTU387, strain 3T5, data not shown). The presence of *Chryseobacterium, Opitutus, Chitinophaga* and *Xanthomonas* in RWS might relate to secondary functions, the nature of which is unclear. In TWS, members of *Stenotrophomonas, Pseudomonas* and *Flavobacterium* can be involved in the degradation of furanic compounds [52,29]. However, in both our consortia, *P. putida*_OTU418 might also act as a sugar cheater. Interestingly, Ronan et al. (2013) [53] reported an aerotolerant bacterial consortium composed of *Clostridium* and *Flavobacterium* that had the ability to produce ethanol. Moreover, the production of hydrogen by a consortium composed of *Clostridium, Klebsiella, Acinetobacter, Bacillus, Pseudomonas, Ruminococcus* and *Bacteroides* retrieved from sludge anaerobic digester has been evaluated [19]. These studies highlight the importance of aerobic bacterial members to deconstruct lignocellulose, such as those belonging to *Flavobacterium, Klebsiella* and *Acinetobacter*.

Concerning fungi, *Acremonium* is considered to be a very important organism for the production of (hemi)cellulases, as compared to *Trichoderma reesii*[54,55]. Moreover, *Penicillium* species have an elaborate enzymatic machinery to deconstruct lignocellulose, such as vanillyl-alcohol oxidases, copper-dependent polysaccharide monooxygenases [56], galactosidases, mannosidases and fucosidases [57]. In our consortia, the *Malassezia* species may have acted as sugar monomer cheaters, and their high abundances in RWS might be related with their high abundance in the SS. *Trichosporon* species are anamorphic basidiomycetous yeasts that are widespread in nature [58,59]. The presence of *Trichosporon* in the gut of xylophagous insects is probably facilitated by their ability to assimilate and transform lignin and various phenolic compounds [60]. Recent results from our group confirm the ability of our *Trichosporon* isolates to produce cellobiohydrolases and β-xylosidases (data not shown). *Trichosporon*, an oil-rich yeast, has high biotechnological potential and has been shown to be tolerant to furanic compounds [61,10]. It has been reported that the use of single fungal strains can be highly efficient to deconstruct specific compounds, such as lignin [62]. However, the breakdown of lignocellulose, for example, for biofuel production, often encounters great recalcitrance which will likely require the synergistic action of multispecies consortia (with higher gene diversity) to overcome it [2]. Some enzymatic transformations might be slow in such communities as a consequence of inter-specific competition or even antagonism. To resolve these issues, enzyme cocktails that come from multispecies consortia may be retrieved and applied directly to the plant waste materials.

On theoretical grounds, one could bring up compelling evidence pointing to the scientific danger of attempting to link phylogeny with function by using PICRUSt, and the arguments extend to the limitation of current databases used in the software. However, the linkage might be regarded in a loose manner, including genes/functions that might be actually "floating" in the horizontal gene pool of the community. Thus, such functions are thought of as being not tightly linked to a phylogenetically determined species. In both microbial consortia, the uncertainty of the prediction as revealed by the NSTI was very reduced compared with that in the SS, thus indicating fair reliability and accuracy in the metagenome reconstruction (Figure 6A). The analysis predicted the enrichment of several genes in our consortia that were potentially involved in lignocellulose degradation, and also showed that TWS was possibly a poorer selector of such genes than RWS (Table 1). It was predicted that some peroxidases (EC:1.11.1-), classified as an "auxiliary activities" (AA2 family) in the CAZy (Carbo-hydrate-Active EnZymes database) [63], were enriched in both consortia by the sequential transfers. Such enzymes oxidize phenolic and non-phenolic aromatic compounds and can modify lignin polymers [56]. These enzymes were more evident in the RWS consortium, supporting its potential to act on lignin. Furthermore, glycolate oxidase (EC:1.1.3.15; an oxidoreductase capable of oxidizing glycolate to glyoxylate, producing reactive oxygen species) was progressively enriched in the TWS consortium, suggesting a correlation with the metabolism of furanic compounds. Glycolate oxidases are classified in CAZy as family AA7. In this family, we found gluco-oligosaccharide oxidases capable of oxidizing a variety of carbohydrates and possibly involved in the biotransformation of lignocellulosic compounds [64].

Concerning (hemi)cellulose bioconversion, genes encoding alpha-L-fucosidase (EC:3.2.1.51) (families GH29 and GH95 in CAZy) and alpha-L-arabinofuranosidases (E.C. 3.2.1.55) (GH51 and GH54) were abundant in RWS (T1) and also in TWS (T10). The alpha-L-arabinofuranosidases and alpha-L-fucosidases are the most important (hemi)cellulosic accessory enzymes that catalyze the hydrolysis of arabinans, arabinoxylans and alpha-l-fucosyl residues in agricultural waste [65-67]. In an anaerobic microbial community decomposing poplar wood chips, high levels of genes for these enzymes were found, especially in Bacteroidetes genomes [30]. In our microbial consortia, such genes may have come from *Sphingobacterium* members.

Endo-1.4-beta-xylanases (EC:3.2.1.8) (GH families 5, 8, 10, 11, 43) perform the endohydrolysis of (1 - 4)-beta-D-xylosidic linkages in xylans. These genes were also predicted to occur in both our consortia. Other genes involved in the deconstruction of xylan were also identified; for instance, the gene that encodes xylan-1,4-beta-xylosidase (EC:3.2.1.37) was enriched in T10 in RWS (Table 1). Beta-xylosidases are exotype glycosidases that hydrolyze short xylo-oligomers to single xylose units [68]. These enzymes were active in the secretomes of both microbial consortia, suggesting the expression of these bacterial genes.

The beta-galactosidases, which hydrolyze beta-galactosidic bonds between galactose and its organic functional group and can act on xyloglucans [69], were highly active in both consortia (Figure 6B). The beta-mannosidases (EC:3.2.1.25), involved in the hydrolysis of terminal, non-reducing beta-D-mannosyl residues in beta-D-mannosides [70], were lowly abundant in our consortia as compared to SS, but such activities were also detected in the secretome. The activities of these last two types of enzymes were higher in TWS1 than in RWS (Figure 6B), suggesting the raised availability of beta-D-galactose and beta-D-mannosyl residues in TWS, possibly released due to the torrefaction.

Conversely, mannan endo-1,4-beta-mannosidase (E.C: 3.2.1.78) (GH5, GH26, GH113 and AA10) related genes were enriched in RWS compared to SS. These enzymes are involved in the random hydrolysis of (1 - 4)-beta-D-mannosidic linkages in mannans, galactomannans and glucomannans. Cellobiohydrolases (endo- and exoglucanases) showed low activity in the secretome of TWS, suggesting the presence of high cellulose levels in the untreated compared with the torrefied wheat tissue. Several genes that encode beta-glucosidases were enriched in both consortia compared with SS, suggesting that the conversion of cellobiose to glucose is an important function in these consortia. Finally, cleavage and further metabolism of di-sugars was represented by several predicted enzymes. For example, alpha-L-rhamnosidase related genes were highly abundant in TWS at T10, compared to SS. These enzymes cleave terminal alpha-l-rhamnose from a large number of natural glycosides, and are relevant for application in citrus fruit juice and wine industries [71].

6.4 CONCLUSION

In this study, the application of DNA-based high-throughput sequencing technology allowed the characterization of novel bacterial-fungal consortia growing on wheat straw. The data, in conformity with our previous work [29], indicate that mixed microbial consortia, encompassing specific biodegradative (mainly affiliated to *Klebsiella, Sphingobacterium, Flavobacterium, Acinetobacter, Penicillium* and *Acremonium*) and cheater types, are selected by the specific lignocellulose substrate. The approach allowed us to identify interesting yeasts, such as *Coniochaeta* and *Trichosporon,* that are possibly involved in plant biomass degradation and/or conversion of furanic compounds. Application of PICRUSt to predict the functional profile (using 16S rRNA sequences), in conjunction with the evaluation of enzymatic activities in the consortial secretomes, allowed the inference of genes/proteins that were presumptively involved in lignocellulose degradation (such as peroxidases, beta-mannases, beta-galactosidases, alpha-L-fucosidases, alpha-L-arabinofuranosidases and beta-glucosidases). Finally, assays of the degradation of other plant waste and quantification of initial and final products (for example, cello-oligosaccharides) might demonstrate the degradative potential that is needed for future biofuel production. A closer analysis of the metagenome and mobilome in our consortia will clarify the enzymatic profile and biotechnological potential present and can also shed light on the potential role of HGT in its evolution. A greater understanding of the ecological interactions between consortium members during plant biomass biodegradation is required for further progress in this area.

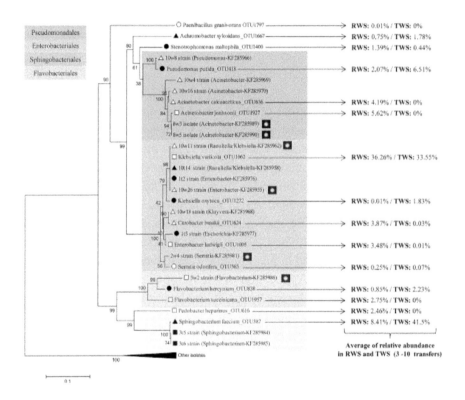

FIGURE 5: Neighbour-joining phylogenetic tree of partial bacterial 16S rRNA gene sequences (274 nucleotides) from bacterial strains and most abundant OTUs in the enriched cultures (RWS and TWS). Dots represent (hemi)cellulolytic activity in agar plates (CMC-ase and xylanase). Taxonomic affiliation and accession numbers of isolates from the GenBank are shown in parentheses. Right side shows average relative abundance of each OTU in the 3 to 10 transfers. Circles, squares and triangles represent sequences retrieved in T1, T3 and T10, respectively in RWS (white) and TWS (black).

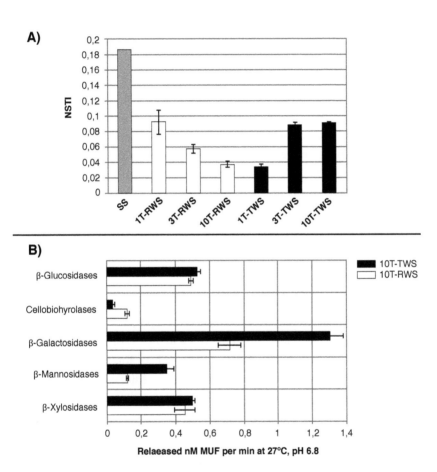

FIGURE 6: NSTI values and quantification of enzymatic activities by methylumbelliferyl (MUF)-substrates. (A) NSTI values in the soil inoculum (SS) and in enriched cultures (RWS and TWS) along the sequential batches. (B) Quantification of specific enzymatic activities by MUF-substrates in the consortial secretome of the last transfer in both enriched cultures.

6.5 METHODS

6.5.1 LIGNOCELLULOLYTIC MICROBIAL CONSORTIA CONSTRUCTION

Soil samples (n=10) were collected and mixed from a forest (top layer, 0 to 10 cm depth) in Groningen, The Netherlands (53.41 N; 6.90 E). Cell soil -suspensions were prepared by adding 10 g of fresh sampled soil to 250-mL flasks containing 10 g of sterile gravel in 90 mL of mineral salt medium (MSM). The flasks were shaken for 20 min at 250 rpm, and aliquots (250 μL) of soil suspension were added to triplicate Erlenmeyer flasks containing 25 mL of MSM with 1% lignocellulose substrate (0.25 g in 25 mL), amended with a trace element and vitamin solution. Two different substrates were thus obtained, to serve as carbon sources: i) "raw" wheat straw (RWS) and ii) heat-treated (torrefied) wheat straw (TWS). The flasks were incubated at 25°C with shaking at 100 rpm. Cultures were monitored at regular time intervals, and once the systems reached high cell density (log 7 to 8 cells/mL), aliquots (25 μL microbial suspension with fibrous material) were transferred to 25 mL of fresh medium. Finally, a sample of soil suspension and duplicate flask samples (selected based on reported DGGE analyses) at the final batches were taken from the RWS and TWS consortia after 1 (T1), 3 (T3) and 10 (T10) transfers (n=13) and used for total DNA extractions and pyrosequencing as described below. Details of the experimental setup, substrate preparation, growth in sequential-batch cultures (cell counts and qPCR) and negative controls have been reported [29].

6.5.2 TOTAL DNA EXTRACTION AND BACTERIAL 16S RRNA/ FUNGAL ITS1 PYROSEQUENCING

The DNA extraction from the cultures and from the soil suspension was performed with the Power Soil DNA extraction kit (MoBio® Laboratories Inc., Carlsbad, CA, USA) according to the manufacturer's instructions. The bacterial 16S rRNA gene amplicons were generated using the universal primer set GM3F (5′-TAGAGTTTGATCMTGGC-3′) and 926R (5′-TCCGTCAATTC-

MTTTGAGTTT-3') [72]. For fungal communities, specific primers ITS1F (5'-CTTGGTCATTTAGAGGAAGTAA-3') and ITS4 (5'-TCCTCCGCT-TATTGATATGC-3') were used to amplify the ITS1, 5.8S rRNA and ITS2 regions of fungal rRNA [73]. The pyrosequencing reactions were performed with the new flow pattern B (software v2.8) and the FLX-Titanium chemistry (Roche/454 Life Sciences) at LGC Genomics (Berlin, Germany). Sequencing of 13 samples resulted in totals of 117,042 and 35,506 raw reads for bacterial 16S rRNA gene and fungal ITS1, respectively.

6.5.3 SEQUENCING PROCESSING AND STATISTICAL ANALYSIS

Pyrosequencing raw data were processed using the Quantitative Insights Into Microbial Ecology (QIIME) toolkit [74]. The sequences were quality trimmed using the following parameters: quality score of >25, sequence length of 300 to 900 bp (for 16S rRNA) and 100 to 900 bp (for ITS1), maximum homopolymer of 6, 0 maximum ambiguous bases and 0 mismatched bases in the primer. In order to select for the same region of each gene, we retrieved sequences with primers GM3F (for the bacterial 16S rRNA) and ITS1F (for fungal ITS1). We identified bacterial and fungal players by grouping highly similar sequences into OTUs (at 97% of nucleotide identity) using UCLUST [75] followed by selection of representative sequences. Subsequently, chimeric sequences were detected using ChimeraSlayer [76] and deleted. Additionally, clusters consisting only of singleton sequences were removed in order to avoid sequencing errors. Analyses of community composition, as well as richness and diversity estimators, were carried out at a depth of 1,400 bacterial and 550 fungal rarefied sequences per sample, to eliminate the effect of sampling effort. QIIME was also used to generate alpha- and beta-diversity metrics, including OTU richness, CRE, SWI, PD and UniFrac distances. Taxonomic classifications at the phylum and order level of each OTU were done using RDP classifier and BLAST algorithms against the Greengenes (16S rRNA), UNITE and GenBank (ITS1) databases. The assignment of each OTU on the genus level was based on the best BLASTn hit against the GenBank database (Additional file 6). Abundant OTUs, more than 10 and 5 sequences per OTU for the 16S rRNA and ITS1 data respectively, were selected to construct the PCA using Canoco software v4.52 (Wageningen, The Netherlands). The 16S rRNA and ITS1sequences were deposited in GenBank with SRA accession numbers [SRP039495].

6.5.4 DETECTION OF ABUNDANT OTUS AS BACTERIAL STRAINS

Isolation of bacterial strains along the experiment and the determination of their taxonomic identification and (hemi)cellulolytic activity in agar plates (with CMC and xylan from birchwood) were previously reported [29]. Partial 16S rRNA gene sequences of these strains were obtained using the same forward primer as used for the 16S rRNA pyrosequencing. To detect which OTUs were possibly recovered as bacterial strains, we constructed a phylogenetic tree using the sequences of the 15 most abundant bacterial OTUs (representing over 72% and 88% of the consortia in RWS and TWS, respectively) in addition to 20 sequences retrieved from the bacterial strains. Sequences were aligned using the ClustalW software, and the phylogenetic analyses (p-distance) were conducted with MEGA v5.1 using the Neighbour-Joining method [77]. The evolutionary distances were computed using the Kimura-2 parameter method and are in the units of the number of base substitutions per site (note scale bar - Figure 5). The branches were tested with bootstrap analyses (1,000 replications). Furthermore, (hemi)cellulolytic activity was linked to the OTUs based on the similarity and clustering with the bacterial strains.

6.5.5 RECONSTRUCTING THE BACTERIAL METAGENOMES WITH PICRUST SOFTWARE

The bacterial metagenomes were reconstructed using the PICRUSt software [27]. A PICRUSt-compatible OTU table was constructed in QIIME (at 97% of nucleotide identity) using the newest available reference closed-reference OTU collection in the Greengenes database [78]. In order to normalize the data, we used 1,000 rarefied sequences of bacterial 16S rRNA per sample as an input. Subsequently, the normalization by 16S rRNA copies number per OTU was performed with the normalize_ by_copy_number.py script and IMG database information. The metagenome inference was done using the predict_metagenomes.py script with the normalized OTU table as an input. We analyzed the average number of annotated genes in each sample and selected the top 40 known genes related with the bioconversion of lignocellulose. PICRUSt also calculated the NSTI, a measure of prediction uncertainty presented here in a comparative way along the sequential batches in both consortia datasets.

6.5.6 QUANTIFICATION OF SPECIFIC ENZYMATIC ACTIVITIES RELATED TO THE (HEMI)CELLULOSE BIOCONVERSION

In order to evaluate the metabolic potential in the degradation of (hemi) cellulose and the expression of selected genes identified by the PICRUSt prediction, we quantified specific enzymatic activities in samples of 2 mL from the enriched cultures after final batch (T10), when the communities are stable. Microbial cells plus wheat substrate were harvested by centrifugation for 3 min at 12,000 rpm, the supernatant (secretome) was recovered and tested for enzymatic activity using MUF-beta-D-xylopyranoside, MUF-beta-D-mannopyranoside, MUF-beta-D-galactopyranoside, MUF-beta-D-cellobioside and MUF-beta-D-glucopyranoside as substrates. The reaction mixture consisted of 10 µl of MUF-substrate (10 mM in dimethyl sulfoxide), 15 µL of McIlvaine buffer (pH 6.8) and 25 µL of each supernatant. The mixture was incubated at 27°C for 45 min in the dark, and the reaction was stopped by adding 150 µL of 0.2 M glycine-NaOH buffer (pH 10.4). Fluorescence was measured at an excitation of 365 nm and emission of 445 nm. We also evaluated the fluorescence without the MUF-substrate as a negative control. Enzyme activities were determined from the fluorescence units using a standard calibration curve and expressed as rates of MUF production (nM MUF per min at 27°C, pH 6.8).

REFERENCES

1. Chandel AK, Singh OV: Weedy lignocellulosic feedstock and microbial metabolic engineering: advancing the generation of 'Biofuel'. Appl Microbiol Biotechnol 2011, 89(5):1289-1303.

2. Zuroff TR, Curtis WR: Developing symbiotic consortia for lignocellulosic biofuel production. Appl Microbiol Biotechnol 2012, 93:1423-1435.

3. Yamada R, Hasunuma T, Kondo A: Endowing non-cellulolytic microorganisms with cellulolytic activity aiming for consolidated bioprocessing. Biotechnol Adv 2013, 31(6):754-763.

4. Liguori R, Amore A, Faraco V: Waste valorization by biotechnological conversion into added value products. Appl Microbiol Biotechnol 2013, 97(14):6129-6147.

5. Hasunuma T, Okazaki F, Okai N, Hara KY, Ishii J, Kondo A: A review of enzymes and microbes for lignocellulosic biorefinery and the possibility of their application to consolidated bioprocessing technology. Bioresour Technol 2013, 135:513-522.

6. Liu G, Qin Y, Li Z, Qu Y: Development of highly efficient, low-cost lignocellulolytic enzyme systems in the post-genomic era. Biotechnol Adv 2013, 31(6):962-975.

7. Kanokratana P, Mhuantong W, Laothanachareon T, Tangphatsornruang S, Eurwilaichitr L, Pootanakit K, Champreda V: Phylogenetic analysis and metabolic potential of microbial communities in an industrial bagasse collection site. Microb Ecol 2013, 66(2):322-334.

8. Berlemont R, Martiny AC: Phylogenetic distribution of potential cellulases in bacteria. Appl Environ Microbiol 2013, 79(5):1545-1554.

9. López MJ, Vargas-Garcia MC, Suarez-Estrella F, Nichols NN, Dien BS, Moreno J: Lignocellulose-degrading enzymes produced by the ascomycete Coniochaeta ligniaria and related species: application for a lignocellulosic substrate treatment. Enzyme Microb Technol 2007, 40:794-800.

10. Johnson EA: Biotechnology of non-Saccharomyces yeasts-the basidiomycetes. Appl Microbiol Biotechnol 2013, 97(17):7563-7577.

11. Gusakov AV: Alternatives to Trichoderma reesei in biofuel production. Trends Biotechnol 2011, 29(9):419-425.

12. Lee DJ, Show KY, Wang A: Unconventional approaches to isolation and enrichment of functional microbial consortium - a review. Bioresour Technol 2013, 136:697-706.

13. Suenaga H: Targeted metagenomics: a high-resolution metagenomics approach for specific gene clusters in complex microbial communities. Environ Microbiol 2012, 14(1):13-22.

14. Neufeld JD, Dumont MG, Vohra J, Murrell JC: Methodological considerations for the use of stable isotope probing in microbial ecology. Microb Ecol 2006, 53(3):435-442.

15. Wongwilaiwalin S, Rattanachomsri U, Laothanachareon T, Eurwilaichitr L, Igarashi Y, Champreda V: Analysis of a thermophilic lignocellulose degrading microbial consortium and multi-species lignocellulolytic enzyme system. Enzyme Microb Technol 2010, 47:283-290.

16. Zhang Q, Tian M, Tang L, Li H, Li W, Zhang J, Zhang H, Mao Z: Exploration of the key microbes involved in the cellulolytic activity of a microbial consortium by serial dilution. Bioresour Technol 2013, 132:395-400.

17. Ho KL, Lee DJ, Su A, Chang JS: Biohydrogen from cellulosic feedstock: dilution-to-stimulation approach. Int J Hydrogen Energy 2012, 37:15582-15587.

18. Cheng JR, Zhu MJ: A novel co-culture strategy for lignocellulosic bioenergy production: a systematic review. Int J Mod Biol Med 2012, 1(3):166-193.

19. Zuroff TR, Xiques SB, Curtis WR: Consortia-mediated bioprocessing of cellulose to ethanol with a symbiotic Clostridium phytofermentans/yeast co-culture. Biotechnol Biofuels 2013, 6(1):59.
20. Wirth R, Kovács E, Maróti G, Bagi Z, Rákhely G, Kovács KL: Characterization of a biogas-producing microbial community by short-read next generation DNA sequencing. Biotechnol Biofuels 2012, 5(1):41.
21. Wongwilaiwalin S, Laothanachareon T, Mhuantong W, Tangphatsornruang S, Eurwilaichitr L, Igarashi Y, Champreda V: Comparative metagenomic analysis of microcosm structures and lignocellulolytic enzyme systems of symbiotic biomass-degrading consortia. Appl Microbiol Biotechnol 2013, 97(20):8941-8954.
22. Štursová M, Zifčáková L, Leigh MB, Burgess R, Baldrian P: Cellulose utilization in forest litter and soil: identification of bacterial and fungal decomposers. FEMS Microbiol Ecol 2012, 80(3):735-746.
23. Ma A, Zhuang X, Wu J, Cui M, Lv D, Liu C, Zhuang G: Ascomycota members dominate fungal communities during straw residue decomposition in arable soil. PLoS One 2013, 8(6):e66146.
24. Trifonova R, Postma J, Ketelaars JJ, van Elsas JD: Thermally treated grass fibers as colonizable substrate for beneficial bacterial inoculum. Microb Ecol 2008, 56(3):561-571.
25. Wang XJ, Yuan XF, Wang H, Li J, Wang XF, Cui ZJ: Characteristics and community diversity of a wheat straw-colonizing microbial community. Afr J Biotechnol 2011, 10(40):7853-7861.
26. Pankratov TA, Ivanova AO, Dedysh SN, Liesack W: Bacterial populations and environmental factors controlling cellulose degradation in an acidic Sphagnum peat. Environ Microbiol 2011, 13:1800-1814.
27. Langille MG, Zaneveld J, Caporaso JG, McDonald D, Knights D, Reyes JA, Clemente JC, Burkepile DE, Vega Thurber RL, Knight R, Beiko RG, Huttenhower C: Predictive functional profiling of microbial communities using 16S rRNA marker gene sequences. Nat Biotechnol 2013, 31(9):814-821.
28. McHardy IH, Goudarzi M, Tong M, Ruegger PM, Schwager E, Weger JR, Graeber TG, Sonnenburg JL, Horvath S, Huttenhower C, McGovern DP, Fornace AJ Jr, Borneman J, Braun J: Integrative analysis of the microbiome and metabolome of the human intestinal mucosal surface reveals exquisite inter-relationships. Microbiome 2013, 1:17.
29. Jiménez DJ, Korenblum E, van Elsas JD: Novel multi-species microbial consortia involved in lignocellulose and 5-hydroxymethylfurfural bioconversion. Appl Microbiol Biotechnol 2014, 98(6):2789-2803.
30. van der Lelie D, Taghavi S, McCorkle SM, Li LL, Malfatti SA, Monteleone D, Donohoe BS, Ding SY, Adney WS, Himmel ME, Tringe SG: The metagenome of an anaerobic microbial community decomposing poplar wood chips. PLoS One 2012, 7(5):e36740.
31. Yang H, Wu H, Wang X, Cui Z, Li Y: Selection and characteristics of a switchgrass-colonizing microbial community to produce extracellular cellulases and xylanases. Bioresour Technol 2011, 102:3546-3550.
32. Nikolopoulos N, Isemin R, Atsonios K, Kourkoumpas D, Kuzmin S, Mikhalev A, Nikolopoulos A, Agraniotis M, Grammelis P, Em K: Modeling of wheat straw tor-

refaction as a preliminary tool for process design. Waste Biomass Valor 2013, 4:409-420.

33. Semenov AV, Silva MCPe, Szturc-Koestsier AE, Schmitt H, Salles JS, van Elsas JD: Impact of incorporated fresh 13C potato tissues on the bacterial and fungal community composition of soil. Soil Biol Biochem 2012, 49:88-95.

34. DeAngelis KM, Allgaier M, Chavarria Y, Fortney JL, Hugenholtz P, Simmons B, Sublette K, Silver WL, Hazen TC: Characterization of trapped lignin-degrading microbes in tropical forest soil. PLoS One 2011, 6(4):e19306.

35. Baldrian P, Kolařík M, Stursová M, Kopecký J, Valášková V, Větrovský T, Zifčáková L, Snajdr J, Rídl J, Vlček C, Voříšková J: Active and total microbial communities in forest soil are largely different and highly stratified during decomposition. ISME J 2012, 6(2):248-258.

36. Romano N, Gioffré A, Sede SM, Campos E, Cataldi A, Talia P: Characterization of cellulolytic activities of environmental bacterial consortia from an Argentinian native forest. Curr Microbiol 2013, 67(2):138-147.

37. Hui W, Jiajia L, Yucai L, Peng G, Xiaofen W, Kazuhiro M, Zongjun C: Bioconversion of un-pretreated lignocellulosic materials by a microbial consortium XDC-2. Bioresour Technol 2013, 136:481-487.

38. Eichorst SA, Kuske CR: Identification of cellulose-responsive bacterial and fungal communities in geographically and edaphically different soils by using stable isotope probing. Appl Environ Microbiol 2012, 78(7):2316-2327.

39. Aylward FO, Burnum KE, Scott JJ, Suen G, Tringe SG, Adams SM, Barry KW, Nicora CD, Piehowski PD, Purvine SO, Starrett GJ, Goodwin LA, Smith RD, Lipton MS, Currie CR: Metagenomic and metaproteomic insights into bacterial communities in leaf-cutter ant fungus gardens. ISME J 2012, 6(9):1688-1701.

40. Wang Y, Liu Q, Yan L, Gao Y, Wang Y, Wang W: A novel lignin degradation bacterial consortium for efficient pulping. Bioresour Technol 2013, 139:113-119.

41. Voříšková J, Baldrian P: Fungal community on decomposing leaf litter undergoes rapid successional changes. ISME J 2013, 7(3):477-486.

42. Okeke BC, Lu J: Characterization of a defined cellulolytic and xylanolytic bacterial consortium for bioprocessing of cellulose and hemicelluloses. Appl Biochem Biotechnol 2011, 163(7):869-881.

43. Suen G, Scott JJ, Aylward FO, Adams SM, Tringe SG, Pinto-Tomás AA, Foster CE, Pauly M, Weimer PJ, Barry KW, Goodwin LA, Bouffard P, Li L, Osterberger J, Harkins TT, Slater SC, Donohue TJ, Currie CR: An insect herbivore microbiome with high plant biomass-degrading capacity. PLoS Genet 2010, 6(9):e1001129.

44. Kog S, Ogawa J, Choi Y, Shimizu S: Novel bacterial peroxidase without catalase activity from Flavobacterium meningosepticum: purification and characterization. Biochim Biophys Acta 1999, 1435:117-126.

45. Chandra R, Abhishek A, Sankhwar M: Bacterial decolorization and detoxification of black liquor from rayon grade pulp manufacturing paper industry and detection of their metabolic products. Bioresour Technol 2011, 102(11):6429-6436.

46. McBride MJ, Xie G, Martens EC, Lapidus A, Henrissat B, Rhodes RG, Goltsman E, Wang W, Xu J, Hunnicutt DW, Staroscik AM, Hoover TR, Cheng YQ, Stein JL: Novel features of the polysaccharide-digesting gliding bacterium Flavobacterium

johnsoniae as revealed by genome sequence analysis. Appl Environ Microbiol 2009, 75(21):6864-6875.

47. Deangelis KM, D'Haeseleer P, Chivian D, Fortney JL, Khudyakov J, Simmons B, Woo H, Arkin AP, Davenport KW, Goodwin L, Chen A, Ivanova N, Kyrpides NC, Mavromatis K, Woyke T, Hazen TC: Complete genome sequence of "Enterobacter lignolyticus" SCF1. Stand Genomic Sci 2011, 5(1):69-85.

48. Talia P, Sede SM, Campos E, Rorig M, Principi D, Tosto D, Hopp HE, Grasso D, Cataldi A: Biodiversity characterization of cellulolytic bacteria present on native Chaco soil by comparison of ribosomal RNA genes. Res Microbiol 2012, 163(3):221-232.

49. Matsuyama H, Katoh H, Ohkushi T, Satoh A, Kawahara K, Yumoto I: Sphingobacterium kitahiroshimense sp. nov., isolated from soil. Int J Syst Evol Microbiol 2008, 58(Pt 7):1576-1579.

50. Tamboli DP, Telke AA, Dawkar VV, Jadhav SB, Govindwar SP: Purification and characterization of bacterial aryl alcohol oxidase from Sphingobacterium sp. ATM and its uses in textile dye decolorization. Biotech Bioprocess Eng 2011, 16:661-668.

51. Zhou J, Huang H, Meng K, Shi P, Wang Y, Luo H, Yang P, Bai Y, Zhou Z, Yao B: Molecular and biochemical characterization of a novel xylanase from the symbiotic Sphingobacterium sp. TN19. Appl Microbiol Biotechnol 2009, 85(2):323-333.

52. López MJ, Nichols NN, Dien BS, Moreno J, Bothast RJ: Isolation of microorganisms for biological detoxification of lignocellulosic hydrolysates. Appl Microbiol Biotechnol 2004, 64(1):125-131.

53. Ronan P, Yeung W, Schellenberg J, Sparling R, Wolfaardt GD, Hausner M: A versatile and robust aerotolerant microbial community capable of cellulosic ethanol production. Bioresour Technol 2013, 129:156-163.

54. Fujii T, Fang X, Inoue H, Murakami K, Sawayama S: Enzymatic hydrolyzing performance of Acremonium cellulolyticus and Trichoderma reesei against three lignocellulosic materials. Biotechnol Biofuels 2009, 2(1):24.

55. Hideno A, Inoue H, Fujii T, Yano S, Tsukahara K, Murakami K, Yunokawa H, Sawayama S: High-coverage gene expression profiling analysis of the cellulase producing fungus Acremonium cellulolyticus cultured using different carbon sources. Appl Microbiol Biotechnol 2013, 97(12):5483-5492.

56. Levasseur A, Drula E, Lombard V, Coutinho PM, Henrissat B: Expansion of the enzymatic repertoire of the CAZy database to integrate auxiliary redox enzymes.

57. Biotechnol Biofuels 2013, 6(1):41.

58. Baldrian P, Voříšková J, Dobiášová P, Merhautová V, Lisá L, Valášková V: Production of extracellular enzymes and degradation of biopolymers by saprotrophic microfungi from the upper layers of forest soil. Plant Soil 2011, 338:111-125.

59. Laitila A, Wilhelmson A, Kotaviita E, Olkku J, Home S, Juvonen R: Yeasts in an industrial malting ecosystem. J Ind Microbiol Biotechnol 2006, 33(11):953-966.

60. Heneberg P, Řezáč M: Two Trichosporon species isolated from Central-European mygalomorph spiders (Araneae: Mygalomorphae). A Van Leeuw J Microb 2013, 103(4):713-721.

61. Gujjari P, Suh SO, Lee CF, Zhou JJ: Trichosporon xylopini sp. nov., a hemicellulose-degrading yeast isolated from the wood-inhibiting beetle Xylopinus saperdioides. Int J Syst Evol Microbiol 2011, 61:2538-2542.

62. Chen X, Li Z, Zhang X, Hu F, Ryu DD, Bao J: Screening of oleaginous yeast strains tolerant to lignocellulose degradation compounds. Appl Biochem Biotechnol 2009, 159(3):591-604.

63. Šnajdr J, Steffen KT, Hofrichter M, Baldrian P: Transformation of 14C-labelled lignin and humic substances in forest soil by the saprobic basidiomycetes Gymnopus erythropus and Hypholoma fasciculare. Soil Biol Biochem 2010, 42:1541-1548.

64. Cantarel BL, Coutinho PM, Rancurel C, Bernard T, Lombard V, Henrissat B: The Carbohydrate-Active EnZymes database (CAZy): an expert resource for Glycogenomics. Nucleic Acids Res 2009, 37(Database issue):D233-D238.

65. van Hellemond EW, Leferink NG, Heuts DP, Fraaije MW, van Berkel WJ: Occurrence and biocatalytic potential of carbohydrate oxidases. Adv Appl Microbiol 2006, 60:17-54.

66. Numan MT, Bhosle NB: Alpha-L-arabinofuranosidases: the potential applications in biotechnology. J Ind Microbiol Biotechnol 2006, 33:247-260.

67. Benesová E, Lipovová P, Dvoráková H, Králová B: α-L-fucosidase from Paenibacillus thiaminolyticus: its hydrolytic and transglycosylation abilities. Glycobiology 2013, 23(9):1052-1065.

68. Wright DW, Moreno-Vargas AJ, Carmona AT, Robina I, Davies GJ: Three dimensional structure of a bacterial α-l-fucosidase with a 5-membered iminocyclitol inhibitor. Bioorg Med Chem 2013, 21(16):4751-4754.

69. Zhou J, Bao L, Chang L, Liu Z, You C, Lu H: β-xylosidase activity of a GH3 glucosidase/xylosidase from yak rumen metagenome promotes the enzymatic degradation of hemicellulosic xylans. Lett Appl Microbiol 2012, 54(2):79-87.

70. Husain Q: Beta galactosidases and their potential applications: a review. Crit Rev Biotechnol 2010, 30(1):41-62.

71. Scully ED, Geib SM, Hoover K, Tien M, Tringe SG, Barry KW, del Rio GT, Chovatia M, Herr JR, Carlson JE: Metagenomic profiling reveals lignocellulose degrading system in a microbial community associated with a wood-feeding beetle. PLoS One 2010, 8(9):e73827.

72. Yadav V, Yadav PK, Yadav S, Yadav KD: α-l-Rhamnosidase: a review. Process Biochem 2010, 45(8):1226-1235.

73. Klindworth A, Pruesse E, Schweer T, Peplies J, Quast C, Horn M, Glöckner FO: Evaluation of general 16S ribosomal RNA gene PCR primers for classical and next-generation sequencing-based diversity studies. Nucleic Acids Res 2013, 41(1):e1.

74. Bokulich NA, Mills DA: Improved selection of internal transcribed spacer-specific primers enables quantitative, ultra-high-throughput profiling of fungal communities. Appl Environ Microbiol 2013, 79(8):2519-2526.

75. Caporaso JG, Kuczynski J, Stombaugh J, Bittinger K, Bushman FD, Costello EK, Fierer N, Peña AG, Goodrich JK, Gordon JI, Huttley GA, Kelley ST, Knights D, Koenig JE, Ley RE, Lozupone CA, McDonald D, Muegge BD, Pirrung M, Reeder J, Sevinsky JR, Turnbaugh PJ, Walters WA, Widmann J, Yatsunenko T, Zaneveld J, Knight R: QIIME allows analysis of high-throughput community sequencing data. Nat Methods 2010, 7(5):335-336.

76. Edgar RC: Search and clustering orders of magnitude faster than BLAST. Bioinformatics 2010, 26:2460-2461.

77. Haas BJ, Gevers D, Earl AM, Feldgarden M, Ward DV, Giannoukos G, Ciulla D, Tabbaa D, Highlander SK, Sodergren E, Methé B, DeSantis TZ, Petrosino JF, Knight R, Birren BW, The Human Microbiome Consortium: Chimeric 16S rRNA sequence formation and detection in Sanger and 454-pyrosequenced PCR amplicons. Genome Res 2011, 21:494-504.

78. Tamura K, Peterson D, Peterson N, Stecher G, Nei M, Kumar S: MEGA5: molecular evolutionary genetics analysis using maximum likelihood, evolutionary distance, and maximum parsimony methods. Mol Biol Evol 2011, 28(10):2731-2739.

79. McDonald D, Price MN, Goodrich J, Nawrocki EP, DeSantis TZ, Probst A, Andersen GL, Knight R, Hugenholtz P: An improved Greengenes taxonomy with explicit ranks for ecological and evolutionary analyses of bacteria and archaea. ISME J 2012, 6(3):610-618.

There are several supplemental files that are not available in this version of the article. To view this additional information, please use the citation on the first page of this chapter.

PART II

PRETREATMENTS

CHAPTER 7

PRETREATMENT OF LIGNOCELLULOSIC BIOMASS USING MICROORGANISMS: APPROACHES, ADVANTAGES, AND LIMITATIONS

THOMAS CANAM, JENNIFER TOWN, KINGSLEY IROBA, LOPE TABIL, AND TIM DUMONCEAUX

7.1 INTRODUCTION

Much of Earth's recent geologic history is dominated by periods of extensive glaciation, with relatively low global mean temperatures and correspondingly low atmospheric CO_2 concentrations [1]. The current interglacial period stands out as an anomaly because the atmospheric CO_2 concentration has risen sharply above the range of approximately 180–280 parts per million by volume that has defined the past 420,000 years to reach levels that are nearly 40% higher than the biosphere has experienced over this time frame [2]. This rapid increase in CO_2 concentration is primarily due to the release of ancient fixed atmospheric CO_2 into the modern atmosphere through the combustion of fossil fuel resources over the past 200 years. Since it is clear from ice core records that atmospheric

CO_2 concentration has a strong positive correlation to global temperature, it is expected that changes to global climate are forthcoming [3]. There are substantial uncertainties regarding the ability of terrestrial and oceanic carbon sinks to absorb this anthropogenic CO_2 on time scales that are relevant to human society [2], so the continued release of ancient CO_2 into the modern atmosphere at current rates carries with it an important risk of inducing climate changes of unknown amplitude along with a host of ancillary changes that are difficult to predict with certainty. This has led to the search for alternatives to fossil fuels to meet a rising global energy demand, and one such option is the use of extant organic matter to produce energy. This resource contains carbon that was fixed from the modern atmosphere, which means it does not result in a net increase in atmospheric CO_2 upon combustion.

Meeting the world's energy demands requires resources that are abundant and inexpensive to produce. Biomass from forestry and agricultural activities is certainly a candidate, as hundreds of millions of tonnes of agricultural waste from rice, wheat, corn, and other crops are produced worldwide, which could generate billions of litres of ethanol [4]. For ethanol, butanol, methane, and other biofuels to be produced economically, however, requires an integrated approach, with a number of value-added co-products produced in addition to the energy—a "biorefinery" that stands in analogy to petroleum refineries that produce both energy and a wide range of petroleum-based chemicals and products [5-7]. The biorefinery concept is hardly new, as the industrial-scale bacterial fermentation of starch to acetone and butanol (A-B) was developed a century ago. These A-B fermentations were done on an industrial scale in the West during World War I and persisted into the 1950's. They continued in Russia until late in the Soviet era, ultimately using corn cobs and other agricultural residues as input [8]. However, releasing the energy and co-product potential of plant-based material requires energy inputs and processing steps, as discussed below; this hinders the ability of biofuels to compete economically with petroleum resources, which have been exposed to millions of years of geological energy input to reach their current biochemical state.

Current paradigms for biofuels production include the production of ethanol by yeast or bacteria from glucose produced from soluble sugars and starch (1st generation ethanol) or from the cellulosic fraction of biomass (2nd generation ethanol). Due to a lack of competition with food production, the latter is typically seen as more sustainable on a long-term basis [9, 10]. An emerging option is the co-production of ethanol and hydrogen via consolidated bioprocessing [11]. In addition to the now little-used anaerobic A-B fermentations discussed above, another scheme for biofuels production from plant biomass involves the anaerobic production of methane by microbial consortia (anaerobic digestion) [12, 13]. The common link for all of these strategies is the exploitation and optimization of natural microbial activity to produce energy-rich molecules for combustion to produce energy. Direct thermochemical conversion of biomass via pyrolysis or gasification is also possible, although these strategies involve a large amount of energy input by heating the biomass to very high temperatures (normally >500°C) and are therefore independent of microbial activity [14].

Regardless of the means by which biofuels are produced by microbial activity from extant plant material, the same essential challenge must be faced: the substrate for biofuels production is the carbohydrate fraction, which must be made available to the microorganisms in order for the biochemical reactions to proceed efficiently. In the case of 1st generation ethanol, soluble sugars and starch are relatively easily converted to glucose that is fermented into ethanol by yeast. Strategies that utilize the non-food portion of crops, however, face a more formidable challenge. The resource from which energy is to be produced consists of three major biopolymers: cellulose (β(1,4)-linked glucose residues with a degree of polymerization up to ~15,000); hemicellulose (a heterogeneous, short-chained, branched carbohydrate with both 5- and 6-carbon sugars); and lignin (a complex aromatic polymer consisting of nonrepeating covalently linked units of coniferyl, sinapyl, and coumaryl alcohols). These polymers exist together in the plant as a composite, tightly interconnected molecule called lignocellulose [15]. Within lignocellulose, the lignin fraction in particular acts as a barrier to enzyme or microbial penetration, which greatly decreases the yields of fermentable sugars and negatively affects the overall process of energy production from these resources to the extent that it is uneconomical [5, 16]. To overcome this limitation, some form of pretreatment of the biomass is required for economical and efficient production of biofuels by any of the strategies described above [13, 16-21].

The purpose of this chapter is to review the various pretreatment options available for lignocellulosic biomass, with particular emphasis on agricultural residues and on strategies that exploit the natural metabolic activity of microbes to increase the processability of the biomass. These microbial-based strategies can be effective pretreatments on their own or, more probably, can be used in combination with thermomechanical pretreatments in order to provide a cost-effective means to make lignocellulosic substrates available for conversion to biofuels by microorganisms. The key advantages and disadvantages of this strategy will be presented along with a vision for how microbial pretreatment can be integrated into an economical biorefinery process for biofuels and co-product production.

7.2 MECHANICAL, THERMOMECHANICAL AND THERMOCHEMICAL PRETREATMENTS

An early, essential mechanical pretreatment step is comminution, or mechanical particle size reduction, to transform the biomass from its native state into a suitable substrate for further pretreatment and energy production [17]. This step is often not considered in the energy balance of biofuels processes, but it is important to keep in mind that particle size reduction involves energy input that can influence the effective energy yield of these processes [22]. While smaller particle sizes are often considered to be more desirable for yields of fermentable sugars, sizes smaller than about 0.4-0.5 mm provide no additional benefit [17, 23], and the process becomes economically unfeasible at even smaller particle sizes [22]. Methods for mechanical size reduction include wet milling, dry milling, ball milling or vibratory ball milling, and other forms of chipping and grinding of biomass [4, 17]. Regardless of the method employed, particle size reduction requires energy input; therefore, strategies that facilitate the production of biomass in the proper size range while minimizing energy input will provide positive benefits to the overall economics of biofuels processes.

A wide range of options is available for preparing ground biomass for further processing. One of the most common and simple technologies for rendering the carbohydrate fraction available for biofuels production is the application of a dilute solution of sulfuric acid (0.5%–2%) at temperatures of 140°C–180°C with residence times of 10-30 minutes [24]. This process leaves a residue that is depleted in hemicellulose but retains most of the cellulose intact, making it an ideal substrate for enzymatic hydrolysis to yield fermentable sugars for ethanol production. There is a range of conditions for acid hydrolysis that will result in more or less carbohydrate remaining in the solid fraction, with the most severe conditions used to completely degrade the carbohydrate fraction for the determination of cell wall carbohydrate composition [25]. Harsher conditions (e.g. higher acid concentration and temperature), while resulting in a substrate that is highly digestible with enzymes to generate fermentable sugars, also result in a higher yield of compounds derived from pentoses (furfural), hexoses (5-hydroxymethylfurfural) and lignin (low molecular weight phenolic compounds) that are inhibitory to subsequent fermentation by ethanologenic yeasts [26]. The mathematical concept of combined severity, which combines the various factors that define acid hydrolysis conditions (e.g. temperature, residence time, pH), allows objective comparisons between different conditions that enables the determination of optimal conditions for a given substrate [26]; however, doubts have been raised about its accuracy [17].

Another highly effective pretreatment strategy is steam explosion, in which biomass is briefly heated to high temperatures (~200°C) under high pressure, then subjected to a rapid pressure drop that renders the biomass more penetrable by enzymes for subsequent hydrolysis [18]. In some cases, steam explosion is enhanced by the addition of an acid catalyst such as sulfuric acid [27]. For lignocellulosic agricultural residues, steam explosion under optimized conditions has been shown to be an effective pretreatment strategy for enzymatic saccharification [28]. Steam explosion has also been successfully used in combination with other physiochemical pretreatments such as acid/water impregnation of cereal straws [29]. Both of the latter studies resulted in the release of hemicellulose-derived pentose oligomers into the liquid fraction, and it was suggested that the use of ethanologenic strains capable of converting these pentoses into ethanol would further improve overall process efficiency [28]. Other assessments have suggested that the hemicellulose fraction would be more efficiently converted to other value-added products rather than ethanol using post-treatment enzyme addition or further acid hydrolysis [30].

Organosolv is a process by which the lignin fraction is chemically modified and essentially removed from biomass using high-temperature extraction with alcohols such as methanol or ethanol or other solvents, sometimes with dilute acid (e.g. hydrochloric or sulfuric acid) as a catalyst [17]. While organosolv processes require a solvent recovery step to be economical and efficient, they provide a robust means of generating three streams of potential products: an extracted, modified lignin component, a hemicellulose-enriched aqueous phase, and a residue that is highly enriched in cellulose and an excellent substrate for the production of biofuels by enzymatic saccharification followed by bacterial or yeast fermentation. Organosolv is one of the pretreatment options that results in a fraction containing chemically modified, low molecular weight lignin components. This stream has a good deal of product potential in addition to its possible use as a fuel for combustion to provide energy to the process [7, 31]. While organosolv is particularly suited to very lignin-rich feedstocks such as wood [32], there is increasing interest in using organosolv extractions for agricultural residues such as wheat straw and dedicated biofuels crops [33]. Goh et al. [34] optimized organosolv conditions for empty palm fruit bunch using combined severity calculations, with excellent results and the ability to accurately predict product stream yields.

Microwave pretreatment of biomass is another option that has been reported to improve subsequent enzymatic saccharification of rice straw [35]. Microwaves have the advantage of combining very rapid heating times with a lower energy input than conventional heating strategies. This irradiative pretreatment creates localized hotspots, which open up the lignocellulose composite molecule, thereby facilitating enzyme access for saccharification and biofuel production by fermentation [4]. A successful combination of microwave and chemical pretreatments in a microwave-acid-alkali-hydrogen peroxide sequence resulted in efficient enzymatic saccharification of rice straw [36]. A related pretreatment option that has been exploited to improve the enzymatic digestibility of switchgrass is the use of radio frequency heating in combination with alkali; this treatment has the key advantage of allowing a much higher solids content than conventional heating [37]. Irradiation of biomass can also enhance methane production by anaerobic digestion [12].

A number of other pretreatment options exist, including ammonia fiber explosion (AFEX), liquid hot water, alkalai/wet oxidative pretreament, and others; several recent reviews discuss these processes and their advantages and disadvantages in detail [4, 17-20, 23, 30, 38, 39]. Regardless of the strategy employed, a common feature of any pretreatment option is that energy input is required. Pretreatment is a major part of the overall operating expense and energy efficiency of any biofuels process, and, while essential, typically accounts for over 30% of the costs of biorefinery operation [40, 41]. Strategies to reduce these costs will have a major impact on the energy balance and economic sustainability of biorefineries.

7.3 BIOLOGICAL PRETREATMENTS

Microorganisms have evolved a capacity to modify and access lignocellulosic biomass to meet their metabolic needs. The exploitation of this capacity offers a natural, low-input means for preparing biomass for biofuels processes. Natural modification and degradation of the lignin component in particular can reduce the severity requirements of subsequent thermochemical pretreatment steps. For example, Itoh et al. [42] used a variety of lignin-degrading white-rot fungi to treat wood chips prior to extracting lignin by an organosolv method, and demonstrated that improved ethanol yields were obtained from the solid fraction along with a 15% savings in electricity use. Similarly, brown-rot fungal species *Coniophora puteana* and *Postia placenta* have been successfully used to improve glucose yields upon enzymatic saccharification of pine, acting as a complete replacement for thermomechanical pretreatments [43]. While it is clear that it is possible to exploit the metabolic capabilities of microorganisms to facilitate biofuels production, the very wide taxonomic array of microorganisms that modify or degrade lignocellulose presents a tremendous variety of choices for implementing such a strategy. Each approach carries its own advantages and challenges.

7.3.1 MICROBIAL CONSORTIA

One approach for applying the power of microbial metabolism to the challenges of biofuel production involves ensiling, which is a commonly used means for enhancing the digestibility of forage and other biomass for ruminants [44, 45]. The process of ensiling exploits the capacity of naturally occurring bacteria, mostly *Lactobacillaceae*, to ferment the sugars within lignocellulosic residues and produce a substrate that is more easily digested by ruminal microorganisms. While these bacterial consortia lack the ability to substantially degrade the lignin component, the changes effected on the biomass can improve yields of fermentable sugars upon subsequent enzymatic hydrolysis. For example, ensiling a variety of agricultural residues, including wheat, barley, and triticale straws along with cotton stocks resulted in significant improvements in fermentable carbohydrate yields upon application of cellulose-degrading enzymes [46]. Due to limitations in the ability of ensilage to substantially modify the lignin component, this method is not normally a suitable stand-alone biological pretreatment. However, ensiling has been exploited as a means to preserve biomass for biofuels production and has been found to be a very effective, on-farm biomass pretreatment. A strain of *Lactobacillus fermentum* was highly effective in preserving sugar beet pulp cellulose and hemicellulose, and ensiling improved enzymatic saccharification by as much as 35% [47]. Ensiling has also been found to improve yields of methane in anaerobic digestion, with the added benefit of facilitating the longer-term storage of biomass (up to 1 year) while retaining the yield improvements [48, 49]. Improvements in methane yields of up to 50% have been observed with hemp and maize residue, while other crops showed little improvement [50]. However, other researchers have cautioned that the total solids loss may be overestimated for certain substrates, which may result in a misleading, apparent improvement in methane yields by ensiling [51]. Furthermore, while some studies noted above have shown that desirable carbohydrates can be preserved through ensiling, others have noted degradation of cellulose and hemicellulose of up to 10% in this relatively uncontrolled, complex process [46]. Nevertheless, ensiling does offer the substantial benefit of biomass preservation and, importantly, it utilizes existing technology and expertise and can be performed on-farm using unmodified farm equipment. Moreover, ensiling is a relatively low-input process that is anaerobic and therefore does not require mixing and aeration. For these reasons, ensiling could easily be incorporated into an overall biorefinery process at the earliest stages of energy production.

7.3.2 LIGNIN-DEGRADING FUNGI

The earliest colonization of land by plants began around 450 million years ago. The evolutionary innovation that facilitated their spread and success in the non-marine environment was lignification, which provided protection from ultraviolet radiation, structural rigidity and eventually protection from coevolved pathogens and herbivores [52]. The complexity of the phenylpropanoid polymer also provided a carbon sink as land plants fixed atmospheric CO_2 into degradation-resistant lignin. The vast coal reserves whose combustion have contributed to the recent spike in atmospheric CO_2 concentrations trace their origins to the Carboniferous period (~350-300 million years ago), when lignin was not effectively decomposed [52]. Near the end of the Carboniferous period, saprophytic fungi of the class Agaricomycetes evolved the ability to degrade the lignin component of plant biomass, which contributed to a substantial decline in organic carbon burial to the extent that little coal formation occurs today [53, 54]. The large majority of fungal species that are capable of wood decay are known as "white-rot" fungi, which degrade all of the major wood polymers. Approximately 6% of wood decay species are "brown-rot" fungi, which evolved from white-rot fungi and selectively degrade the cellulose and hemicellulose fraction of wood, leaving a lignin-rich residue that is a major contributor to soil carbon in forest ecosystems [55].

The taxonomically broadly distributed white- and brown-rot fungi have developed a variety of means to access and degrade lignocellulose over their long evolutionary history, and their powerful metabolism has been exploited for industrial applications in recent decades. For example, lignin-degrading fungi were noted to have a brightening effect on kraft pulp derived from hardwoods, with savings in bleaching chemicals and potentially decreased environmental impact on paper mill operations [56]. This "biobleaching" was developed further using well-known fungi, such as *Trametes versicolor* [57, 58] and *Phanerochaete chrysosporium* [59, 60]. Similar approaches were used to decolorize and detoxify pulp mill effluent and black liquor [61-63]. In addition, white-rot fungi have been exploited for their ability to decrease energy requirements in pulp manufacturing. This process, known as biopulping, softens the woody substrate and substantially decreases mill electricity requirements for mechanical pulp manufacture [64, 65]. The required scale of industrial pulp manufacture and the applicability of white-rot fungi in providing manufacturing benefits led to the development of feasible means of applying white-rot fungi to biomass on an industrially-relevant scale [66]. This two-auger system featured a wood chip decontamination step and an inoculation step, followed by incubation at ambient temperatures in large chip piles with forced aeration. A series of outdoor trials of this method each featured the treatment of ~36 tonnes of softwood chips with the biopulping fungus *Ceriporiopsis subvermispora* for two weeks. The results were energy savings of around 30% in subsequent pulping, which is slightly higher than was observed in bench-scale trails [66].

7.3.2.1 SPECIES AND SYSTEMS INVESTIGATED

More recently, wood-degrading fungi have been investigated for their ability to assist in processing biomass for biofuels production. Again, with the tremendous variety of wood-rotting species and feedstocks available, there is a wide array of strategies reported for biological pretreatment. One very promising approach used rice straw as feedstock, treated with the white-rot fungus *Pleurotus ostreatus* (oyster mushroom) followed by AFEX [67]. This strategy resulted in significant reductions in the severity of the required pretreatment along with improved glucose yields upon enzyme treatment and produced edible mushrooms as a by-product. Another study found that the incubation time required for *Pleurotus ostreatus* to improve enzymatic saccharification with rice hulls was decreased from 60 days to 18 days by pretreating the rice hulls with hydrogen peroxide prior to fungal inoculation [68]. Similarly, preconditioning of softwood using various white-rot fungi resulted in degradation and modification of the lignin, although significant cellulose loss was also observed [69]. Nevertheless, improved glucose yields were observed by enzymatic saccharification of softwood treated with *Stereum hirsutum* compared to untreated controls, which was attributed to an increase in the pore size of the substrate [69]. Other studies have exploited the selective lignin degradation ability of the white-rot fungus *Echinodontium taxodii* to enhance enzymatic saccharification of water hyacinth in combination with dilute acid pretreatment [70], or of woody substrates without subsequent thermochemical pretreatment [71]. Biological pretreament has also been shown to improve biogas yields from agricultural residues via anaerobic digestion [72]. A tremendous variety of other approaches to biological pretreatment has been reported to be successful on many different lignocellulosic substrates [73, 74].

Exploitation of fungal metabolic activity for industrial purposes can take a variety of forms. For white- and brown-rot fungi, the mode of cultivation can have an effect on the results obtained, and the choice of cultivation conditions depends on the desired outcomes. In general, fungi can be cultivated under solid-state conditions (solid-state fermentation, or SSF), or using submerged fermentation (SmF). SSF involves culturing the fungus on the substrate under relatively low moisture conditions (~60-70%), while SmF uses liquid cultures of the fungus co-incubated with the normally insoluble substrate. Early pulp biobleaching experiments used SmF of white-rot fungi such as *Trametes versicolor*, which featured the advantage of shorter incubation times than SSF [75], but suffered the drawback that very large fermentation vessels would be required for industrial-scale treatments. Many white-rot fungi grow well and perform the desired metabolism under solid-state conditions. For example, species of the genera *Trametes, Phanerochaete*, and *Pycnoporus* preferentially removed color and chemical oxygen demand from olive mill wastewaters and pulp mill black liquors under SSF cultivation conditions [61, 76, 77]. SSF using white-rot fungi has also been used to modify the lignin in agricultural residues, such as wheat straw, for biofuels processes [78].

Despite relatively long incubation times, SSF offers an inexpensive and effective means of fungal cultivation that can also be used for the production of potentially valuable fungal enzymes [79-81]. Fungal enzymes produced by SSF have been used to enhance methane production by anaerobic digestion [82]. Alternatively, fungal lignocellulose modifying enzymes produced by SSF have been used to improve the ruminal digestibility of agricultural residues [83]. However, for SSF to work efficiently with white- or brown-rot fungi requires a decontamination step to allow the fungi to establish on the residues. In lab-scale studies, this is usually accomplished by autoclaving the residues prior to inoculation [84, 85]. While this is necessary at the research scale to establish with certainty the effects of the inoculated fungus on the substrate, autoclaving is in itself a form of pretreatment and is not feasible on an industrial scale. This is a limitation of SSF for application on the large scale that would be required for biological pretreatment of agricultural residues for biofuels production.

7.3.2.2 ENZYMATIC MECHANISMS OF FUNGAL LIGNOCELLULOSE DEGRADATION

The mechanisms that saprophytic wood degrading fungi have evolved to access their difficult growth substrate can be divided into two categories: oxidative mechanisms and hydrolytic mechanisms. These two groups of enzymes and chemicals act together in various combinations to effect the degradation of lignocellulose by different organisms.

7.3.2.2.1 Oxidative Mechanisms

Due to the highly compact, complex nature of lignocellulose, enzymes cannot effectively penetrate this molecule to interact with their substrates. To overcome this limitation, wood-degrading fungi use chemical means to access the recalcitrant substrate. The production of reactive oxygen species (ROS) is a recurring theme in fungal lignocellulose degradation [86]. Specifically, since wood contains sufficient redox-active iron, fungal production of hydrogen peroxide will produce hydroxyl radicals via the Fenton reaction [86]. Hydroxyl radicals (•OH) are extremely powerful oxidizing agents that can catalyze highly non-specific reactions leading to the cleavage of covalent bonds in both lignin and cellulose [86]. Hydrogen peroxide is commonly produced through the action of fungal redox enzymes, such as glyoxal oxidase, pyranose-2 oxidase, and aryl-alcohol oxidase [15].

Another redox enzyme produced by a wide variety of wood-degrading fungi (as well as plants) is laccase, a multicopper oxidase. Laccase acts by removing a single electron from its substrate, which is typically a low-molecular weight compound (mediator) that can diffuse into the densely packed lignocellulose molecule and initiate free radical-mediated reactions leading to the depolymerisation of the substrate. The white-rot fungus *Pycnoporus cinnabarinus* uses laccase in combination with a secondary metabolite, 3-hydroxyanthranilic acid, to effect lignin depolymerisation [87, 88]. Laccase has been used in combination with a wide variety of chemical mediators to effect lignin degradation in wood pulp, with excellent results [89-91].

The presence of manganese in woody substrates is exploited by lignin-degrading fungi through the production of the enzyme manganese peroxidase (MnP). The importance of MnP in lignin degradation is illustrated by its presence in the genomes of white-rot fungi and absence in the non-lignin-degrading brown-rot fungi [92], as well as by the inability of MnP-deficient mutants of *Trametes versicolor* to delignify hardwood-derived kraft pulp [93]. MnP is a heme-containing enzyme with a catalytic cycle that is typical of heme peroxidases, but is uniquely selective for Mn^{2+} as its preferred electron donor [94]. The oxidation of Mn^{2+}, which is accompanied by the reduction of hydrogen peroxide to water, results in the formation of Mn^{3+}. The latter ion is a powerful, diffusible oxidant that is chelated by organic acids such as oxalate produced as a secondary metabolite of the fungus [94]. This highly reactive ion interacts with a wide variety of substrates, including phenols, non-phenolic aromatics, carboxylic acids, and unsaturated fatty acids, producing further ROS and resulting in lignocellulose bond cleavage through oxidative mechanisms [94]. Like laccase, MnP has found application as a delignifying enzyme for pine wood [95] as well as kraft pulp [96, 97]. Peroxidases related to MnP, including lignin peroxidase (LiP) and versatile peroxidase (VP) are also produced by a variety of wood-degrading fungi and play an important role in lignin degradation [98].

Cellobiose dehydrogenase (CDH) is a unique enzyme containing both a heme and a flavin cofactor [99]. CDH is produced by a wide range of fungal species, including both lignin-degrading organisms and fungi that are incapable of degrading lignin [100, 101]. CDH catalyzes the two-electron oxidation of a narrow range of $\beta(1,4)$-linked sugar molecules, principally cellobiose, and transfers these electrons to a very wide array of substrates, including metals such as ferric, cupric, or manganic ions, iron-containing proteins (e.g. cytochrome c), quinones, and other large and small molecules [102, 103]. The diversity of reduced substrates has led to much speculation regarding the role of CDH in lignocellulose degradation; roles have been postulated in the degradation of both cellulose [104] and lignin [105]. The reduction of cupric and ferric ions by CDH and the production of hydrogen peroxide by lignin-degrading fungi suggests that CDH may be involved in sustaining hydroxyl radical-based Fenton's chemistry, with many possible secondary reactions leading to lignocellulose bond cleavage [106]. The role of CDH in lignin-degrading basidiomycetes was addressed by generating mutants of *Trametes versicolor* that did not produce the enzyme, suggesting that CDH plays a role in cellulose degradation, with a more minor role in lignin degradation [107, 108]. Similarly, a recent study with the non- lignin-degrading ascomycete Neurospora crassa revealed that deletion of the gene encoding CDH resulted in vastly decreased cellulase activity, and that the oxidation of cellobiose was coupled to the reductive activation of copper-containing polysaccharide monooxygenases [109]. These studies strongly suggest a role for CDH in supporting cellulose catabolism by fungi, with the latter study in particular providing a highly plausible mechanism for the in vivo function of CDH.

7.3.2.2.2 Hydrolytic Mechanisms

Complementing the degradative power of the redox chemistry catalyzed by the enzymes produced by lignocellulose-degrading fungi is a suite of enzymes that act by adding a water molecule to glycosidic bonds, resulting in bond cleavage and depolymerization. In contrast to the redox enzymes, these hydrolytic enzymes recognize and act on specific glycosidic linkages, releasing sugar molecules that can be utilized as an energy source to support fungal metabolism. Cellulose degradation is catalyzed by the synergistic action of three classes of hydrolytic cellulase enzymes: endo-(1,4)-β-glucanase (endocellulase), cellobiohydrolase (exocellulase), and β-glucosidase [110]. Endocellulases catalyze the cleavage of cellulose chains internally at amorphous regions, while exocellulases remove cellobiose units from the ends of cellulose chains. β-glucosidases are extracellular, cell wall-associated or intracellular enzymes that cleave cellobiose into glucose, which also supports exocellulase activity by relieving end-product inhibition [110]. The redundancy in cellulase genes in fungi is at least partially explained by the fact that different exocellulase enzymes preferentially attack the reducing or non-reducing end of a cellulose chain. This has the effect of exposing new sites for exocellulases of the opposite specificity and also generates new amorphous regions to be acted upon by endocellulases [110, 111]. Hemicellulose degradation is effected by the activity of a wide range of hydrolytic enzymes, including endo-xylanases; endo-α-L-arabinase; endo-mannanase, β-galactosidase, and an array of corresponding β-glucosidases [112]. In addition, covalent bonds within lignocellulose are hydrolyzed by cinnamoyl or feruloyl esterases, which cleave the ester bond between polymerized lignin subunits and the hemicellulose within the composite molecule [113, 114]. Complementary cellulase activity by these various "accessory enzymes" is shown on complex substrates by the improvement in enzymatic saccharification observed when enzymes such as xylanase, pectinase, and feruloyl esterase are added to cellulase cocktails [115, 116].

7.3.2.2.3 Fungal Enzyme Discovery, Production, and Application

Tremendous progress has been made in the last decade concerning the genetic mechanisms underlying plant biomass degradation and modification by microbes, specifically ascomycetous and basidiomycetous fungi. Key to these advancements was the complete genome sequencing of several biomass-degrading fungi, including *Phanerochaete chrysosporium, Phanerochaete carnosa, Postia placenta* and *Trametes versicolor.* The first basidiomycete genome to be sequenced and analyzed, *Phanerochaete chrysosporium*, revealed a tremendous diversity of genes encoding enzymes involved in wood degradation [117]. Among these genes were approximately 240 carbohydrate-active enzymes and several lignin and manganese-dependent peroxidases, which function to degrade the cellulosic/hemicellulosic and lignin components of the cell wall, respectively. This research provided the groundwork for more comprehensive analyses of the genome [118], transcriptome and secretome of *Phanerochaete chrysosporium* [119-122]. These studies highlighted hundreds of wood-degrading genes that were upregulated when *P. chrysosporium* was grown in cellulose-rich medium, including almost 200 genes encoding enzymes of unknown function [122].

Complementary to this research on white-rot fungi was a genome/transcriptome/proteome study on the brown-rot fungus *Postia placenta* [123]. Despite an abundance of similarities between *P. chrysosporium* and *P. placenta*, there were notably fewer glycoside hydrolases expressed by *P. placenta*, such as extracellular cellulases (e.g. endo-(1,4)-β-glucanases), highlighting the mechanistic differences between white- and brown-rot fungi. This work was followed by transcriptomic and proteomic studies investigating the biomass-degrading activity of *Phanerochaete carnosa* [124, 125]. Despite the overall similarity of the transcriptome composition among *P. carnosa* and *P. chrysosporium*, the most abundant transcripts in *P. carnosa* grown on wood substrates (hardwood and softwood) were peroxidases and oxidases involved in lignin degradation [124], whereas *P. chrysosporium* grown only on hardwood revealed only a few highly expressed lignin-degrading enzymes [121]. The differing expression of lignocellulosic enzymes in response to different woody substrates was also explored by examining gene expression patterns in both *P. placenta* and *P. chrysosporium* on hardwood and softwood species [126]. The results of this study strongly suggest that both species of fungi alter their gene expression patterns to degrade wood with different structural characteristics.

In addition to helping uncover the fundamental biochemical machinery involved in biomass degradation, these genomic, transcriptomic and proteomic studies of biomass-degrading fungi have also identified hundreds of target enzymes that could be utilized industrially for bioenergy production, with unique enzyme cocktails suited for specific substrates (e.g. hardwoods vs. softwoods). Several commercial enzymes are commonly used to degrade lignocellulosic residue into fermentable sugars (e.g. Celluclast and Novozyme 188). These 'omic' studies have identified hundreds of fungal glycoside hydrolases that may supplement or completely replace these industry standards. Pre-treatment strategies may also take advantage of the numerous lignin-modifying enzymes identified from biomass-degrading fungi, including lignin and manganese-dependent peroxidases, which have the potential to reduce the severity of thermomechanical and thermochemical pretreatment processes.

7.3.2.2.4 Exploiting Fungal Mutants For Biological Pretreatment

The explosion of 'omic' data for a wide variety of lignocellulose-degrading fungi [117, 123-125] along with the development of sophisticated tools for annotating fungal genomes [127] will continue to add to our understanding of the mechanisms of fungal decay of lignocellulose. Furthermore, increased knowledge of fungal decay mechanisms can aid in the development of strains with improved characteristics. For example, a major limitation to the application of fungal strains for biological pretreatment is the degradation of the desired carbohydrates (cellulose and hemicellulose) for fungal metabolism [73, 74]. Creating or selecting strains that lack the ability to degrade these carbohydrates while retaining the ability to degrade and modify lignin would provide a means to avoid this drawback of fungal pretreatment. Early studies with strains that were deficient in the production of cellulase met with only moderate success, with substantial degradation of cellulose observed [128, 129]. This is probably attributable to the high degree of redundancy in fungal cellulases, with large numbers of genes contributing to the hydrolytic degradation of cellulose and hemicellulose in various species [15]. More recently, we have applied a strain of *Trametes versicolor* that is unable to produce cellobiose dehydrogenase (CDH) to the pretreatment of canola residue, and found that the strain was proficient in lignin degradation but was unable to catabolize the cellulose [107]. Xylose within the substrate appeared to have been utilized to support the greatly decreased fungal growth compared to the wild-type strain. Furthermore, we found that the application of a fungal cell wall-degrading enzyme cocktail (glucanex; a concentrated supernatant of a SmF culture of *Trichoderma harzianum*) to the fungus-treated biomass resulted in the release of fungal cell wall-associated glucose [107]. Biological pretreatment with *T. versicolor* therefore had the overall effect of converting some of the xylose within the substrate to glucose, which is more easily fermented by ethanologenic yeasts.

Studies such as these also provide biological data regarding the role of the genes that are down-regulated in the mutant strains. This reverse genetics approach is a powerful method for investigating gene function, and in the current genomic era reverse genetics tools can often be applied in the known context of the entire genome of the fungus. Gene silencing by RNA interference (RNAi) is a common method for down-regulating

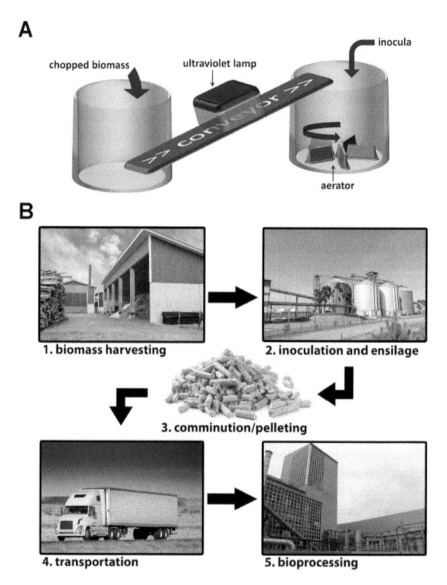

FIGURE 1: A possible scheme for incorporating biological pretreatment into biofuels manufacturing. A. Treatment to suppress the growth of endogenous microorganisms to allow establishment of the inoculated fungal culture. A variety of treatments could be utilized, including ultraviolet, microwave, or radio frequency treatment. This scheme is based on the successful biopulping inoculation strategy described by Scott et al. [66] B. Overall scheme for biofuels production including biological pretreatment. (B.5. photo courtesy of Jay Grabiec, Eastern Illinois University).

genes in a variety of model systems [130, 131], and the recent demonstration of RNAi mechanisms in the model white-rot fungus *Phanerochaete chrysosporium* [132] suggests that RNAi could be used for targeted down-regulation of specific genes in species that are useful for biological pretreatment. The availability of convenient gene silencing transformation vectors for ascomycetes such as pSilent [133] and pTroya [134] as well as pHg/pSILBAγ for basidiomycetes [135] will greatly facilitate the investigation of gene function and may also result in the development of modified strains featuring enhanced properties for biological pretreatment of lignocellulosic substrates for biofuels production.

7.4 CONCLUSIONS AND FUTURE OUTLOOK

Pretreatment of lignocellulosic materials with white- or brown-rot fungi can be incorporated into any strategy for the production of biofuels and bioproducts, with significant advantages including decreased energy requirements for subsequent steps, production of fewer fermentation-inhibiting substances, and the potential for the production of value-added co-product streams [73, 74]. With the wide variety of potential strains and substrates available, and the possibility to create or select new strains with more desirable properties, it seems likely that biological pretreatment can be used on nearly any biomass that is currently produced. One of the most important benefits of biological pretreatment is the resultant reduction in the severity of the subsequent thermomechanical or thermochemical pretreatment step that is required for efficient enzymatic saccharification. While this is a very important benefit, Keller et al. [136] identified six criteria for strains to be selected for biological pretreatment of agricultural waste: little carbohydrate degradation, low costs for nutrients, a reasonable storage time, ability to compete with endogenous microbiota, decreased thermomechanical pretreatment severity, improved yields of glucose upon enzymatic saccharification, and a lack of production of compounds inhibitory to fermenting organisms. These criteria underscore the major limitations of biological pretreatment, the most important of which are the propensity of the organisms to degrade the carbohydrate component, their inability to establish growth on unsterilized biomass, and the relatively long incubation times that are required. These limitations are related to the ecological niche that these saprophytic fungi fill in nature. That is, they have evolved to access and utilize those plant carbohydrates that are difficult for other microorganisms to access. For this reason, these fungi typically appear at the end of an ecological succession of organisms that degrade decaying wood and are often ill equipped to compete with the faster-growing molds and bacteria that access the more easily degraded plant carbohydrates [137]. While it may be possible using reverse genetics tools and/or strain selection to limit carbohydrate degradation by pretreatment fungi [107], it is likely that such strains will be even less able to compete with endogenous microorganisms; therefore, establishment on recently harvested biomass will remain a challenge. Some sort of treatment of the biomass to suppress the growth of endogenous molds prior to inoculation with the pretreatment fungi will likely be necessary.

The unavoidable expense of the pre-inoculation treatment can be compensated by taking advantage of a potential benefit of biological pretreatment that has received very little attention: wood-degrading fungi may modify the lignin component sufficiently to provide positive benefits for particle compression of agricultural biomass during densification. Densification (briquetting or pelleting) of biomass aims to increase the bulk density of agricultural residues far beyond what is achievable by baling, and it is an essential step for providing biomass with sufficient caloric density for efficient transportation [138]. The production of biomass pellets provides a substrate that is suitable for conversion into biofuels through microbial processes or gasification [139, 140], or can be combusted directly to produce energy [141]. A wide variety of agricultural feedstocks is suitable for pelleting [142]; however, untreated biomass is very difficult to densify and, without pretreatment, produces weak, powdery pellets that are expensive to produce and cannot withstand the physical rigors of transportation. Lignin acts as a natural binder that provides strength and durability to biomass pellets, and pretreatment of the biomass is required in order to release lignin fragments during compaction and produce pellets with the desired characteristics [139, 143]. A number of options are available to prepare biomass for pelleting, with two very promising methods being microwave heating and radio frequency heating. Both of these methods provide a number of advantages over conventional heating, particularly regarding treatment times [144-146]. We have found that a very brief microwave treatment of a variety of agricultural feedstocks suppresses the growth of endogenous molds and bacteria sufficiently for inoculated white-rot fungi to establish growth on these substrates. Moreover, canola residue treated with *Trametes versicolor* produces pellets with excellent compaction characteristics and durability (Canam, Town, and Dumonceaux, unpublished). Such pellets would retain the thermochemical pretreatment benefits afforded by the fungal pretreatment in terms of enzymatic saccharification [107], but would offer vastly increased transportation efficiency in a full-scale biorefinery scenario.

We can therefore envision a means by which fungal pretreatment might be incorporated into an overall process for producing energy from biomass by a variety of strategies (Figure 1). Biological pretreatment should be included at the earliest stages in order to take maximum advantage of its beneficial effects. Building on the successful bio-pulping model described by Scott et al. [66], chopped biomass would be briefly decontaminated by microwave or radio frequency heating within a conveyor; the objective of this is not sterilization or complete thermal pretreatment of the biomass, but primarily growth suppression of endogenous microbiota. The lightly treated biomass would then be inoculated with a fungal suspension or formulation and transferred via auger to a pile analogous to a silage pile, but with aeration. The inoculated biomass would be incubated at ambient temperatures for several weeks to allow fungal growth. The fully infested biomass would then be milled to an appropriate size using standard equipment. After cooling and/or drying, pellets formed from the milled, pretreated biomass would be suitable for transport to a biorefinery for biofuels/bioproducts manufacture. In the absence of a viable product stream or a biorefinery, biomass pellets could be burned in a high-efficiency oven to exploit their calorific value [139]. Biological pretreatment would provide an array of benefits along this production chain, including decreased milling energy, decreased compression energy requirements, improved densification characteristics and the consequent reduction in transportation cost, decreased severity of thermochemical pretreatments, decreased production of fermentation inhibitors, improved yield of fermentable sugars upon enzymatic saccharification, and possibly co-products derived from the more easily extractable lignin phase. All of these benefits would be realized with a fairly minor energy input, and although it is difficult to avoid the long incubation times, SSF can be performed on time scales only slightly longer than the common on-farm practice of ensilage. Biological pretreatments can therefore, in theory, be performed on-farm or nearby, offering significant logistical and technical advantages when incorporated into an overall process for biofuels manufacture.

REFERENCES

1. Wolff EW. Greenhouse gases in the Earth system: A palaeoclimate perspective. Philosophical Transactions of the Royal Society A: Mathematical, Physical and Engineering Sciences. 2011;369(1943):2133-47.
2. Falkowski P, Scholes RJ, Boyle E, Canadell J, Canfield D, Elser J, et al. The Global Carbon Cycle: A Test of Our Knowledge of Earth as a System. Science. 2000;290(5490):291-6.
3. Crowley TJ. Causes of climate change over the past 1000 years. Science. 2000;289(5477):270-7.
4. Sarkar N, Ghosh SK, Bannerjee S, Aikat K. Bioethanol production from agricultural wastes: An overview. Renewable Energy. 2012;37(1):19-27.
5. Menon V, Rao M. Trends in bioconversion of lignocellulose: Biofuels, platform chemicals & biorefinery concept. Progress in Energy and Combustion Science. 2012;38(4):522-50.
6. Ragauskas AJ, Williams CK, Davison BH, Britovsek G, Cairney J, Eckert CA, et al. The path forward for biofuels and biomaterials. Science. 2006;311(5760):484-9.
7. Sannigrahi P, Pu Y, Ragauskas A. Cellulosic biorefineries-unleashing lignin opportunities. Curr Opin Environ Sustain. 2010;2(5-6):383-93.
8. Zverlov VV, Berezina O, Velikodvorskaya GA, Schwarz WH. Bacterial acetone and butanol production by industrial fermentation in the Soviet Union: Use of hydrolyzed agricultural waste for biorefinery. Appl Microbiol Biotechnol. 2006;71(5):587-97.
9. Rude MA, Schirmer A. New microbial fuels: a biotech perspective. Curr Opin Microbiol. 2009;12(3):274-81.
10. Khanna M, Hochman G, Rajagopal D, Sexton S, Zilberman D. Sustainability of food, energy and environment with biofuels. CAB Reviews: Perspectives in Agriculture, Veterinary Science, Nutrition and Natural Resources. 2009;4(28).
11. Carere CR, Sparling R, Cicek N, Levin DB. Third generation biofuels via direct cellulose fermentation. Int J Molec Sci. 2008;9(7):1342-60.
12. Chandra R, Takeuchi H, Hasegawa T. Methane production from lignocellulosic agricultural crop wastes: A review in context to second generation of biofuel production. Renewable and Sustainable Energy Reviews. 2012;16(3):1462-76.
13. Frigon JC, Guiot SR. Biomethane production from starch and lignocellulosic crops: A comparative review. Biofuels, Bioproducts and Biorefining. 2010;4(4):447-58.

14. Bahng MK, Mukarakate C, Robichaud DJ, Nimlos MR. Current technologies for analysis of biomass thermochemical processing: A review. Analytica Chimica Acta. 2009;651(2):117-38.

15. Martinez AT, Ruiz-Duenas FJ, Martinez MJ, Del Rio JC, Gutierrez A. Enzymatic delignification of plant cell wall: from nature to mill. Curr Opin Biotechnol. 2009;20(3):348-57.

16. Margeot A, Hahn-Hagerdal B, Edlund M, Slade R, Monot F. New improvements for lignocellulosic ethanol. Curr Opin Biotechnol. 2009;20(3):372-80.

17. Agbor VB, Cicek N, Sparling R, Berlin A, Levin DB. Biomass pretreatment: Fundamentals toward application. Biotechnol Adv. 2011;29(6):675-85.

18. Chandra RP, Bura R, Mabee WE, Berlin A, Pan X, Saddler JN. Substrate pretreatment: the key to effective enzymatic hydrolysis of lignocellulosics? Adv Biochem Eng Biotechnol. 2007;108:67-93.

19. Conde-Mejía C, Jiménez-Gutiérrez A, El-Halwagi M. A comparison of pretreatment methods for bioethanol production from lignocellulosic materials. Process Safety and Environmental Protection.

20. Hendriks ATWM, Zeeman G. Pretreatments to enhance the digestibility of lignocellulosic biomass. Bioresour Technol. 2009;100(1):10-8.

21. Kumar P, Barrett DM, Delwiche MJ, Stroeve P. Methods for pretreatment of lignocellulosic biomass for efficient hydrolysis and biofuel production. Industrial and Engineering Chemistry Research. 2009;48(8):3713-29.

22. da Costa Sousa L, Chundawat SP, Balan V, Dale BE. 'Cradle-to-grave' assessment of existing lignocellulose pretreatment technologies. Curr Opin Biotechnol. 2009;20(3):339-47.

23. Chang VS, Burr B, Holtzapple MT. Lime pretreatment of switchgrass. Appl Biochem Biotechnol. 1997;63-65:3-19.

24. Yang B, Lu Y, Wyman CE. Cellulosic ethanol from agricultural residues. In: Balscheck HP, Ezeji TC, Scheffran J, editors. Biofuels from Agricultural Wastes and Byproducts. Ames, Iowa: Wiley-Blackwell; 2010. p. 175-200.

25. Sluiter A, Hames B, Ruiz R, Scarlata C, Sluiter J, Templeton D, et al. Determination of structural carbohydrates and lignin in biomass. NREL/TP-510-4261. 2008.

26. Larsson S, Palmqvist E, Hahn-Hägerdal B, Tengborg C, Stenberg K, Zacchi G, et al. The generation of fermentation inhibitors during dilute acid hydrolysis of softwood. Enz Microb Technol. 1999;24(3-4):151-9.

27. Ballesteros I, Negro MJ, Oliva JM, Cabanas A, Manzanares P, Ballesteros M. Ethanol production from steam-explosion pretreated wheat straw. Appl Biochem Biotechnol. 2006;129-132:496-508.

28. Ruiz E, Cara C, Manzanares P, Ballesteros M, Castro E. Evaluation of steam explosion pre-treatment for enzymatic hydrolysis of sunflower stalks. Enz Microb Technol. 2008;42(2):160-6.

29. Rosgaard L, Pedersen S, Meyer AS. Comparison of different pretreatment strategies for enzymatic hydrolysis of wheat and barley straw. Appl Biochem Biotechnol. 2007;143(3):284-96.

30. Carvalheiro F, Duarte LC, GÃrio FM. Hemicellulose biorefineries: A review on biomass pretreatments. Journal of Scientific and Industrial Research. 2008;67(11):849-64.

31. Doherty WOS, Mousavioun P, Fellows CM. Value-adding to cellulosic ethanol: Lignin polymers. Industrial Crops and Products. 2011;33(2):259-76.

32. Bozell JJ, Black SK, Myers M, Cahill D, Miller WP, Park S. Solvent fractionation of renewable woody feedstocks: Organosolv generation of biorefinery process streams for the production of biobased chemicals. Biomass Bioenergy. 2011;35:4197-208.

33. Buranov AU, Mazza G. Lignin in straw of herbaceous crops. Industrial Crops and Products. 2008;28(3):237-59.

34. Goh CS, Tan HT, Lee KT, Brosse N. Evaluation and optimization of organosolv pretreatment using combined severity factors and response surface methodology. Biomass Bioenergy. 2011;35(9):4025-33.

35. Ma H, Liu WW, Chen X, Wu YJ, Yu ZL. Enhanced enzymatic saccharification of rice straw by microwave pretreatment. Bioresour Technol. 2009;100(3):1279-84.

36. Zhu S, Wu Y, Yu Z, Wang C, Yu F, Jin S, et al. Comparison of three microwave/chemical pretreatment processes for enzymatic hydrolysis of rice straw. Biosys Engin. 2006;93(3):279-83.

37. Hu Z, Wang Y, Wen Z. Alkali (NaOH) pretreatment of switchgrass by radio frequency-based dielectric heating. Appl Biochem Biotechnol. 2008;148(1-3):71-81.

38. Khanal SK, Rasmussen M, Shrestha P, Van Leeuwen H, Visvanathan C, Liu H. Bioenergy and biofuel production from wastes/residues of emerging biofuel industries. Wat Envrion Res. 2008;80(10):1625-47.

39. Zhang X, Tu M, Paice MG. Routes to potential bioproducts from lignocellulosic biomass lignin and hemicelluloses. Bioenergy Res. 2011;4(4):246-57.

40. Lynd LR. Overview and evaluation of fuel ethanol from cellulosic biomass: Technology, economics, the environment, and policy. Annu Rev Energy Environ. 1996;21(1):403-65.

41. Lynd LR, Elander RT, Wyman CE. Likely features and costs of mature biomass ethanol technology. Appl Biochem Biotechnol. 1996;57-58:741-61.

42. Itoh H, Wada M, Honda Y, Kuwahara M, Watanabe T. Bioorganosolve pretreatments for simultaneous saccharification and fermentation of beech wood by ethanolysis and white rot fungi. J Biotechnol. 2003;103(3):273-80.

43. Ray MJ, Leak DJ, Spanu PD, Murphy RJ. Brown rot fungal early stage decay mechanism as a biological pretreatment for softwood biomass in biofuel production. Biomass Bioenergy. 2010;34(8):1257-62.

44. McEniry J, O'Kiely P, Clipson NJW, Forristal PD, Doyle EM. Bacterial community dynamics during the ensilage of wilted grass. J Appl Microbiol. 2008;105(2):359-71.

45. Thompson DN, Barnes JM, Houghton TP. Effect of additions on ensiling and microbial community of senesced wheat straw. Appl Biochem Biotechnol. 2005;121(1-3):21-46.

46. Chen Y, Sharma-Shivappa RR, Chen C. Ensiling agricultural residues for bioethanol production. Appl Biochem Biotechnol. 2007;143(1):80-92.

47. Zheng Y, Yu C, Cheng YS, Zhang R, Jenkins B, VanderGheynst JS. Effects of ensilage on storage and enzymatic degradability of sugar beet pulp. Bioresour Technol. 2011;102(2):1489-95.

48. Herrmann C, Heiermann M, Idler C. Effects of ensiling, silage additives and storage period on methane formation of biogas crops. Bioresour Technol. 2011;102(8):5153-61.

49. Nizami AS, Korres NE, Murphy JD. Review of the integrated process for the production of grass biomethane. Environmental Science and Technology. 2009;43(22):8496-508.

50. Pakarinen A, Maijala P, Jaakkola S, Stoddard F, Kymäläinen M, Viikari L. Evaluation of preservation methods for improving biogas production and enzymatic conversion yields of annual crops. Biotechnol Biofuels. 2011;4(20).

51. Kreuger E, Nges I, Björnsson L. Ensiling of crops for biogas production: Effects on methane yield and total solids determination. Biotechnol Biofuels. 2011;4(44).

52. Weng JK, Chapple C. The origin and evolution of lignin biosynthesis. New Phytologist. 2010;187(2):273-85.

53. Floudas D, Binder M, Riley R, Barry K, Blanchette RA, Henrissat B, et al. The paleozoic origin of enzymatic lignin decomposition reconstructed from 31 fungal genomes. Science. 2012;336(6089):1715-9.

54. Hittinger CT. Evolution: Endless rots most beautiful. Science. 2012;336(6089):1649-50.

55. Eastwood DC, Floudas D, Binder M, Majcherczyk A, Schneider P, Aerts A, et al. The plant cell wall-decomposing machinery underlies the functional diversity of forest fungi. Science. 2011;333(6043):762-5.

56. Kirk TK, Yang HH. Partial delignification of unbleached kraft pulp with ligninolytic fungi. Biotechnol Lett. 1979;1(9):347-52.

57. Roy BP, Archibald F. Effects of kraft pulp and lignin on Trametes versicolor carbon metabolism. Appl Environ Microbiol. 1993;59(6):1855-63.

58. Selvam K, Saritha KP, Swaminathan K, Manikandan M, Rasappan K, Chinnaswamy P. Pretreatment of wood chips and pulp with Fomes lividus and Trametes versicolor to reduce chemical consumption in paper industries. Asian J Microbiol Biotechnol Environ Sci. 2006;8(4):771-6.

59. Jiménez L, López F, Martínez C. Biological pretreatments for bleaching wheat-straw pulp. Process Biochem. 1994;29(7):595-9.

60. Jiménez L, Martínez C, Pérez I, López F. Biobleaching procedures for pulp from agricultural residues using Phanerochaete chrysosporium and enzymes. Process Biochem. 1997;32(4):297-304.

61. Da Re V, Papinutti L. Black liquor decolorization by selected white-rot fungi. Appl Biochem Biotechnol. 2011;165(2):406-15.

62. Kumar A, Shrivastava V, Pathak M, Singh RS. Biobleaching of pulp and paper mill effluent using mixed white rot fungi. Biosci Biotechnol Res Asia. 2010;7(2):925-31.

63. Mittar D, Khanna PK, Marwaha SS, Kennedy JF. Biobleaching of pulp and paper mill effluents by Phanerochaete chrysosporium. J Chem Technol Biotechnol. 1992;53(1):81-92.

64. Scott GM, Akhtar M, Kirk TK, editors. An Update on Biopulping Commericialization. Proceedings of the 2000 TAPPI Pulping/Process and Product Quality Process; 2000; Boston, MA.

65. Sena-Martins G, Almeida-Vara E, Moreira PR, Polónia I, Duarte JC, editors. Biopulping of pine wood chips for production of kraft paper board. Proceedings of the 2000 TAPPI Pulping/Process and Product Quality Process; 2000; Boston, MA.

66. Scott GM, Akhtar M, Lentz MJ, Kirk TK, Swaney R. New technology for papermaking: Commercializing biopulping. Tappi J. 1998;81(11):220-5.

67. Balan V, Da Costa Sousa L, Chundawat SPS, Vismeh R, Jones AD, Dale BE. Mushroom spent straw: A potential substrate for an ethanol-based biorefinery. J Ind Microbiol Biotechnol. 2008;35(5):293-301.

68. Yu J, Zhang J, He J, Liu Z, Yu Z. Combinations of mild physical or chemical pretreatment with biological pretreatment for enzymatic hydrolysis of rice hull. Bioresour Technol. 2009;100(2):903-8.

69. Lee JW, Gwak KS, Park JY, Park MJ, Choi DH, Kwon M, et al. Biological pretreatment of softwood Pinus densiflora by three white rot fungi. J Microbiol. 2007;45(6):485-91.

70. Ma F, Yang N, Xu C, Yu H, Wu J, Zhang X. Combination of biological pretreatment with mild acid pretreatment for enzymatic hydrolysis and ethanol production from water hyacinth. Bioresour Technol. 2010;101(24):9600-4.

71. Yu H, Guo G, Zhang X, Yan K, Xu C. The effect of biological pretreatment with the selective white-rot fungus Echinodontium taxodii on enzymatic hydrolysis of softwoods and hardwoods. Bioresour Technol. 2009;100(21):5170-5.

72. Muthangya M, Mshandete AM, Kivaisi AK. Two-stage fungal pre-treatment for improved biogas production from sisal leaf decortication residues. Int J Molec Sci. 2009;10(11):4805-15.

73. Isroi, Millati R, Syamsiah S, Niklasson C, Cahyanto MN, Lundquist K, et al. Biological pretreatment of lignocelluloses with white-rot fungi and its applications: A review. BioResources. 2011;6(4):5224-59.

74. Chen S, Zhang X, Singh D, Yu H, Yang X. Biological pretreatment of lignocellulosics: potential, progress and challenges. Biofuels. 2010;1(1):177-99.

75. Paice MG, Archibald FS, Bourbonnais R, Jurasek L, Reid ID, Charles T, et al. Enzymology of kraft pulp bleaching by Trametes versicolor. In: Jeffries TW, Viikari L, editors. Enzymes for Pulp and Paper Processing. Washington: American Chemical Society; 1996. p. 151-64.

76. Alaoui SM, Merzouki M, Penninckx MJ, Benlemlih M. Relationship between cultivation mode of white rot fungi and their efficiency for olive oil mill wastewaters treatment. Elec J Biotechnol. 2008;11(4).

77. Aloui F, Abid N, Roussos S, Sayadi S. Decolorization of semisolid olive residues of "alperujo" during the solid state fermentation by Phanerochaete chrysosporium, Trametes versicolor, Pycnoporus cinnabarinus and Aspergillus niger. Biochem Engin J. 2007;35(2):120-5.

78. Dinis MJ, Bezerra RMF, Nunes F, Dias AA, Guedes CV, Ferreira LMM, et al. Modification of wheat straw lignin by solid state fermentation with white-rot fungi. Bioresour Technol. 2009;100(20):4829-35.

79. Rodríguez Couto S, Moldes D, Liébanas A, Sanromán A. Investigation of several bioreactor configurations for laccase production by Trametes versicolor operating in solid-state conditions. Biochem Engin J. 2003;15(1):21-6.

80. Rodríguez Couto S, Sanromán MA. Application of solid-state fermentation to lignolytic enzyme production. Biochem Engin J. 2005;22(3):211-9.

81. Winquist E, Moilanen U, Mettälä, A., Leisola M, Hatakka A. Production of lignin modifying enzymes on industrial waste material by solid-state cultivation of fungi. Biochem Engin J. 2008;42(2):128-32.

82. Bochmann G, Herfellner T, Susanto F, Kreuter F, Pesta G. Application of enzymes in anaerobic digestion. 2007. p. 29-35.

83. Graminha EBN, GonÃ§alves AZL, Pirota RDPB, Balsalobre MAA, Da Silva R, Gomes E. Enzyme production by solid-state fermentation: Application to animal nutrition. Anim Feed Sci Technol. 2008;144(1-2):1-22.

84. Kuhar S, Nair LM, Kuhad RC. Pretreatment of lignocellulosic material with fungi capable of higher lignin degradation and lower carbohydrate degradation improves substrate acid hydrolysis and the eventual conversion to ethanol. Can J Microbiol. 2008;54(4):305-13.

85. Rasmussen ML, Shrestha P, Khanal SK, Pometto Iii AL, van Leeuwen J. Sequential saccharification of corn fiber and ethanol production by the brown rot fungus Gloeophyllum trabeum. Bioresour Technol. 2010;101(10):3526-33.

86. Hammel KE, Kapich AN, Jensen Jr KA, Ryan ZC. Reactive oxygen species as agents of wood decay by fungi. Enz Microb Technol. 2002;30(4):445-53.

87. Eggert C, Temp U, Dean JF, Eriksson KE. A fungal metabolite mediates degradation of non-phenolic lignin structures and synthetic lignin by laccase. FEBS Lett. 1996;391(1-2):144-8.

88. Eggert C, Temp U, Eriksson KE. Laccase is essential for lignin degradation by the white-rot fungus Pycnoporus cinnabarinus. FEBS Lett. 1997 Apr 21;407(1):89-92.

89. Bajpai P, Anand A, Bajpai PK. Bleaching with lignin-oxidizing enzymes. Biotechnology Annual Review2006. p. 349-78.

90. Bourbonnais R, Paice MG. Enzymatic delignification of kraft pulp using laccase and a mediator. Tappi J. 1996;79(6):199-204.

91. Call HP, Mucke I. History, overview and applications of mediated lignolytic systems, especially laccase-mediator-systems (Lignozym®-process). J Biotechnol. 1997;53(2-3):163-202.

92. Ruiz-Duenas F, Martınez A. Microbial degradation of lignin: How a bulky recalcitrant polymer is efficiently recycled in nature and how we can take advantage of this. Microb Biotechnol. 2009;2:164-77.

93. Addleman K, Dumonceaux T, Paice MG, Bourbonnais R, Archibald FS. Production and characterization of Trametes versicolor mutants unable To bleach hardwood kraft pulp. Appl Environ Microbiol. 1995 10;61(10):3687-94.

94. Hofrichter M. Review: Lignin conversion by manganese peroxidase (MnP). Enz Microb Technol. 2002;30(4):454-66.

95. Hofrichter M, Lundell T, Hatakka A. Conversion of milled pine wood by manganese peroxidase from Phlebia radiata. Appl Environ Microbiol. 2001;67(10):4588-93.

96. Feijoo G, Moreira MT, Alvarez P, Lú-Chau TA, Lema JM. Evaluation of the enzyme manganese peroxidase in an industrial sequence for the lignin oxidation and bleaching or eucalyptus kraft pulp. J Appl Polymer Sci. 2008;109(2):1319-27.

97. Paice MG, Reid ID, Bourbonnais R, Archibald FS, Jurasek L. Manganese peroxidase, produced by Trametes versicolor during pulp bleaching, demethylates and delignifies kraft pulp. Appl Environ Microbiol. 1993;59(1):260-5.

98. Hammel KE, Cullen D. Role of fungal peroxidases in biological ligninolysis. Curr Opin Plant Biol. 2008;11(3):349-55.

99. Henriksson G, Johansson G, Pettersson G. A critical review of cellobiose dehydrogenases. J Biotechnol. 2000;78(2):93-113.

100. Harreither W, Sygmund C, Augustin M, Narciso M, Rabinovich ML, Gorton L, et al. Catalytic properties and classification of cellobiose dehydrogenases from Ascomycetes. Appl Environ Microbiol. 2011;77(5):1804-15.

101. Zamocky M, Ludwig R, Peterbauer C, Hallberg BM, Divne C, Nicholls P, et al. Cellobiose dehydrogenase - A flavocytochrome from wood-degrading, phytopathogenic and saprotropic fungi. Curr Protein Peptide Sci. 2006;7(3):255-80.

102. Henriksson G, Ander P, Pettersson B, Pettersson G. Cellobiose dehydrogenase (cellbiose oxidase) from Phanerochaete chrysosporium as a wood degrading enzyme. Studies on cellulose, xylan and synthetic lignin. Appl Microbiol Biotechnol. 1995;42(5):790-6.

103. Roy BP, Dumonceaux T, Koukoulas AA, Archibald FS. Purification and characterization of cellobiose dehydrogenases from the white rot fungus Trametes versicolor. Appl Environ Microbiol. 1996 12;62(12):4417-27.

104. Mansfield SD, De Jong E, Saddler JN. Cellobiose dehydrogenase, an active agent in cellulose depolymerization. Appl Environ Microbiol. 1997;63(10):3804-9.

105. Roy BP, Paice MG, Archibald FS, Misra SK, Misiak LE. Creation of metal-complexing agents, reduction of manganese dioxide, and promotion of manganese peroxidase-mediated Mn(III) production by cellobiose:quinone oxidoreductase from Trametes versicolor. J Biol Chem. 1994;269(31):19745-50.

106. Mason MG, Nicholls P, Wilson MT. Rotting by radicals--the role of cellobiose oxidoreductase? Biochem Soc Trans. 2003;31(Pt 6):1335-6.

107. Canam T, Town JR, Tsang A, McAllister TA, Dumonceaux TJ. Biological pretreatment with a cellobiose dehydrogenase-deficient strain of Trametes versicolor enhances the biofuel potential of canola straw. Bioresour Technol. 2011;102:10020–7.

108. Dumonceaux T, Bartholomew K, Valeanu L, Charles T, Archibald F. Cellobiose dehydrogenase is essential for wood invasion and nonessential for kraft pulp delignification by Trametes versicolor. Enz Microb Technol. 2001;29(8-9):478-89.

109. Phillips CM, Beeson WT, Cate JH, Marletta MA. Cellobiose dehydrogenase and a copper-dependent polysaccharide monooxygenase potentiate cellulose degradation by Neurospora crassa. ACS Chem Biol. 2011 Dec 16;6(12):1399-406.

110. Baldrian P, Valaskova V. Degradation of cellulose by basidiomycetous fungi. FEMS Microbiol Rev. 2008;32(3):501-21.

111. Gilkes NR, Kwan E, Kilburn DG, Miller RC, Warren RAJ. Attack of carboxymethylcellulose at opposite ends by two cellobiohydrolases from Cellulomonas fimi. J Biotechnol. 1997;57(1-3):83-90.

112. Shallom D, Shoham Y. Microbial hemicellulases. Curr Opin Microbiol. 2003;6(3):219-28.

113. Crepin VF, Faulds CB, Connerton IF. Functional classification of the microbial feruloyl esterases. Appl Microbiol Biotechnol. 2004;63(6):647-52.

114. Mathew S, Abraham TE. Ferulic acid: An antioxidant found naturally in plant cell walls and feruloyl esterases involved in its release and their applications. Crit Rev Biotechnol. 2004;24(2-3):59-83.

115. Berlin A, Gilkes N, Kilburn D, Bura R, Markov A, Skomarovsky A, et al. Evaluation of novel fungal cellulase preparations for ability to hydrolyze softwood substrates - Evidence for the role of accessory enzymes. Enz Microb Technol. 2005;37(2):175-84.

116. Berlin A, Maximenko V, Gilkes N, Saddler J. Optimization of enzyme complexes for lignocellulose hydrolysis. Biotechnol Bioengin. 2007;97(2):287-96.

117. Martinez D, Larrondo LF, Putnam N, Gelpke MD, Huang K, Chapman J, et al. Genome sequence of the lignocellulose degrading fungus Phanerochaete chrysosporium strain RP78. Nat Biotechnol. 2004 Jun;22(6):695-700.

118. Vanden Wymelenberg A, Minges P, Sabat G, Martinez D, Aerts A, Salamov A, et al. Computational analysis of the Phanerochaete chrysosporium v2.0 genome database and mass spectrometry identification of peptides in ligninolytic cultures reveal complex mixtures of secreted proteins. Fungal Genet Biol. 2006;43(5):343-56.

119. Abbas A, Koc H, Liu F, Tien M. Fungal degradation of wood: initial proteomic analysis of extracellular proteins of Phanerochaete chrysosporium grown on oak substrate. Curr Genet. 2005;47(1):49-56.

120. Ravalason H, Jan G, Mollé D, Pasco M, Coutinho P, Lapierre C, et al. Secretome analysis of Phanerochaete chrysosporium strain CIRM-BRFM41 grown on softwood. Appl Microbiol Biotechnol. 2008;80(4):719-33.

121. Sato S, Feltus F, Iyer P, Tien M. The first genome-level transcriptome of the wood-degrading fungus Phanerochaete chrysosporium grown on red oak. Curr Genet. 2009;55(3):273-86.

122. Vanden Wymelenberg A, Gaskell J, Mozuch M, Kersten P, Sabat G, Martinez D, et al. Transcriptome and secretome analyses of Phanerochaete chrysosporium reveal complex patterns of gene expression. Appl Environ Microbiol. 2009 June 15, 2009;75(12):4058-68.

123. Martinez D, Challacombe J, Morgenstern I, Hibbett D, Schmoll M, Kubicek CP, et al. Genome, transcriptome, and secretome analysis of wood decay fungus Postia placenta supports unique mechanisms of lignocellulose conversion. Proc Natl Acad Sci USA. 2009 February 10, 2009;106(6):1954-9.

124. MacDonald J, Doering M, Canam T, Gong Y, Guttman DS, Campbell MM, et al. Transcriptomic responses of the softwood-degrading white-rot fungus Phanerochaete carnosa during growth on coniferous and deciduous wood. Appl Environ Microbiol. 2011;77(10):3211-8.

125. MacDonald J, Master ER. Time-dependent profiles of transcripts encoding lignocellulose-modifying enzymes of the white rot fungus Phanerochaete carnosa grown on multiple wood substrates. Appl Environ Microbiol. 2012 March 1, 2012;78(5):1596-600.

126. Vanden Wymelenberg A, Gaskell J, Mozuch M, Splinter BonDurant S, Sabat G, Ralph J, et al. Significant alteration of gene expression in wood decay fungi Postia placenta and Phanerochaete chrysosporium by plant species. Appl Environ Microbiol. 2011;77(13):4499-507.

127. Levasseur A, Piumi F, Coutinho PM, Rancurel C, Asther M, Delattre M, et al. FOLy: an integrated database for the classification and functional annotation of fungal oxidoreductases potentially involved in the degradation of lignin and related aromatic compounds. Fungal Genet Biol. 2008 May;45(5):638-45.

128. Karunanandaa K, Fales SL, Varga GA, Royse DJ. Chemical composition and biodegradability of crop residues colonized by white-rot fungi. J Sci Food Agric. 1992;60(1):105-12.

129. Akin DE, Sethuraman A, Morrison WH, III, Martin SA, Eriksson KEL. Microbial delignification with white rot fungi improves forage digestibility. Appl Environ Microbiol. 1993;59(12):4274-82.

130. Mahmood ur R, Ali I, Husnain T, Riazuddin S. RNA interference: The story of gene silencing in plants and humans. Biotechnol Adv. 2008;26(3):202-9.

131. Nakayashiki H, Nguyen QB. RNA interference: roles in fungal biology. Curr Opin Microbiol. 2008;11(6):494-502.

132. Matityahu A, Hadar Y, Dosoretz CG, Belinky PA. Gene silencing by RNA interference in the white rot fungus Phanerochaete chrysosporium. Appl Environ Microbiol. 2008;74(17):5359-65.

133. Nakayashiki H, Hanada S, Nguyen BQ, Kadotani N, Tosa Y, Mayama S. RNA silencing as a tool for exploring gene function in ascomycete fungi. Fungal Genet Biol. 2005;42(4):275-83.

134. Shafran H, Miyara I, Eshed R, Prusky D, Sherman A. Development of new tools for studying gene function in fungi based on the Gateway system. Fungal Genet Biol. 2008;45(8):1147-54.

135. Kemppainen MJ, Pardo AG. pHg/pSILBAγ vector system for efficient gene silencing in homobasidiomycetes: Optimization of ihpRNA - Triggering in the mycorrhizal fungus Laccaria bicolor. Microb Biotechnol. 2010;3(2):178-200.

136. Keller FA, Hamilton JE, Nguyen QA. Microbial pretreatment of biomass: potential for reducing severity of thermochemical biomass pretreatment. Appl Biochem Biotechnol. 2003;105 -108:27-41.

137. Lakovlev A, Stenlid J. Spatiotemporal patterns of laccase activity in interacting mycelia of wood-decaying basidiomycete fungi. Microb Ecol. 2000;39(3):236-45.

138. Panwar V, Prasad B, Wasewar KL. Biomass residue briquetting and characterization. J Eng Engineering. 2011;137(2):108-14.

139. Granada E, López González LM, Míguez JL, Moran J. Fuel lignocellulosic briquettes, die design and products study. Renewable Energy. 2002;27(4):561-73.

140. Kaliyan N, Morey RV. Natural binders and solid bridge type binding mechanisms in briquettes and pellets made from corn stover and switchgrass. Bioresour Technol. 2010;101(3):1082-90.

141. Alaru M, Kukk L, Olt J, Menind A, Lauk R, Vollmer E, et al. Lignin content and briquette quality of different fibre hemp plant types and energy sunflower. Field Crops Res. 2011;124(3):332-9.

142. Karunanithy C, Wang Y, Muthukumarappan K, Pugalendhi S. Physiochemical Characterization of Briquettes Made from Different Feedstocks. Biotechnol Res Int. 2012;2012:165202.

143. Adapa PK, Tabil LG, Schoenau GJ. Compression characteristics of non-treated and steam-exploded barley, canola, oat, and wheat straw grinds. Appl Engin Agric. 2010;26(4):617-32.

144. Ramaswamy H, Tang J. Microwave and radio frequency heating. Food Sci Technol International. 2008;14(5):423-7.

145. Iroba K, Tabil L, editors. Densification of radio frequency pretreated lignocellulosic biomass barley straw. ASABE Annual International Meeting; 2012; St. Joseph, MI, USA: American Society of Agricultural and Biological Engineers.

146. Kashaninejad M, Tabil LG. Effect of microwave-chemical pre-treatment on compression characteristics of biomass grinds. Biosys Engin. 2011;108(1):36-45.

CHAPTER 8

DISCOVERY AND CHARACTERIZATION OF IONIC LIQUID-TOLERANT THERMOPHILIC CELLULASES FROM A SWITCHGRASS-ADAPTED MICROBIAL COMMUNITY

JOHN M. GLADDEN, JOSHUA I. PARK, JESSICA BERGMANN, VIMALIER REYES-ORTIZ, PATRIK D'HAESELEER, BETANIA F. QUIRINO, KENNETH L. SALE, BLAKE A. SIMMONS, AND STEVEN W. SINGER

8.1 BACKGROUND

With global energy demands rising rapidly, new technologies need to be developed that utilize new resources for transportation fuels. Lignocellulosic biomass is one promising resource, where an estimated one billion tons will be available annually by 2030 in the US alone [1]. Lignocellulosic biomass is primarily composed of plant cell-wall polysaccharides, such as cellulose and hemicelluloses, which together constitute 60 to 70% of the biomass by weight for potential energy crops such as switchgrass [2]. These polymers are composed of hexose and pentose sugars that can

Discovery and Characterization of Ionic Liquid-Tolerant Thermophilic Cellulases from a Switchgrass-Adapted Microbial Community. © Gladden JM, Park JI, Bergmann J, Reyes-Ortiz V, D'haeseleer P, Quirino BF, Sale KL, Simmons BA, and Singer SW.; licensee BioMed Central Ltd. Biotechnology for Biofuels *7,15 (2014), doi:10.1186/1754-6834-7-15. Licensed under Creative Commons Attribution 2.0 Generic License, http://creativecommons.org/licenses/by/2.0/.*

be fermented into substitutes for gasoline, diesel and jet fuel [3-7], augmenting or partially displacing current petroleum-based sources of liquid transportation fuels. One of the challenges of using lignocellulosic biomass for production of biofuels is the recalcitrance of plant biomass to deconstruction, a property that necessitates some form of chemical or physical pretreatment to permit enzymes or chemicals to gain access to and hydrolyze the plant polymers into fermentable sugars [4,6,8]. This study focuses on this challenge and discloses the discovery and characterization of biomass-deconstructing enzymes that are more compatible with certain forms of biomass pretreatment solvents than the current commercially available enzyme cocktails.

The recalcitrance of lignocellulosic biomass has been a difficult hurdle to overcome, but promising new technologies using certain ionic liquids (ILs) that have come about in the last decade indicate that we are well on our way to moving past this barrier [9,10]. Pretreating biomass with certain classes of ILs, most notably those with imidazolium-based cations, can be more efficient and tunable than other existing forms of pretreatment, and technoeconomic analysis of IL-pretreatment suggests that there are potential routes to economic viability [8,11]. One remaining issue with this technology that needs to be addressed to maximize efficiency and reduce capital costs is the incompatibility of ILs with cellulase cocktails derived from filamentous fungi. These enzyme cocktails can be strongly inhibited by certain ILs, such as 1-ethyl-3 methylimidazolium acetate [C2mim] [OAc], necessitating expensive and inefficient washing steps to remove residual IL from the biomass prior to addition of enzymes [8,12-14]. One solution to this issue is to develop enzyme cocktails that are tolerant to ILs. Fortunately, it has been shown that certain thermophilic bacterial cellulase enzymes can tolerate high levels of the IL [C2mim][OA]], and in fact these enzymes have been used to develop an IL-tolerant cellulase cocktail called JTherm [13-17]. It has been further demonstrated that JTherm can be used in a one-pot IL pretreatment and saccharification bioprocessing scheme that eliminates the need to wash the pretreated biomass with water, significantly reducing the number of process steps [18].

The next step toward an economically viable IL-based bioprocessing scheme for the conversion of lignocellulosic biomass to biofuels will be to further integrate and improve all components of the process. For IL-tolerant cellulase cocktails, this includes reducing enzyme loadings by reformulating the cocktails to achieve greater saccharification efficiency; a modification predicted by technoeconomic modeling to substantially reduce overall costs in a biorefinery [11]. Enzyme cocktail reformulation will require screening through an expansive list of IL-tolerant cellulase enzymes to identify those that enhance saccharification efficiency under conditions likely to be found in a biorefinery. However, the number of known IL-tolerant cellulase enzymes, specifically those tolerant to [C2mim] [OAc], is quite small, an issue that hampers cocktail reformulation efforts. The goal of this study was therefore to discover and characterize an expanded set of [C2mim][OAc]-tolerant cellulase enzymes to enable future development of highly efficient IL-tolerant biomass-deconstructing enzyme cocktails. This study focused on thermophilic organisms based on clues provided by previous studies that indicate that thermotolerance may be positively correlated with IL-tolerance. Hence, we can leverage a naturally evolved physiological characteristic of an enzyme and use it as a proxy to discover enzymes with a non-natural industrially relevant characteristic, such as IL-tolerance.

This concept drove recent work where complex compost-derived microbial communities were cultivated on switchgrass under thermophilic conditions to enrich for organisms that produce mixtures of IL-tolerant cellulases and xylanases [14]. The community was composed of several abundant bacterial populations related to *Thermus thermophilus*, *Rhodothermus marinus*, *Paenibacillus*, *Thermobacillus* and an uncultivated lineage in the *Gemmatimonadetes phylum* [19]. The glycoside hydrolases from this community were found to have high optimum temperatures (approximately 80°C) and tolerated relatively high levels of [C2mim][OAc] compared to commercial cellulase cocktails (>50% activity in the presence of 30% (v/v) [C2mim][OAc]). Therefore, these communities provide a rich reservoir of potential enzyme targets to develop thermophilic and IL-tolerant cellulase cocktails to be used in lignocellulosic biofuel production platforms. To discover the genes that encode these IL- and thermo-tolerant enzymes, metagenomic and proteomic analysis was conducted on the community [14,19]. The analysis identified a variety of genes encoding potential cellulose and hemicellulose-degrading enzymes, a subset of which were assembled into complete open reading frames (ORFs) from the metagenome. To validate the concept that thermotolerance can be used as an engine for discovery of IL-tolerant enzymes, this study expressed and characterized 37 of these predicted cellulase genes from the metagenome using both cell-free and in vivo *Escherichia coli* (*E. coli*) expression systems. Both expression methods were employed to determine which method is most suitable for rapid and efficient screening of metagenome-derived gene sets. We found that several of the ORFs encode IL- and thermo-tolerant cellulase enzymes, including enzymes with activities that are stimulated in the presence of ILs.

8.2 RESULTS

8.2.1 IDENTIFICATION OF CELLULASES IN A SWITCHGRASS-ADAPTED METAGENOME

The metagenome of a thermophilic switchgrass-degrading bacterial community was curated for genes with cellulase-related annotations or homology to sequences for cellulase enzymes deposited in the CAZy database (http://www.cazy.org/), including β-glucosidases (BG), cellobiohydrolases (CBH), and endoglucanases (Endo). A total of nineteen predicted BGs, two CBHs, and sixteen Endos were identified that appeared to be complete ORFs (Table 1; see Methods). The top BLASTP hit for each identified cellulase is indicated in Table 1, including the maximum identity and source organism of the top hit in GenBank. Many of the ORFs are homologous to those found in isolates that cluster with abundant community members, such as *Rhodothermus marinus, Paenibacillus, Thermobacillus* and *Gemmatimonadetes*. Many of the ORFs fall into sequence bins assigned to these organisms in the metagenome that are consistent with the phylogenetic affiliation predicted by the BLASTP search (Table 1, Additional file 1, and D'Haeseleer et. al. 2013 [19]). Several of the ORFs in Table 1 contained sequencing errors or were identified as fragments and were manually corrected/assembled (see Methods for details). For J08/09 and J38/39, the manual assembly resulted in two closely related proteins, and therefore both versions were tested.

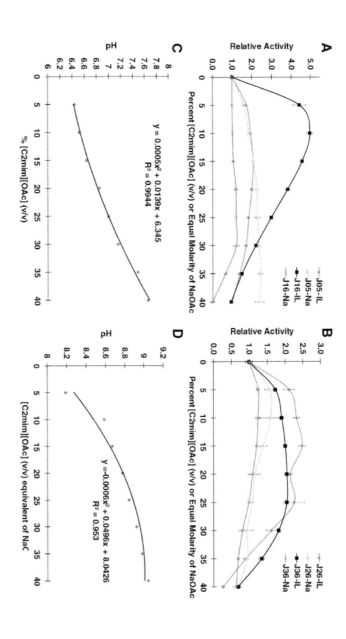

FIGURE 1: Plot of enzyme activity in the presence of 0 to 40% 1-ethyl-3 methylimidazolium acetate [C2mim][OAc] or an equal molarity of sodium acetate (NaOAc). Relative activity is based on activity in water (0% ionic liquid (IL) value). (A) Two IL-tolerant endoglucanases were profiled. The pH was determined at each concentration of (C) [C2mim][OAc] and (D) NaOAc. Error bars represent one standard deviation (they are too small to be visualized on C and D).

8.2.2 CELL-FREE AND E. COLI EXPRESSION AND SCREENING OF PREDICTED CELLULASE GENES

Each of the 37 predicted metagenome-derived cellulase genes were synthesized and cloned into a custom vector for in vitro cell-free expression using a T7 promoter/terminator-based system [20]. Each gene was expressed in vitro and screened for Endo, CBH and BG activity (Table 2). For comparison to the cell-free system, each gene was then cloned into the pDEST17 vector for expression in *E. coli* and screened for the same activities (Table 2). There was a large degree of overlap in terms of active genes detected between the two expression methods, but the *E. coli*-based screen detected activity from a larger subset of genes than the cell-free screen (26 versus 19). BG activity was detected for 15 of the 19 predicted BGs, and none of these enzymes showed Endo activity, consistent with their annotation assigned by the JGI and D'haeseleer et al. [19]. Furthermore, 12 of these 15 positive candidates exhibited CBH activity on, indicating that these enzymes have activity on glucose oligomers with n >2. For the predicted Endos, activity was detected for 11 of the 16 candidates. In addition to Endo activity, 7 of the 11 Endos also had BG and/or CBH activity. No activity was detected for the two predicted CBH genes.

8.2.3 ACTIVITY PROFILE OF CELLULASES

Of the 37 enzymes in the initial screen, 15 of the 19 BGs and 6 of the 16 Endos were expressed at sufficient quantities to profile in greater detail. The activity of each enzyme was measured at temperatures ranging from 45 to 99°C, pH between 4.0 and 8.0, and IL concentrations ranging from 0 to 40% [C2mim][OAc] (v/v). These data were then plotted and optimal temperature/pH and IL-tolerance was determined for each enzyme (Table 3). To illustrate the dynamic activity range of each enzyme, the temperature, pH and IL concentration ranges that gave greater than 80 or 50% activity compared to the optimal activity are also reported in Table 3. All of the enzymes were active at elevated temperature, but the range of optimum temperatures (T_{opt}) was broad, ranging from 45 to 95°C. The enzymes were divided into two groups: seven enzymes with a T_{opt} within

5° of 70°C and another seven near 90°C. Of the remaining enzymes, five had a T_{opt} below 70°C and two had an intermediate T_{opt} of 80°C. The enzymes also showed a similar clustering around optimal pH values (pH$_{opt}$), with fourteen enzymes having a slightly acidic pH$_{opt}$ between 5.0 and 6.0 and the remaining seven enzymes having a pHopt between 6.5 and 7.5. However, many of these enzymes were active over a broad pH range, and all but J16 retained ≥50% activity at pH 7.0. Five of the enzymes were more than 80% active at the highest pH tested of 8.0, indicating that these enzymes also tolerate slightly alkaline conditions.

Enzyme activity was profiled at temperatures (Temp) between 45 and 95°C, pH between 4 and 8, and IL concentrations between 0 and 40% (v/v) of 1-ethyl-3 methylimidazolium acetate [C2mim][OAc]. The temperature and pH that elicited the highest activity is indicated in the row for optimum temperature (T_{opt}) and optimal PH values (pH$_{opt}$), respectively. Temperature and pH ranges that permitted greater than 80% and 50% activity are indicated below the optimum value. Ionic liquid (IL)-tolerance is indicated as the maximum concentration of [C2mim][OAc] that permits at least 80% and 50% enzyme activity (that is, a value of 15 in the 80% row would indicate that 15% (v/v) of [C2mim][OAc] is the maximum concentration of [C2mim][OAc] that can be used to retain at least 80% enzyme activity). Most enzymes showed a steady decline in activity with increasing IL concentrations. *Maximum (Max) activity in IL is reported as the highest fold change of activity in the presence of IL compared to water and the values in brackets are the IL concentrations (v/v) in which that highest activity was achieved. Values less than 1 indicate the enzyme is less active in IL than in water, and values greater than 1 indicate the enzyme had increased activity in the presence of IL.

Surprisingly, most of the enzymes (16 of the 21 tested) showed an initial increase in activity in the presence of [C2mim][OAc] compared to water (0% IL), with a 15 to 500% enhancement in activity that eventually declined at higher [C2mim][OAc] concentrations (Table 3). This phenomenon is illustrated in the row labeled "Max activity in IL" in Table 3 that lists the highest fold change in activity in the presence of [C2mim][OAc]. For example, enzyme J16 was found to be five times more active in 10% (v/v) [C2mim][OAc] than in water. The majority of the enzymes were active in at least 20% (v/v) [C2mim][OAc] and maintained greater than

50% activity. Six of the enzymes (J03, J05, J16, J25, J26 and J36) maintained more than 80% activity in 35 to 40% [C2mim][OAc]. Only a single enzyme, J15, lost activity at low [C2mim][OAc] concentrations. The BG enzymes J5 and J16 and Endo enzymes J26 and J36 showed the highest increase in activity in the presence of [C2mim][OAc]. To examine the relationship of IL-tolerance to potential halo-tolerance, their activity was measured in equal molar concentrations of [C2mim][OAc] and sodium acetate (NaOAc) (Figure 1A-B). Each of these enzymes also showed greater or equal activity in the presence of NaOAc, despite this salt buffering the solution at a more basic pH, which tends to be outside the optimal activity range for these enzymes (in water), especially J16 (Figure 1C-D).

The T_{opt} and pH_{opt} values of these enzymes were compared to their IL-tolerance to determine whether either of these properties positively correlates with high IL-tolerance. A plot of the optimum temperature or pH of the enzyme versus the highest concentration of [C2mim][OAc] in which the enzyme retains $\geq$80% of its activity was examined for clustering of values that would indicate that a particular range of pH or temperature positively correlates with high IL-tolerance. Of these two properties, only the T_{opt} showed any discernible correlation with high IL-tolerance (Figure 2). It appears that a T_{opt} >70°C is a positive indicator of high IL-tolerance. Enzymes with a T_{opt} $\leq$70°C have only an 18% probability of being highly tolerant to [C2mim][OAc], whereas enzymes with a T_{opt} >70°C have a 78% chance of being highly IL-tolerant (see Figure 2 legend for details).

8.3 DISCUSSION

Developing IL-tolerant enzymatic mixtures for cellulose hydrolysis will permit the advancement of technologies that combine IL-based pretreatment using [C2mim][OAc] with enzymatic hydrolysis. This type of process intensification will be critical for the development of cost-competitive lignocellulosic biofuel technologies [11]. However, there are few IL-tolerant enzymes known and more must be discovered before these technologies can be matured to the point of large-scale implementation in a biorefinery. This study was based on the hypothesis that thermotolerance and IL-tolerance are correlated, and therefore sought to expand the list of known IL-tolerant enzymes by identifying, expressing, and characterizing multiple thermophilic biomass deconstructing enzymes sourced from a single compost-derived microbial community that was previously used as a test bed for comparing IL and thermotolerance [14,19]. In the course of this study, we compared cell-free and in vivo *E. coli* expression methods for rapidly (and with high fidelity) screening through predicted enzyme candidates to narrow down the list of targets to functional and properly annotated enzymes. Results from this study elicit several interesting conclusions regarding the utility of in vitro versus in vivo screening methods, the activity of the recombinant enzymes versus the native enzymes from the parent microbial community, and the hypothesis that thermotolerance and IL-tolerance are correlated.

Comparison of the cell-free and in vivo *E. coli* screens yielded several observations: 1) both screens work well at quickly screening through candidate genes to identify functional genes; 2) the screens produce similar results in regards to predicted annotation and 3) the cell-free screen is more rapid (24 hours) compared to the in vivo screen (5 days); however, 4) the cell-free screen missed about 27% of the positive candidates (19 versus 26), and 5) the cell-free screen will eventually require porting into an in vivo expression system to conduct more detailed enzyme profiling. In light of these observations, the cell-free screen would be advantageous if the number of candidates to screen is large, as it is more rapid and less labor-intensive than the in vivo screen, whereas the in vivo screen would be more advantageous in smaller screens as it provides greater returns and enables more detailed characterization. Overall, the assigned annotation of each enzyme accurately reflected its measured activity. Several enzymes showed activity on multiple substrates, but in most cases the highest measured activity matched the annotation of the enzyme.

After the initial screening, there were 21 promising enzyme targets (15 BG and 6 Endo) to profile in more detail for optimum temperature, pH and IL-tolerance. The profiles revealed that the enzymes are indeed thermotolerant (T_{opt} between 45 and 95°C), and the two clusters of optimum temperatures observed for these enzymes (70 and 90°C) mirror the pattern seen in the profile of the native enzymes produced by the parent community from which these genes were isolated, except that the native enzymes had their had two T_{opt} peaks 10° lower than the heterologous enzymes (60 and 80°C) [14]. It is unclear why this may be. Perhaps the community produces a complex mixture of enzymes, the average of which results in observed T_{opt} at around 60 and 80°C, or the community only expresses a complement of enzymes with T_{opt} near 60 and 80°C.

The native enzymes produced by the parent microbial community were also [C2mim][OAc]-tolerant, a trait mirrored by the majority of enzymes profiled in this study. However, unlike the recombinant enzymes in this study, the native cellulase enzymes were not observed to have an increase in activity in the presence of ILs [14]. Many of the enzymes in this study showed an increase in activity in the lower range of [C2mim][OAc] concentrations tested (0.3 to 0.9 M), some several fold higher than the activity in water. The fact that several of these enzymes also showed increased activity in the presence of NaOAc suggests that these enzymes may require the presence of salt for optimal activity. The increase in activity with NaOAc was not as high for enzyme J16 as in the corresponding concentration of IL, which is likely due to the more basic pH of NaOAc and the lower pH optimum of J16 (pH 5.0). This phenomenon was less apparent for the other enzymes tested, but generally the enzymes demonstrated relatively higher levels of activity in the presence of [C2mim][OAc] compared to NaOAc. This apparent IL- and salt-tolerance is not surprising, considering that these enzymes are similar to those derived from thermotolerant and slightly halo-tolerant organisms like *Rhodothermus marinus*, which requires salt and grows optimally in about 0.3 M NaCl [21]. Unlike many fungal enzymes, these cellulases tend to prefer more neutral pH (6.0 or 7.0), and many retained more than 80% activity at the highest pH tested of 8.0. [C2mim][OAc] buffers around neutral pH in the range of concentrations tested, a property that may further ameliorate tolerance to this IL by several of the enzymes tested. The affinity of these enzymes for more neutral pH may reflect their origin; for example, *R. marinus* grows optimally at pH 7.0 [21].

The mechanisms of IL-tolerance are not well understood; few enzymes have been investigated for IL-tolerance in general and there are no studies that have looked at a large enough set of enzymes with a single type of IL, such as [C2mim][OAc], to do any type of thorough comparative analysis. The 21 enzymes characterized in this study had varying degrees of [C2mim][OAc]-tolerance and therefore provide an opportunity to look for correlations between IL-tolerance and other characteristics of the enzymes, that is, T_{opt} and pH ranges. Of those two properties, there only appears to be a correlation between IL-tolerance and T_{opt}, consistent with the conclusion from studies of other thermotolerant enzymes. A comparison of the IL-tolerance and T_{opt} revealed that the enzymes with a $T_{opt} > 70°C$ tend to have a higher probability of tolerating high concentrations of [C2mim] [OAc]. This indicates that evolution towards higher T_{opt} frequently alters the properties of an enzyme in a manner that also promotes tolerance to [C2mim][OAc]. The data from this study also indicates that the correlation is not simply between thermotolerance and IL-tolerance but more specifically hyperthermotolerance and IL-tolerance. Only a single enzyme studied with a $T_{opt} < 70°C$ displayed appreciable levels of IL-tolerance. This observation helps explain why enzymes from filamentous fungi used in commercial cellulase cocktails do not display IL-tolerance; there are no known hyperthermophilic filamentous fungi. Furthermore, future studies aimed at studying the mechanisms of [C2mim][OAc]-tolerance may benefit from a refined hypothesis that hyperthermotolerance (>70°C) is correlated with IL-tolerance.

The results presented here can also be used to comment on the general strategy used to identify enzymes with a particular set of characteristics, in this case IL-tolerance. The microbial community from which these enzymes were derived was originally established under the premise that organisms endowed with a particular functionality could be selectively enriched in abundance from a complex microbial community by cultivation under defined conditions. This selective enrichment could then help researchers target organisms and genes with a desired set of characteristics. In this case, the desired functionality was production of cellulase enzymes and the desired characteristic was thermo- and IL-tolerant cellulase enzymes. This strategy was implemented by cultivating a microbial community derived from green-waste compost under thermophilic conditions with plant biomass as a sole carbon source [14]. The native enzymes produced by this community were both thermo- and IL-tolerant and so were the recombinant enzymes derived from this community, suggesting that selective cultivation is a good method for discovering enzymes that function under a desired set of conditions.

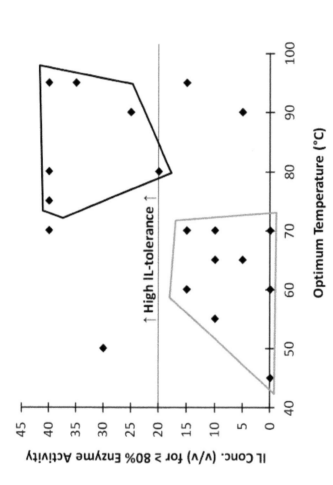

FIGURE 2: A plot highlighting the correlation between thermotolerance and ionic liquid (IL)-tolerance of the enzymes shown in Table3. The plot shows the maximum [C2mim][OAc] concentration that permits $\geq$80% enzyme activity compared to water versus the optimum temperature (Topt) of the enzyme. There are two overlapping data points at 95°C, 35% IL. Enzymes with high IL-tolerance are defined as the enzymes that can tolerate 20% (v/v) [C2mim][OAc] or greater (above horizontal line). The enzymes fall into two clusters: the black polygon where 78% (7/9) of the enzymes with a Topt >70°C have high IL-tolerance, and the grey polygon where 82% (9/11) of the enzymes with a Topt $\leq$70°C have high IL-tolerance. Only 18% (2/11) of the enzymes with a Topt $\leq$70°C have low or no IL-tolerance.

8.4 CONCLUSIONS

The enzymes characterized in this report are some of the most tolerant to [C2mim][OAc] reported to date [12,14,15,17]. Tolerance to this particular IL is of increasing interest as it is currently one of the most effective and well-studied ILs for pretreatment of lignocellulosic biomass [22]. Recent efforts to develop IL-tolerant cellulase cocktails and to incorporate these cocktails into one-pot pretreatment and saccharification bioprocessing schemes show that IL-tolerant enzymes can be used to develop new technologies to deconstruct biomass, and open up the technological landscape for lignocellulosic biorefineries [18]. The enzymes described in this report can be used to further those technologies.

8.5 METHODS

8.5.1 MANUAL CELLULASE GENE ASSEMBLY

Although most of the full-length ORFs in Table 1 were taken directly from the metagenome, several were manually reconstructed from fragmented genes identified in the assembly of the metagenomic dataset. The following ORFs were manually assembled: J03 had an incorrectly predicted start codon. The start of this ORF was moved 5' to match the start of its top BLAST hit. J08/09 are two versions of a single ORF composed of four gene fragments from the metagenome (IMG gene IDs 2061981261, 2062002762, 2062037967, 2061992858), which all have very high homology with a predicted BG from *Thermobaculum terrenum* ATCC BAA-798 [GenBank: ACZ42845.1]. J08 is an assembly of 2061981261 (N-terminus), 2062002762 (C-terminus), and ACZ42845.1 (sequence that encodes AAIVITENGAAYPDE inserted between the two sequences), and J09 is a compilation of 2062037967, 2061992858, and the same fragment from ACZ42845.1 assembled in the same order as J08. Overall, J08 and J09 differ by 5 AA. The same situation applies to J10, which is assembled from 2062002992 (N-terminus), 2062002993 (C-terminus), and a middle fragment (sequence encoding NAVKVTAAA) from ACX65411.1, a glycoside hydrolase family 3 protein from *Geobacillus sp.* Y412MC1. J11 was also assembled in the same manner; two consecutive ORFs (2062005533 and 2062005534) were merged with a fragment encoding (YVR) derived from a glycoside hydrolase family 3 protein from *Ktedonobacter racemifer* DSM 44963 (EFH83601.1). J38/39 are two versions of two consecutive orfs (2062019305, and 2062019306), which may be separated by a single base pair frame-shift or a larger deletion. J38 is a merger of the two orfs by inserting a single base pair to encode a leucine codon at residue 103. J39 is a merger of the two ORFs with a 316 base pair insertion at the same location derived from a BG from *Paenibacillus sp.* JDR-2 (ACT00588.1), to repair the glycoside hydrolase family 3 N-terminal domain.

8.5.2 GENE SYNTHESIS AND CLONING

Each gene was codon-optimized for expression in *E. coli* and synthesized by Genscript (Piscataway, NJ, USA). They were then cloned into a modified pUC57 vector constructed at Genscript, pUC57CFv1, with an added T7 promoter and terminator, as well as gateway attB1/attB2 sequences flanking the ORF, and a 8 × C-terminal 8 × His and Strep-tag II dual tag. There was an in-frame NheI-XhoI cloning site added between the attB1/attB2 sequences to place the ORFs into the pUC57CFv1 vector. The added vector sequences were cloned into the pUC57 vector at the EcoRI and SacI sites. Synthesized ORFs were then cloned into the pUC57CFv1 vector at the NheI-XhoI sites. The synthesized genes in the pUC57CFE1 vector were transformed in to TOP10 *E. coli* for storage at -80°C.

The T7, Gateway attB1/attB2 and His tag sequences added to pUC57 are:

```
GAATTCTAAATTAATACGACTCACTATAGGGAGAC-
CACAACGGTTTCCCTCTAGAAATAATTTTGTTTA-
ACTTTAAGAAGGAGATATACATATGACAAGTTT-
GTACAAAAAAGCAGGCTTCGCTAGCCCAATCC
AATCTCGAGGACCCAGCTTTCTTGTACAAAGTG-
GTCCATCATCACCATCACCATTAACAATAACTAG-
```

CATAACCCCTTGGGGCCTCTAAACGGGTCTT-
GAGGGGTTTTTTGGAGCTC

8.5.3 IN VITRO AND IN VIVO EXPRESSION OF CELLULASES

Each of the 37 cellulases was expressed in vitro using the RTS 100 *E. coli* 100 Hy cell-free expression Kit (Roche Diagnostics, Mannheim, Germany, Catalogue Number 03 186 148 001), using 0.5 µg of vector and following the manufacturer's instructions. The lyophilized plasmids were dissolved in DNase/RNase-free water before use. The in vitro protein expression was performed at 30°C for six hours. The expression products were used immediately for enzyme assay reactions.

To validate the enzyme activity results of in vitro protein expression and assays, the cellulase genes were cloned into the low-copy bacterial expression plasmid pDEST17 by Gateway cloning techniques following the manufacturer's instructions (Invitrogen). The sequences of all cloned genes in the pDONR221 and pDEST17 vectors were verified by DNA sequencing (Quintara Biosciences; Albany, CA, USA). All cellulase genes in the pDEST17 vector, except J24 and J29, were transformed into BL21(DE3)Star *E. coli* (Invitrogen, Carlsbad, CA, USA). The J24 and and J29 genes in the pDEST17 vector were transformed into the T7 Express I^q*E. coli* strain (New England BioLabs, Ipswich, Massachusetts, USA) to attenuate the basal level of cellulase expression during the growth phase prior to induction of protein expression. This was done because the expression vectors containing J24 and J29 were toxic to TOP10 and BL21(DE3)Star strains of *E. coli*, presumably due to the leaky activation of the T7 promoter. Bacterial cultures were grown in 96-deep well-plates in 800 µL of Luria-Bertani (LB) Miller broth containing carbenicillin (50 µg/ml) in each well. The overnight cultures of *E. coli* were inoculated to fresh LB medium containing Overnight Express Autoduction System 1 (Calbiochem, San Diego, CA, USA) reagent and carbenicillin. In the autoinduction medium, the bacterial cultures were incubated at 37°C with constant shaking at 200 rpm for the first four hours. Then the cultures were grown at 30°C for 18 hours with constant shaking at 200 rpm. The cell pellets were harvested by centrifugation at 6,000 g for 30 minutes, and then stored at -20°C. Each of the frozen cell pellets was thawed and resuspended in 0.1 mL of BugBuster containing lysozyme (1 mg/mL), benzonase (25 U/ml) and phenylmethanesulfonylfluoride (PMSF) (1 mM). After 30 minutes of incubation at room temperature, the cell lysates were centrifuged at 4,000 g for 30 minutes at 4°C. The soluble protein extracts (supernatants) were filtered through 0.45-μm syringe filters, and then used for enzymatic assays.

8.5.4 ENZYME ASSAYS FOR IN VITRO AND IN VIVO SCREENS

The enzyme activities of the in vitro protein expression products from the pUC57CFE1 vector were screened on the following substrates: 4-nitrophenyl-β-D-glucopyranoside (pNPG, 5 mM), 4-nitrophenyl-β-D-cellobioside (pNPC, 5 mM), and 1% carboxymethyl cellulose (Sigma Aldrich). Each enzyme reaction mixture containing one of these substrates and 5 μL of in vitro expression product or soluble extract from *E. coli* cell lysates (before or after induction) was done in 50 mM sodium acetate buffer at pH 5 in a total volume of 50 μL. The final concentration of 4-nitrophenol-labeled substrate (pNPC, or pNPG) was 5 mM, and that of carboxymethylcellulose (CMC) was 1% in each reaction. The enzymatic reaction was done at 50°C for 16 hours. For the reaction mixtures containing CMC, a 3,5-dinitrosalicylic acid (DNS) assay was used to quantify hydrolyzed products. For the reaction mixtures containing pNPG, or pNPC, an equal volume of 2% sodium carbonate (Na_2CO_3) was added prior to measuring absorbance at 420 nm to detect hydrolyzed 4-nitrophenol.

8.5.5 ENZYME ASSAYS FOR ACTIVITY PROFILING OF CELLULASES

To profile the enzyme activity of positive cellulases in the screen, each enzyme was expressed in vivo as described above, except the culture volume was scaled to 50 ml. For each enzyme assay, 5 to 20 ul of lysate was used to ensure that each enzyme had an activity that fell within the linear range of the activity assay. Enzymes J1 to J19 were screened using pNPG (5 mM final concentration) and enzymes J21 to J39 were screened using CMC (1% w/v final concentration) in a 100-ul reaction volume. Each value reported in Table 3 is from the average of triplicate reactions. For the temperature profile, the reaction was set up using 50 mM 2-(N-morpholino)ethanesulfonic acid (MES) buffer pH 6.5, and reactions were run for 15 to 60 minutes, depending on enzyme activity, at 5° increments from 45 to 99°C. For the pH profile, the reactions were run at approximately 10°C below the optimal temperature of each enzyme in 100 mM NaOAc 50 mM MES and 50 mM 4-(2-hydroxyethyl)-1-piperazineethanesulfonic acid (HEPES) buffers between pH 4.0 and 8.0. The buffers were made by mixing two aliquots of the aforementioned buffer set to either pH 4.0 (buffer A) or 8.0 (buffer B) in 10% increments, starting from 0% B to 100% B, giving 11 points total between pH 4.0 and 8.0. For IL-tolerance profiles, the reactions were run without added buffer in IL concentrations between 0 and 40% w/v [C2mim][OAc] at approximately 10°C below the optimal temperature of each enzyme. Reaction times were set to keep the values within the linear range of detection. For some enzymes, the same reaction was set up substituting an equal molar amount of NaOAc for [C2mim][OAc]. Figure 1C-D shows the pH at each concentration of IL and molar equivalent concentrations of NaOAc.

REFERENCES

1. Laboratory ORN: U.S. Billion-Ton Update: Biomass Supply for a Bioenergy and Bioproducts Industry. US DOE Energy Efficiency and Renewable Energy web site. 2011. http://www1.eere.energy.gov/bioenergy/pdfs/billion_ton_update.pdf

2. Wiselogel AE, Agblevor FA, Johnson DK, Deutch S, Fennell JA, Sanderson MA: Compositional changes during storage of large round switchgrass bales. Bioresource Technol 1996, 56:103-109.

3. Wald ML: U.S. Backs Project to Produce Fuel From Corn Waste. July 7th 2011, B10. [The New York Times, New York edition]

4. U.S. DOE: Using Fermentation and Catalysis to Make Fuels and Products: BIOCHEMICAL CONVERSION. US DOE Energy Efficiency and Renewable Energy web site. 2010. http://www1.eere.energy.gov/bioenergy/pdfs/biochemical_four_pager.pdf

5. Steen EJ, Kang Y, Bokinsky G, Hu Z, Schirmer A, McClure A, Del Cardayre SB, Keasling JD: Microbial production of fatty-acid-derived fuels and chemicals from plant biomass. Nature 2010, 463:559-562.
6. Peralta-Yahya PP, Keasling JD: Advanced biofuel production in microbes. Biotechnol J 2010, 5:147-162.
7. Nakayama S, Kiyoshi K, Kadokura T, Nakazato A: Butanol production from crystalline cellulose by cocultured clostridium thermocellum and clostridium saccharoperbutylacetonicum N1-4. Appl Environ Microb 2011, 77:6470-6475.
8. Li C, Knierim B, Manisseri C, Arora R, Scheller HV, Auer M, Vogel KP, Simmons BA, Singh S: Comparison of dilute acid and ionic liquid pretreatment of switchgrass: biomass recalcitrance, delignification and enzymatic saccharification. Bioresource Technol 2010, 101:4900-4906.
9. Zhao H, Jones CIL, Baker GA, Xia S, Olubajo O, Person VN: Regenerating cellulose from ionic liquids for an accelerated enzymatic hydrolysis. J Biotechnol 2009, 139:47-54.
10. Tadesse H, Luque R: Advances on biomass pretreatment using ionic liquids: An overview. Energ Environ Sci 2011, 4:3913-3929.
11. Klein-Marcuschamer D, Simmons BA, Blanch HW: Techno-economic analysis of a lignocellulosic ethanol biorefinery with ionic liquid pre-treatment. Biofuels Bioprod Biorefin 2011, 5:562-569.
12. Turner MB, Spear SK, Huddleston JG, Holbrey JD, Rogers RD: Ionic liquid salt-induced inactivation and unfolding of cellulase from Trichoderma reesei. Green Chem 2003, 5:443-447.
13. Park JI, Steen EJ, Burd H, Evans SS, Redding-Johnson AM, Batth T, Benke PI, D'Haeseleer P, Sun N, Sale KL, Keasling JD, Lee TS, Petzold CJ, Mukhopadhyay A, Singer SW, Simmons BA, Gladden JM: A thermophilic ionic liquid-tolerant cellulase cocktail for the production of cellulosic biofuels. PLoS One 2012, 7:e37010.
14. Gladden JM, Allgaier M, Miller CS, Hazen TC, VanderGheynst JS, Hugenholtz P, Simmons BA, Singer SW: Glycoside hydrolase activities of thermophilic bacterial consortia adapted to switchgrass. Appl Environ Microbiol 2011, 77:5804-5812.
15. Datta S, Holmes B, Park JI, Chen ZW, Dibble DC, Hadi M, Blanch HW, Simmons BA, Sapra R: Ionic liquid tolerant hyperthermophilic cellulases for biomass pretreatment and hydrolysis. Green Chem 2010, 12:338-345.
16. Gladden JM, Eichorst SA, Hazen TC, Simmons BA, Singer SW: Substrate perturbation alters the glycoside hydrolase activities and community composition of switchgrass-adapted bacterial consortia. Biotechnol Bioeng 2012, 109:1140-1145.
17. Zhang T, Datta S, Eichler J, Ivanova N, Axen SD, Kerfeld CA, Chen F, Kyrpides N, Hugenholtz P, Cheng J-F, Sale KL, Simmons BA, Rubin E: Identification of a haloalkaliphilic and thermostable cellulase with improved ionic liquid tolerance. Green Chem 2011, 13:2083-2090.
18. Shi J, Gladden JM, Sathitsuksanoh N, Kambam P, Sandoval L, Mitra D, Zhang S, George A, Singer SW, Simmons BA, Singh S: One-pot ionic liquid pretreatment and saccharification of switchgrass. Green Chem 2013, 15:2579-2589.
19. D'haeseleer P, Gladden JM, Allgaier M, Chain PSG, Tringe SG, Malfatti SA, Aldrich JT, Nicora CD, Robinson EW, Paša-Tolić L, Hugenholtz P, Simmons BA, Singer

SW: Proteogenomic analysis of a thermophilic bacterial consortium adapted to deconstruct switchgrass. PLoS ONE 2013, 8:e68465.

20. Rosenberg AH, Lade BN, Chui DS, Lin SW, Dunn JJ, Studier FW: Vectors for selective expression of cloned DNAs by T7 RNA polymerase. Gene 1987, 56:125-135.

21. Bjornsdottir SH, Blondal T, Hreggvidsson GO, Eggertsson G, Petursdottir S, Hjorleifsdottir S, Thorbjarnardottir SH, Kristjansson JK: Rhodothermus marinus: physiology and molecular biology. Extremophiles 2006, 10:1-16.

22. Sathitsuksanoh N, George A, Zhang YHP: New lignocellulose pretreatments using cellulose solvents: a review. J Chem Technol Biotechnol 2013, 88:169-180.

There are several tables that are not available in this version of the article. To view this additional information, please use the citation on the first page of this chapter.

CHAPTER 9

ULTRASONIC DISINTEGRATION OF MICROALGAL BIOMASS AND CONSEQUENT IMPROVEMENT OF BIOACCESSIBILITY/BIOAVAILABILITY IN MICROBIAL FERMENTATION

BYONG-HUN JEON, JEONG-A CHOI, HYUN-CHUL KIM, JAE-HOON HWANG, REDA A.I. ABOU-SHANAB, BRIAN A. DEMPSEY, JOHN M. REGAN, AND JUNG RAE KIM

9.1 BACKGROUND

There has been an increasing interest in use of the renewable and sustainable biomass, namely the third generation feedstock for bioenergy production. Microalgae have gained considerable attention as an alternative biofuel feedstock [1,2] as recent discoveries indicate that most algal biomass is exceedingly rich in carbohydrate and oil [3], which can be converted to biofuels using existing technology. Especially high levels of biogas and biofuel can be produced using carbohydrate of algal biomass via fermentation process [4]. Glucose in algal biomass is the most important monosaccharide affecting the fermentative ethanol production that is greatly dependent upon the composition of carbohydrate components in organic substrates. Bioenergy production from biomass generally requires three sequential processes, i.e., hydrolysis, acidification, and bioenergy generation. Numerous studies using algal biomass have reported that hydrolysis is often the rate-limiting step due to rigid cell walls and cytoplasmic membranes that inhibit or delay subsequent biodegradation in the fermentation processes.

Several methods for algal cell disruption have been evaluated including ultrasonication, bead beating, microwave (at 100°C), osmotic shock (with NaCl) and autoclaving (at 121°C) with varied results [5,6]. Sonication has the advantage of being able to disrupt the cells at relatively low temperatures when compared to microwave and autoclave. In addition, sonication does not require the addition of beads or chemicals, thus decreasing processing cost. Ultrasonication has been commonly used for cell lysis and homogenization, and could be an effective treatment for breaking up the rigid cell envelopes of microalgae [7]. During ultrasonication, sonic waves are transmitted to the microalgal culture. The waves create a series of microbubble cavitations which imparted kinetic energy into the surface of the cells and eventually ruptured the cell walls facilitating the release of carbohydrates and lipids from the cell into the exocellular medium [8]. Acoustic streaming is the other mechanism for using ultrasonication that facilitates the mixing of solution. Such homogenization of algal suspension can improve enzymatic and/or bacterial access to substrates and therefore facilitates the subsequent fermentation process [9,10].

Despite the wide use of algal biomass as a feedstock to produce bioenergy, there are only a few studies that quantitatively determine the compositional distribution of carbohydrate components which affects the productivity of fermentative bioenergy. We previously investigated the feasibility of using ultrasonication as a pretreatment prior to bacterial fermentation of microalgal biomass. Ultrasonication resulted in physical disintegration of microalga cell walls and consequently enhanced fermentative production of hydrogen and ethanol [4]. However, the earlier study did not determine the qualitative and quantitative characteristics of the lysed biochemicals. The main objective of this study therefore was to quantitatively evaluate the performance of sonication on microbial fermentation process and to systematically characterize the biochemical compositions and properties of microalgal biomass before and after the sonication pretreatment in the following ways: (1) investigate the composition of carbohydrate components in both dissolved and solid phases; (2) determine the abiotic conversion of total carbohydrates to the dissolved phase through sonication; (3) measure both surface hydrophobicity and electrical stability of microalgal cell; and (4) evaluate bioenergy productivity via microbial fermentation of microalgal biomass (compared with soluble starch) under different thermal conditions.

9.2 RESULTS AND DISCUSSION

9.2.1 COMPOSITION AND BIOAVAILABILITY OF CARBOHYDRATE COMPONENTS FROM ALGAL BIOMASS

Microalgal suspensions were exposed to four different sonication durations of 0 (non- sonication), 10 (short-term treatment), 15 and 60 min (long-term treatment) at 45°C. No change was found in the concentration of dissolved carbohydrates after 10 min sonication compared to non-sonication (data not shown). A 15 min sonication treatment increased the dissolved fraction of total carbohydrates from 3% to 32%. Further increase of sonication up to 60 min resulted in an insignificant increase of the dissolved fraction, accounting for <1% of total carbohydrates. This result implies that effective algal cell lysis had occurred within 15 min by sonication, resulting in rupture of the cell walls and intracellular materials release to the medium. Ultrasonication of microbes can result in much more hydroxyl groups of carbohydrates and/or lipids on the inner and outer cell surfaces due to extensive cell disintegration and lysis [11]. The hydrophilic nature of saccharide-like substances can also make it possible to increase the solubility of organic materials in culture broth because of the electro-negativity of oxygen atoms in hydroxyl ions [12].

Table 1 shows that total carbohydrates accounted for 37% of the non-sonicated algal biomass. Microalgae such as *Chlorella, Chlamydomonas, Dunaliella, Scendesmus,* and *Tetraselmis* have been shown to accumulate a large amount of carbohydrate (>40% of the dry weight) [13]. Cell wall of the green algae such as *Scenedesmus, Chlorella, Monoraphidium,* and

Ankistrodesmus contains 24–74% of neutral sugars, 1–24% uronic acid, 2–16% proteins, and 0–15% glucosamine [14]. The major fractions of sugars are either mannose/glucose or rhamnose/galactose [14]. Glucose and mannose were the major constituents among the quantified monosaccharides, and accounted for 66.4% (which decreased to 60.4% by a 60 min sonication) and 21.7% (which increased to 25.6% after the sonication) of total monomeric sugars, respectively. The increased portion of mannose might be derived from glucose due to sonication [15,16]. Sonication also resulted in small changes in the concentrations of galactose and glucosamine as minor constituents.

TABLE 1: Effect of sonication treatment on composition and bioavailability of total carbohydrate components

Sonication time, min	0		15		60	
	Non-sonicated	Fermented	Sonicated	Sonicated/fermented	Sonicated	Sonicated/fermented
Total carbohydrate, g g[-1]	0.37 (0.01)[a]	0.22	0.37 (0.12)	0.08	0.39 (0.13)	0.09
Glucose (%)	66.38	31.90	63.67	8.03	60.40	7.62
Mannose (%)	21.76	18.81	24.33	2.78	25.60	2.67
Galactose (%)	5.94	5.68	8.66	8.04	10.22	10.12
Glucosamine (%)	5.92	3.06	3.34	2.78	3.78	2.66

[a]*The values in the parentheses show the concentrations of dissolved carbohydrates (g-carbohydrate g-1-biomass).*
Sonication pretreatment of algal biomass was conducted at 45°C for 15 or 60 min. Both non-sonicated and sonicated algal biomass were mixed with an equivalent volume of fermenting bacteria and subsequently fermented at 35°C for 23 days. Other common monomeric dissolved carbohydrates (such as fucose, rhamnose, and galactosamine) were not detected.

Long-term sonication resulted in large increases in dissolved carbohydrates and in fermentative utilization of the carbohydrates. Table 1 shows that dissolved carbohydrates increased from 1% of biomass in non-sonicated microalga to 12% of biomass after 15 min sonication. A 15 min sonication pretreatment also significantly increased total carbohydrates consumption during the following fermentation stage. The amount of total carbohydrates sharply decreased to 89% for the first 16 days of fermentation, while extremely small amounts of residual carbohydrates consistently decreased in the fermentor up to 23 days. Fermentation of fresh algal biomass at 35°C for 23 days resulted in a decrease in total carbohydrates from 0.37 g-carbohydrate g^{-1}-biomass to 0.22 g-carbohydrate g^{-1}-biomass, while the fermentation after 15 min sonication significantly reduced the total carbohydrates to 0.08 g-carbohydrate g^{-1}-biomass. The improved utilization of carbohydrates by sonication was attributed to both increased dissolved carbohydrate concentrations and increased microbial access to carbohydrates was available on the ruptured cell walls.

9.2.2 ELECTRICAL STABILITY AND FUNCTIONAL PROPERTY OF MICROALGAL CELL

Several previous studies reported that hydrophobic aggregation decreased bioaccessibility of fermenting bacteria to substrate [17,18]. The surface charge represented by zeta potential is important for a better understanding on the nature of the particle stability [19], and it has also been well-known that colloidal particles with higher absolute values of the zeta potential tend to be less aggregated due to high electrical repulsion among the particles [20,21]. The pH of the mixture of microalgal biomass and inoculum (fermenting microbes) ranged from 5.7 to 6.8 at the beginning of fermentation, and the final pH values after the 23 days of fermentation were nearly identical (5.5 ± 0.2) in all sets of tested samples. Figure 1 shows that the zeta potential of algal biomass was slightly negative at pH 5, became more negative as the pH increased to 9, and was sharply decreased by a 15 min sonication regardless of the initial pH values. Our result indicates that algal cell became more electrostatically stable in aqueous solution at pH 5 and 9 as a result of sonication treatment for 15 min or longer due to exposed functional group as well as the release of negatively charged organic constituents upon cell lysis. This result also coincided with a significant increase of the dissolved carbohydrate fraction in the suspension upon sonication [12].

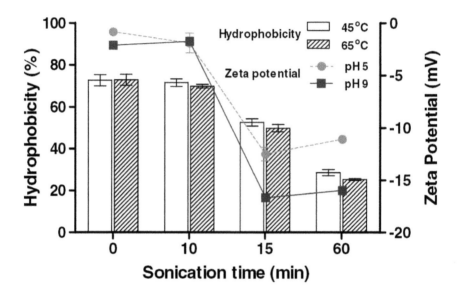

FIGURE 1: The cell surface hydrophobicity (left y-axis) and zeta potential (right y-axis) of algal biomass as a function of sonication duration. Sonication of algal biomass was conducted at two different temperatures of 45°C and 65°C for up to 60 min. The pH of non-sonicated and sonicated algal biomass suspensions was adjusted to either 5 or 9 for zeta potential measurement. Three independent biological replicates were used for the measurements. Error bars indicate standard deviation values from the average.

Sonication treatment for up to 10 min did not significantly changed the algal surface hydrophobicity accounting for $75\pm3\%$, but which was decreased to $54\pm2\%$ or $28\pm3\%$ by a 15 min or 60 min sonication, respectively, at 45°C. It should be noted that almost identical hydrophobicity was observed in sonication of algal biomass at two tested temperatures. The results reveal that algal biomass was significantly hydrophilized by sonication treatment at 45°C or higher for 15 min. Chen et al. (2004) showed that sonication changed the molecular structure of organic matter, especially resulting in decreases in aromaticity, molecular weight, and specific UV absorbance that has been used as an indicator of organic hydrophobicity [22]. These changes to organic properties were consistent with our observation showing a significant decrease in the surface hydrophobicity of microalga, possibly due to hydroxyl radical production resulting in a substantial degradation of organic compounds and thus an increase in the assimilable organic carbon fraction of the total organic carbon pool [23]. The destruction of algal cell structures during sonication pretreatment also released more algal cell fragment to the aqueous phase, which was observed by increased residual turbidity by 24%. The functionality of organic matter is an important factor in determining how efficient heterotrophic bacteria can assimilate the organic substrates. Numerous studies have demonstrated that heterotrophic microorganisms preferably assimilate hydrophilic organic substrates to a much greater extent than hydrophobic organics [24,25].

Our result showed that the sonication decreased hydrophobicity of cell fragments compared to the control, possibly due to a decrease in the number of conjugated bonds in chain structures [22,26]. Review of the literature demonstrates that algal cells are significantly hydrophilized by ultrasonication [27], which was demonstrated with improved availability of algal biomass as a fermentable substrate for the fermenting bacteria in this study. Hydrophilic functional groups of particle surface also contribute to increasing colloidal stability as shown in zeta potential in this study along with an increase of the specific surface area, resulting in improved bacterial accessibility and metabolic activity [17,18]. These previous findings were consistent with our observation showing an increase of carbohydrate consumption in the microbial fermentation of algal biomass after sonication treatment.

9.2.3 EFFECTS OF SONICATION DURATION AND FERMENTING TEMPERATURE ON BIOENERGY PRODUCTION

Figure 2 shows the compositional distribution and concentrations of soluble metabolite product (SMP) after the 23 day fermentation of algae biomass under different thermal conditions (35°C and 55°C). The concentrations of ethanol and volatile fatty acid (VFAs) were increased as the sonication duration was increased to 15 min, while the production of those materials was rarely improved even when the sonication duration was increased up to 60 min regardless of the fermentation temperature.

Figure 2 also compares algal biomass (both sonicated and non-sonicated) as a substrate with starch as a control for microbial fermentation in terms of the bioenergy productivity. The use of algal biomass sonicated for 15 min or longer was comparable to soluble starch as a feedstock for the production of ethanol/VFAs throughout the 23 day fermentation period. In case of sonication pretreatment for 15 min or longer, butyric acid was the dominant form of VFAs followed by acetic acid and propionic acid. Fermentative biofuel production from organic substances results in incomplete decomposition of substrate into organic acids such as acetate and butyrate. Butyrate is more dominant because of its lower Gibbs free energy ($\Delta G = -257.1$ kJ) compared to acetate ($\Delta G = -184.2$ kJ) and its production involves enzyme activity [28]. This observation was consistent with previously reported work in which the butyrate type fermentation process was employed [4,27,29]. Ethanol has also been reported as the major SMP during anaerobic degradation of saccharide-like substances because monomeric sugars can be utilized easily by heterotrophic microbes (e.g., fermentative bacteria) [30].

The results of microbial fermentations with non-sonicated and sonicated algal biomass are shown in Figure 3 in which cumulative hydrogen and ethanol production are plotted according to time. Production of ethanol resulted in decreased hydrogen production, which was consistent with an earlier report that hydrogen production was decreased when ethanol production was initiated [31]. This was due to a shift of metabolic pathway from butyrate fermentation to ethanol fermentation. The maximum production of hydrogen (H_{max}) via microbial fermentation of algal biomass at 55°C was 10% higher than achieved at 35°C, but the fermenting thermal conditions rarely affected the energy conversion efficiency (H_2 yield). Table 2 shows that the maximum hydrogen production rate (R_{max}) was independent of the temperatures between 45°C and 65°C in ultrasonic pretreatment of algal biomass. As the volumetric ratio of algal biomass to fermentative bacteria (AB:FB) was increased from 0.2 to 1.0, the maximum accumulative hydrogen production increased from 0.72 to 2.51 L L^{-1} and from 0.87 to 2.72 L L^{-1} at two different fermenting temperatures of 35°C (mesophilic) and 55°C (thermophilic), respectively. This might be attributed to the improved fermentative bacteria activity at the higher organic loading rate under themophilic conditions. The dissolved carbohydrates concentration increased by sonication was also correlated with the increased H_2 production, and the remarked production of hydrogen was thus due to improved bioavailability of algal biomass for the fermenting bacteria. The λ (average lag time) values calculated from Gompertz equation are close to those observed in the experiments. The λ value prior to exponential hydrogen production was 11 h under mesophilic conditions, while thermophilic showed much shorter λ (2 to 3 h) and relatively higher R_{max} (up to 0.3 L L^{-1} h^{-1}) compared to observed for mesophilic. Our observations were consistent with previously reported work in which anaerobic fermentation of glucose increased the production of hydrogen when operating the fermenter with high organic loading rates under thermophilic conditions [29]. Many factors such as substrates, their concentration, pH and temperature can influence on the fermentative hydrogen production [32,33]. Among them, temperature is a key factor because it can affect the activity of hydrogen producing bacteria (HPB) by influencing the activity of essential enzymes such as hydrogenases for fermentative hydrogen production [34,35].

Ethanol production from algal biomass was increased from 0.9 to 5.6 g L^{-1} with increasing the AB/FB ratio from 0.2 to 1.0 under mesophilic condition. The highest ethanol production among the experimental variations using algal biomass was comparable to the fermentation of equal amount of carbohydrate in starch (see Table 2 and Figure 3b and d). Carbohydrates accounting for 37% of the algal biomass were not only especially beneficial components for heterotrophic bacterial activity, but also a valuable source for fermentative bacteria leading to enhanced energy production.

9.2.4 SEM IMAGES OF THE SONICATED ALGA

Extensive cell wall damage was observed after a 15 or 60 min sonication which allowed fermentative bacteria to access the inner space of the ruptured algal cell (Figure 4c and d), while an external shape of alga sonicated for 10 min did not look much different from the intact surface of algal biomass on which fermentative bacteria worked (Figure 4a and b). SEM images clearly show greater accessibility of fermentative bacteria to surface-bound carbohydrates of algal debris after sonication $\geq$15 min compared to control and 10 min sonication. Further some of the nucleus materials in the sonicated alga presumably spread outside the cell due to complete cell lysis, coincided with a significant increase in the dissolved fraction of total carbohydrates after sonication for 15 min or longer (Table 1). Therefore the long-term sonication pretreatment resulted in enhanced bioaccessibility and bioavailability of algal biomass, which led to the increases in carbohydrate consumption and subsequent bioenergy (hydrogen/ethanol) production.

9.3 CONCLUSIONS

This study has characterized the carbohydrate components in algal suspension upon sonication that could result in significant changes in the physicochemical properties of algal cell and subsequent enhancement of biodegradability/bioaccessibility for microbial fermentation. Sonication pretreatment for 15 min or longer on algal biomass (*S. obliquus* YSW15) resulted in a large increase in dissolved carbohydrates (composed mainly of glucose), likely due to release from the cell wall and the periplasm. Sonication enhanced fermentative bioenergy (hydrogen/ethanol) production, and resulted in comparable bioenergy production as compared to using soluble starch. Algal surface hydrophobicity was substantially decreased and electrostatic repulsion among algal debris dispersed in aqueous solution was significantly increased by a 15 min sonication treatment, which provided more facile access of the fermentative bacteria to algal biomass for assimilating carbohydrates of the algal cell fragments. A substantial uptake of the carbohydrate by the fermenting bacteria occurred during thermophilic fermentation of algal biomass sonicated for 15 min or longer, coincided with the high bioenergy production (e.g., ethanol 5.6 g L^{-1} and hydrogen 2.5 mL L^{-1}). The bioenergy productivity increased with increasing the organic substrate loading rate on the microbial fermentation regardless of the thermal conditions examined in this study. The economic evaluation of using the renewable carbon sources for promoting microbial fermentation concurrent with bioenergy production should be further investigated.

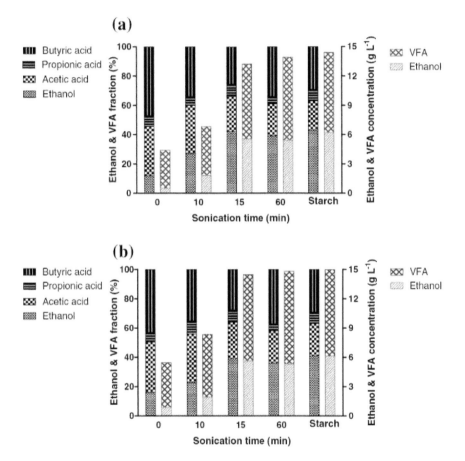

FIGURE 2: Effects of sonication duration and fermentation temperature on the production of ethanol and VFA from microbial fermentation of algal biomass. Either non-sonicated or sonicated (at 45°C) algal biomass was mixed with an equivalent volume of fermentative bacteria, and the mixture was then fermented at a temperature of (a) 35°C or (b) 55°C for 23 days.

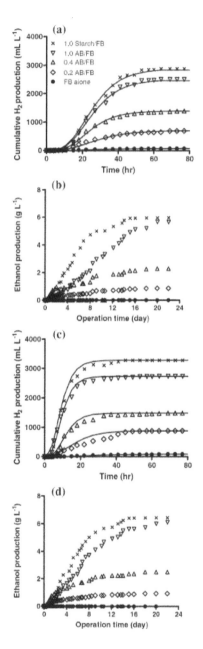

FIGURE 3: Effects of the ratio of algal biomass to fermentative bacteria (AB:FB) on the production of hydrogen and ethanol during the 23 day fermentation at a temperature of 35°C (a and b) or 55°C (c and d). The algal biomass was sonicated at 45°C for 15 min.

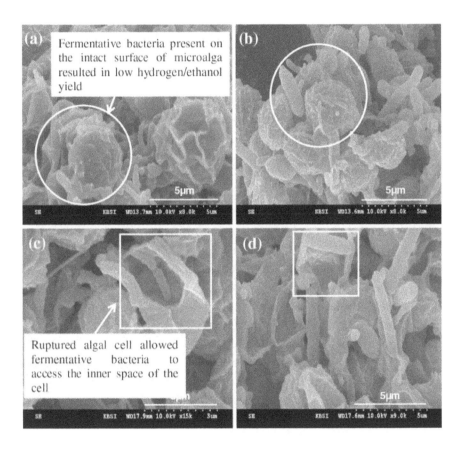

FIGURE 4: SEM images of fermentative bacteria along with non-sonicated or sonicated algal cells. Non-sonicated algal cells (a and b): (a) non- or (b) 10 min sonication. Sonicated algal cells (c and d): (c) 15 min or (d) 60 min sonication. The circles indicate fermentative bacteria present on the intact surface of algae, while the squares show ruptured algal cells allowed fermentative bacteria to access the inner space of the algal cells, resulting in high hydrogen/ethanol production.

9.4 METHODS

9.4.1 CULTIVATION OF MICROALGA

Scenedesmus obliquus YSW15 used in this study was isolated from the effluent of a municipal wastewater treatment plant at Wonju Water Supply and Drainage Center, South Korea [36]. The algal strain was grown in 1 L Erlenmeyer flask containing 0.5 L Bold Basal Medium (BBM) [37]. The culture was incubated on a rotary shaker (SH-804, Seyoung Scientific) at 27°C and 150 rpm under continuous fluorescent illumination with an intensity of 40 μmol m^{-2} s^{-1} for three weeks.

9.4.2 ULTRASONICATION

The harvested microalga biomass (34 mg mL^{-1}) was placed in a glass bottle and sonicated for 10, 15, or 60 min in a Branson 8510-DTH sonicator (Danbury, Connecticut, USA) at two different temperatures (45 and 65°C). The bath-type sonicator was used in this study due to the localized cavitation produced by horn-type sonicators [38,39]. The sonication was conducted at a constant frequency of 40 kHz and an output power of 2.2 kW for which the ultrasonic energy was applied in continuous (non-pulsed) mode with constant amplitude of 40% with specific supplied energy (Es) 70.6 MJ Kg^{-1}.

9.4.3 ANAEROBIC INOCULUMS

Seed sludge used in this study was collected from the anaerobic digesters of a municipal wastewater treatment plant (Wonju Water Supply and Drainage Center, Wonju, South Korea). The microbial sludge was heated at 90°C for 30 min to inactivate methanogenic bacteria and to enhance the activity of H$_2$-producing bacteria (HPB). The HPB was acclimatized to a synthetic medium in an anaerobic chemostat reactor for 1 month [40]. The medium was prepared daily and stored in a substrate reservoir maintained at 4°C. An anaerobic reactor (2 L of capacity with 1 L working volume) was filled with a mixture of pretreated sludge and synthetic medium, and operated with a hydraulic retention time of 12 h at 35°C.

9.4.4 FERMENTATION OF MICROALGAE BIOMASS

Microbial fermentation was performed to evaluate the bioenergy production from the non-sonicated and sonicated algal biomasses under both mesophilic (35°C) and thermophilic (55°C) conditions, and which was also compared with soluble starch as a substrate for the fermenting bacteria. The fermentation was carried out in triplicate using 150 mL serum bottles with a working volume of 100 mL. The volume ratio of algae biomass to fermenting bacteria (AB/FB) ranged from 0 (fermenting bacteria alone) to 1.0, and 50 mL fermenting bacteria (dry biomass = 3.4 g L^{-1}) was used for each of the experimental variations. The headspace of each bottle was flushed with N_2 for 15 min to provide an anaerobic environment and then sealed tightly using a butyl rubber stopper and an aluminum crimp. The bottles were placed in a water bath (SH-502S, Seyoung Scientific) kept at a temperature of 35 or 55°C, and gas/liquid samples were periodically collected from the bottles for measurement of SMP throughout the 23 day fermentation period.

9.4.5 ANALYTICAL PROCEDURES

The non-sonicated and sonicated algal cells were primarily-fixed using 4% glutaraldehyde for 2 h, secondarily-fixed using 1% OsO_4 for 1 h, and rinsed with 0.1 M cacodylate buffer (pH 7.4). The resulting samples were then dehydrated with different concentrations of ethanol, sputter-coated with Au-Pd immediately, and examined with a Low-Vacuum Scanning Electron Microscope (LV-SEM, S-3500 N, Hitachi). An ELS-8000 Electrophoretic Light Scattering Spectrophotometer (Ostuka, Japan) was used to determine zeta potential of microalgal cell. Estimation of microalga cell surface hydrophobicity was also performed with the microbial adhesion to hydrocarbon (MATH) test [41].

The concentrations of carbohydrates were determined using an ICS-5000 bio-liquid chromatography (Dionex, USA) with CarboPac PA1 column [42]. The volatile fatty acids (VFAs) were analyzed by a GC-8A Gas Chromatography (Shimadzu, Japan) equipped with a flame ionization detector (FID) and a glass column packed with 10% Reoplex 400. Ethanol was analyzed by a DS 6200 Gas Chromatography (Do-Nam Ins., Korea) equipped with a FID and a DB-624 column (Agilent, USA). Hydrogen was measured using a gas chromatograph (Shimadzu GC-14, Japan) equipped with a thermal conductivity detector and a molecular sieve 5A. The concentrations of total and volatile suspended solids were determined using the Standard Methods [43]. The pH was also measured by a pH meter (Orion 290A).

REFERENCES

1. Lü J, Sheahan C, Fu P: Metabolic engineering of algae for fourth generation biofuels production. Energy Environ Sci 2011, 4:2451-2466.
2. John RP, Anisha GS, Nampoothiri KM, Pandey A: Micro and macroalgal biomass: a renewable source for bioethanol. Bioresour Technol 2011, 102:186-193.
3. Chisti Y: Biodiesel from microalgae. Biotech Adv 2007, 3:294-306.
4. Choi JA, Hwang JH, Dempsey BA, Abou-Shanab RAI, Min B, Song H, Lee DS, Kim JR, Cho Y, Hong S, Jeon BH: Enhancement of fermentative bioenergy (ethanol/hydrogen) production using ultrasonication of Scenedesmus obliquus YSW15 cultivated in swine wastewater effluent. Energy Environ Sci 2011, 4:3513-3520.
5. Lee JY, Yoo C, Jun SY, Ahn CY, Oh HM: Comparison of several methods for effective lipid extraction from microalgae. Bioresour Technol 2010, 101:S75-S77.
6. Prabakaran P, Ravindran AD: A comparative study on effective cell disruption methods for lipid extraction from microalgae. Let Appl Microbiol 2011, 53:150-154.
7. González-Fernández C, Sialve B, Bernet N, Steyer JP: Comparison of ultrasound and thermal pretreatment of Scenedesmus biomass on methane production. Bioresour Technol 2012, 110:610-616.
8. Tiehm A, Nickel K, Zellhorn M, Neis U: Ultrasonic waste activated sludge disintegration for improving anaerobic stabilization. Water Res 2001, 35:2003-2009.
9. Mason TJ: Industrial sonochemistry: potential and practicality. Ultrasonics 1992, 30:192-196.
10. Mason T, Paniwnyk L, Lorimer J: The uses of ultrasound in food technology. Ultrason Sonochem 1996, 3:S253-S260.
11. Somaglino L, Bouchoux G, Mestas J, Lafon C: Validation of an acoustic cavitation dose with hydroxyl radical production generated by inertial cavitation in pulsed mode: application to in vitro drug release from liposomes. Ultrason Sonochem 2011, 18:577-588.

12. Zakrzewsk ME, Bogel-yukasik E, Bogel-yukasik R: Solubility of carbohydrates in ionic liquids. Energy Fuel 2010, 24:737-745.

13. Becker EW: Micro-algae as a source of protein. Biotech Adv 2007, 25:207-210.

14. Blumreisinger M, Meindl D, Loos E: Cell wall composition of chlorococcal algae. Phytochem 1983, 22:1603-1604.

15. Panneerselvam K, Etchison JR, Freeze HH: Human fibroblasts prefer mannose over glucose as a source of mannose for N-glycosylation. J Biol Chem 1997, 272:23123-23129.

16. Zhang M, Zhang L, Cheung PCK, Ooi VEC: Molecular weight and anti-tumor activity of the water-soluble polysaccharides isolated by hot water and ultrasonic treatment from the sclerotia and mycelia of Pleurotus tuber-regium. Carbohyd Polym 2004, 56:123-128.

17. Oomen AG, Rompelberg CJM, Bruil MA, Dobbe CJG, Pereboom DPKH, Sips AJAM: Development of an In vitro digestion model for estimating the bioaccessibility of soil contaminants. Arch Environ Contam Toxicol 2003, 44:281-287.

18. Mahdy AM, Elkhatib EA, Fathi NO, Lin ZQ: Use of drinking water treatment residuals in reducing bioavailability of metals in biosolid-amended alkaline soils. Commun Soil Sci Plant Anal 2012, 43:1216-1236.

19. Rossi L, ten Hoorn JWM S, Melnikov SM, Velikov KP: Colloidal phytosterols: synthesis, characterization and bioaccessibility. Soft Matter 2010, 6:928-936.

20. Klitzke S, Lang F: Mobilization of soluble and dispersible lead, arsenic, and antimony in a polluted, organic-rich soil - effects of pH increase and counterion valency. J Environ Qual 2009, 38:933-939.

21. Elimelech M, Nagai M, Ko CH, Ryan J: Relative insignificance of mineral grain zeta potential to colloid transport in geochemically heterogeneous porous media. Environ Sci Technol 2000, 34:2143-2148.

22. Chen D, He Z, Weavers LK, Chin YP, Walker H, Hatcher PG: Sonochemical reactions of dissolved organic matter. Res Chem Intermed 2004, 30:735-753.

23. Schechter DS, Singer PC: Formation of aldehydes during ozonation. Ozone Sci Eng 1995, 17:53-69.

24. Volk C, Bell MK, Ibrahim E, Verges D, Amy G, Allier ML: Impact of enhanced and optimized coagulation on removal of organic matter and its biodegradable fraction in drinking water. Water Res 2000, 34:3247-3257.

25. Bosma TP, Middeldorp PJM, Schraa G, Zehnder AJB: Mass transfer limitation of biotransformation:quantifying bioavailability. Environ Sci Technol 1997, 31:248-252.

26. Chin YP, Aiken G, O'Loughlin E: Molecular weight, polydispersity, and spectroscopic properties of aquatic humic substances. Environ Sci Technol 1994, 28:1853-1858.

27. Yin X, Han P, Lu X, Wang Y: A review on the dewaterability of bio-sludge and ultrasound pretreatment. Ultrason Sonochem 2004, 11:337-348.

28. Nandi R, Sengupta S: Microbial production of hydrogen: an overview. Crit Rev Microbiol 1998, 24:61-84.

29. Zhang Y, Liu G, Shen J: Hydrogen production in batch culture of mixed bacteria with sucrose under different iron concentrations. Int J Hydrogen Energy 2005, 30:855-860.

30. Lin CY, Chen HP: Sulfate effect on fermentative hydrogen production using anaerobic mixed microflora. Int J Hydrogen Energy 2006, 31:953-960.
31. Hawkes FR, Dinsdale R, Hawkes DL, Hussy I: Sustainable fermentative hydrogen production: challenges for process optimization. Int J Hydrogen Energy 2002, 27:1339-1347.
32. Li YF, Ren NQ, Chen Y, Zheng GX: Ecological mechanism of fermentative hydrogen production by bacteria. Int J Hydrogen Energy 2007, 32:755-760.
33. Levin DB, Islam R, Cicek N, Sparling R: Hydrogen production by Clostridium thermocellum 27405 from cellulosic biomass substrates. Int J Hydrogen Energy 2006, 31:1496-1503.
34. Lee KS, Lin PJ, Chang JS: Temperature effects on biohydrogen production in a granular sludge bed induced by activated carbon carriers. Int J Hydrogen Energy 2006, 31:465-472.
35. Valdez-Vazquez I, Ríos-Leal E, Esparza-García F, Cecchi F, Poggi-Varaldo HM: Semi-continuous solid substrate anaerobic reactors for H2 production from organic waste: mesophilic versus thermophilic regime. Int J Hydrogen Energy 2005, 30:1383-1391.
36. Abou-Shanab RAI, Hwang JH, Cho Y, Min B, Jeon BH: Characterization of microalgal species isolated from fresh water bodies as a potential source for biodiesel production. Appl Energy 2011, 88:3300-3306.
37. Bischoff HW, Bold HC: in Phycological Sudies IV. Univ Texas Publ 1963, 6318:1-95.
38. Shirgaonkar IZ, Pandit AB: Sonophotochemical destruction of aqueous solution of 2,4,6-trichlorophenol. Utrason Sonochem 1998, 5:53-61.
39. Joseph CG, Puma GL, Bono A, Krishnaiah D: Sonophotocatalysis in advanced oxidation process: a short review. Utrason Sonochem 2009, 16:583-589.
40. Hwang JH, Choi JA, Abou-Shanab RAI, Bhatnagar A, Min B, Song H, Kumar E, Choi J, Lee ES, Kim YJ, Um S, Lee DS, Jeon BH: Effect of pH and sulfate concentration on hydrogen production using anaerobic mixed microflora. Int J Hydrogen Energy 2009, 34:9702-9710.
41. Rosenberg M, Gutnick D, Rosenberg E: Adherence of bacteria to hydrocarbons: a simple method for measuring cell-surface hydrophobicity. FEMS Microbiol Lett 1980, 9:29-33.
42. Ajisaka K, Fujimoto H, Miyasato M: An α-L-fucosidase from Penicillium multicolor as a candidate enzyme for the synthesis of α $(1 \rightarrow 3)$-linked fucosyl oligosaccharides by transglycosylation. Carbohydr Res 1998, 309:125-129.
43. APHA, American Public Health Association: Methods for biomass production. In Standard methods for the examination of water and wastewater. Baltimore (MD, USA); 1998.

There are several tables that are not available in this version of the article. To view this additional information, please use the citation on the first page of this chapter.

CHAPTER 10

RAPID AND EFFECTIVE OXIDATIVE PRETREATMENT OF WOODY BIOMASS AT MILD REACTION CONDITIONS AND LOW OXIDANT LOADINGS

ZHENGLUN LI, CHARLES H. CHEN, ERIC L. HEGG, AND DAVID B. HODGE

10.1 INTRODUCTION

As the global demand for energy grows, the need for a sustainable fuel supply as a supplement or replacement for fossil fuels is becoming imperative [1]. Among possible technology options, the biochemical conversion of plant-derived sugars to biofuels has the potential to displace a substantial fraction of gasoline. This biochemical route involves the enzymatic hydrolysis of plant-derived polysaccharides to monomeric sugars, followed by fermentation of these sugars to biofuels such as ethanol. Starch from corn grain has been a major source of sugars for ethanol production in the U.S., but significant future growth of the corn ethanol industry is limited by the growing demand for both food and animal feed, as well as the recent achievement of maximum production limits on starch-based ethanol set by the Renewable Fuel Standard in the Energy Independence and Security Act of 2007 [2]. Thus, cellulosic biomass (i.e. plant cell wall material) is envisioned as an important feedstock for producing biofuel sustainably in the future as well as meeting renewable transportation fuel mandates.

Woody biomass is an especially attractive alternative to corn as a feedstock for biofuels. In particular, short-rotation woody bioenergy crops such as willow (*Salix* spp.) and hybrid poplar (*Populus* spp.) that are currently grown in temperate regions for combined heat and power bioenergy applications represent important feedstocks for liquid transportation fuels with agronomic and logistical advantages. Specifically, it has been shown that hybrid poplar can be grown on marginal agricultural lands with low energy and chemical input and produce biomass with high energy density at moderately high productivities [3,4], thereby providing significant motivation for developing effective and economic conversion technologies that can be coupled with woody feedstocks.

Due to the higher order structures in the plant cell wall, a chemical, thermal, or physical pretreatment step is necessary to facilitate the biochemical production of biofuels from plant cell wall polysaccharides. This need for pretreatment is primarily a consequence of cell wall lignin that limits cellulolytic enzyme accessibility to polysaccharides, with this cell wall recalcitrance to conversion especially problematic for the cell walls of woody plants. A wide range of pretreatments are known that differ in chemistry and mechanism but share the same outcome of increasing the accessibility of cell wall polysaccharides to cellulolytic enzymes [5]. Alkaline hydrogen peroxide (AHP) pretreatment is one such approach that has been studied since the 1980s [6-8]. AHP results in significant improvement in the enzymatic digestibility of commelinid monocots including corn stover and wheat straw [9,10] and can generate hydrolysates with more than 100 g/L of monomeric sugars that are fermentable without detoxification (manuscript in preparation). The mechanisms by which AHP pretreatment reduces recalcitrance include the fragmentation and solubilization of ester-linked xylan and lignin [11,12], as well as the oxidation, solubilization, and at high H_2O_2 loadings, mineralization of lignin [7]. While effective at increasing enzymatic digestibility, however, the majority of previous work employed high H_2O_2 loadings (i.e. >250 mg/g biomass) that would be economically challenging to implement industrially due to the high cost of H_2O_2 [6,7], and it is well-established that oxidative delignification using H_2O_2 treatment under alkaline conditions requires high H_2O_2 loadings to realize significant lignin removal from wood pulps [13].

Relative to herbaceous monocots, woody biomass such as hybrid poplar presents special challenges for AHP because of its thicker cell walls, denser vascular structure, and higher lignin content [14,15]. As a result, the improvement in enzymatic digestibility after AHP pretreatment is limited [16,17], and this lack of efficacy on woody biomass is a ubiquitous challenge faced by many pretreatment methods [5,18,19]. Although a few methods including organosolv, dilute acid, and SPORL (a sulfite pretreatment combined with mechanical size reduction) have been reported to be effective pretreatments for hybrid poplar [20,21], all of these methods still have drawbacks such as a high consumption of chemicals and the generation of fermentation inhibitors [22]. As a result, there is great interest in identifying effective pretreatments for woody biomass.

In nature, plant cell walls can be biologically altered, catabolized, or degraded by plant pathogens and saprophytes including basidiomycetes [23], ascomycetes [24], and bacteria [25,26]. The strategies used by these organisms include the release of reactive oxygen species produced by redox-active metals and metalloenzymes [27] as well as the excretion of species-dependent combinations of lignin-modifying oxidoreductases [28-31], monooxygenases [32], and glycan-acting hydrolases, esterases, and lyases. Several abiotic catalytic oxidative treatments that mimic certain features of these successful biological approaches have recently been investigated as technologies for pulp bleaching or delignification [33,34] and the pretreatment of cellulosic biomass for the production of biofuels [35,36]. Leskelä and co-workers developed a pressurized O_2-dependent strategy catalyzed by copper-diimine complexes that is effective in both pretreatment processes and pulp bleaching [36-38]. Another effective biomimetic approach reported by Lucas et al. uses oxidation by H_2O_2 catalyzed by manganese acetate to improve the hydrolysis yield from poplar sawdust [39].

We previously reported that AHP pretreatment catalyzed by copper(II) 2,2'-bipyridine complexes (Cu(bpy)) can result in significant improvements in the enzymatic digestibility of a number of biomass feedstocks including switchgrass, silver birch, and most notably hybrid poplar [35]. The improved cellulose digestibility correlated with increased lignin removal as well as modifications to cellulose that included oxidative depolymerization, the introduction of carboxylate groups, and the solubilization and/or oxidative degradation of only up to 5% of the glucan in the original biomass. In the present manuscript we describe our investigation into the key parameters that impact the effectiveness of Cu(bpy)-catalyzed AHP pretreatment on hybrid poplar heartwood as quantified by glucose and xylose release during the subsequent enzymatic digestion of the pretreated biomass. Importantly, we report that the presence of catalytic amounts of Cu(bpy) during AHP pretreatment greatly improves process performance and decreases the required H_2O_2 loading, pretreatment time, enzyme loading, and hydrolysis time. Together, these reduced inputs result in significantly lower pretreatment costs and provide a compelling strategy for an improved pretreatment process for woody biomass.

10.2 RESULTS AND DISCUSSION

The key factors that influence the effectiveness of Cu(bpy)-catalyzed AHP pretreatment are reactant concentrations, catalyst concentrations, and reaction times. These variables not only strongly influence the extent of sugar release, but also greatly affect the overall economics of the pretreatment process. For example, lower H_2O_2 and enzyme concentrations result in decreased raw material costs while increased reaction rates lower capital costs associated with reactor volume due to decreased residence time. The initial pH for uncatalyzed AHP is 11.5 (the approximate pKa of H_2O_2), thereby largely defining the required NaOH loading. Therefore, in this work we focused on H_2O_2, biomass, and Cu(bpy) loadings, and how these variables affect pretreatment times, the required enzyme loading, and subsequent sugar release during saccharification.

To identify a potential lower limit for H_2O_2 during Cu(bpy)-catalyzed AHP, the effect of H_2O_2 loading on the subsequent enzymatic glucose and xylose yield of pretreated hybrid poplar was tested (Figure 1). These results demonstrate that while there is only minimal improvement in glucose and xylose yields with increasing H_2O_2 loadings for uncatalyzed AHP, the presence of catalytic amounts of Cu(bpy) results in monomeric glucose yields of more than 80% and monomeric xylose yields of more than 70% at the highest H_2O_2 loading (100 mg/g biomass) after 72 h of hydrolysis. Importantly, these results demonstrate that the H_2O_2 loading can be halved (from 100 to 50 mg/g biomass) with less than a 4% decrease in the 72 h glucose and xylose yields. Additionally, the trend predicts that the H_2O_2 loading could be further decreased to as low as 35 mg/g biomass (comparable to loadings used in commercial pulp bleaching sequences [40,41]) and still result in more than 70% glucose yields for 72 h of hydrolysis. Considering that the cost of H_2O_2 would likely be one of the primary contributions to the raw materials costs, (along with biomass feedstock, enzyme, and catalyst cost) this 50-65% decrease in the H_2O_2 loading is substantial. For instance, based on a H_2O_2 cost of $1000/ton, this 65% reduction in peroxide loading would decrease the H_2O_2 cost from $0.305/kg of total sugar generated to only $0.081/kg.

The concentration of the Cu(bpy) catalyst utilized during pretreatment is another variable that can be optimized. This water-soluble metal complex has many advantages including its ease of synthesis from $CuSO_4$ and 2,2'-bipyridine [42] and the fact that it is small enough to diffuse into nanoscale pores within plant cell walls to perform catalysis in situ. Reducing Cu(bpy) loadings would be advantageous because this would reduce input costs, alleviate potential copper inhibition to the fermentation microorganism, and diminish environmental concerns about the fate of the catalyst in process water treatment streams. To this end, we tested the effect of catalyst loading on the enzymatic digestibility of pretreated hybrid poplar (Figure 2). Our results demonstrate that after 24 h pretreatment at 20% solids loading, the glucose and xylose yields both essentially saturate at a Cu(bpy) concentration of 2.0 mM (corresponding to a catalyst loading of 10 μmol/g biomass) regardless of the hydrolysis time. In addition, the catalyst concentration can be further halved to 1.0 mM (5.0 μmol/g biomass) with only a 10% loss in the 72 h glucose yield (Figure 2A) and essentially no loss in the xylose yield (Figure 2B).

Importantly, the loading of catalyst on the basis of biomass can be decreased still further by increasing the solids concentrations during pretreatment while keeping the catalyst concentration constant (Table 1, Figure 3). Performing pretreatment and hydrolysis at high solids concentrations with no subsequent washing imparts a number of process benefits, including a decrease of process water usage, a decrease in required reactor volumes, and an increase in sugar titers from hydrolysis and subsequent ethanol titers from fermentation. Intriguingly, uncatalyzed AHP pretreatment shows a noticeable increase in glucan and xylan digestibilities as the hybrid poplar solids are increased from 10% to 20% (w/v), with further modest increases continuing even up to 50% (w/v) solids concentration. The catalyzed AHP pretreatment shows a different trend in that the maximum enzymatic digestibility of hybrid poplar is achieved for solids concentrations in the range of 10% to 20% (w/v) solids with pretreatment efficacy decreasing above 30% solids (w/v) concentration. We suspect that at higher solids concentrations (>20% w/v), the efficacy of the catalyzed pretreatment may be affected by limited mass transfer due to the lack of free water [43], decreased selectivity of H_2O_2 for the biomass versus disproportionation due to the change in reactant concentrations, and/or the decrease in catalyst loading on the biomass (decreasing to 4.0 µmol/g biomass).

Pretreatment reaction kinetics is important for the economics of a process since reactor volume and hence capital equipment requirement is proportional to the residence time of the reactor. An advantage of Cu(bpy)-catalyzed AHP pretreatment is that the pretreatment time is significantly shorter than for uncatalyzed pretreatment. In fact, the enzymatic glucan yield of pretreated poplar rapidly increases to approach a near maximum value within only 10–30 min pretreatment time at 10% solids (w/v) concentrations, while increasing the solids to 20% (w/v) results in achieving the maximum value in less than 10 min (Figure 4A). Comparable increases in the xylan yields can also be achieved within the same short period of time (Figure 4B). Conversely, uncatalyzed AHP pretreatment results in considerably lower yield improvements and requires significantly longer pretreatment time for maximum efficacy.

In addition to employing enzymatic hydrolysis as a screen for determining the differences in pretreatment conditions, we also investigated both enzyme loading and xylanase supplementation for their impact on glucan and xylan yields. While an extensive optimization of enzyme cocktail was not performed, several enzyme loadings and cellulase:xylanase ratios were tested, with the results presented in Figure 5. This shows that the improvement in sugar yield between catalyzed and uncatalyzed pretreatment is significant for all enzyme loadings tested, but the greatest absolute difference is at the higher enzyme loadings. Furthermore, it can be observed that substantially less enzyme is needed to achieve higher digestibilities (i.e. less mass enzyme protein per mass sugar generated) using Cu(bpy)-catalyzed AHP treated poplar relative to the AHP pre-treated material. Another observation is that the xylanase supplementation provides improvement in both the glucose and xylose yields with the synergy between xylanases and cellulases increased at limiting enzyme loadings (Figure 5). This indicates that, like other xylan-retaining pretreatments, xylanase leveraging is possible [44], although the optimal ratios cannot be established from this data. While this was not a complete enzyme cocktail optimization, these results indicate that for the given pretreatment conditions, glucan and xylan yields nearly saturate at their maximum achievable levels with respect to enzyme loading. Additionally, the enzyme dosage can be decreased by at least a 50% total enzyme loading of 30 mg protein/g glucan (at a 15:15 cellulase:xylanase ratio) with only minor losses in glucose and xylose yields. This relatively high enzyme loading could likely be decreased further if a full optimization were performed. This decrease is important considering that enzyme costs are anticipated to be one of largest contributions to cellulosic biofuels costs [45].

The kinetics of the enzymatic hydrolysis for catalyzed and uncatalyzed pretreatment during pretreatment was also investigated (Figure 6), highlighting a number of important outcomes of the pretreatments. As demonstrated above, both the rate and extent of enzymatic hydrolysis are significantly improved following Cu(bpy)-catalyzed AHP treatment relative to uncatalyzed treatment. After 3 hours of hydrolysis, the enzymatic yields of glucose in Cu(bpy)-catalyzed AHP pretreated poplar is approximately two-fold higher than that in hybrid poplar after uncatalyzed AHP pretreatment, and this ratio increases even further with longer hydrolysis time. Another key finding is that while longer pretreatment times result in higher monomeric glucose yields for both catalyzed and uncatalyzed AHP pretreatment, the majority of the glucose yield improvement by pretreatment takes place within the first 30 minutes. Additionally, the differences in sugar yield between 1 h and 24 h pretreatment times nearly disappear at 20% solids, in agreement with previous results (Figure 4).

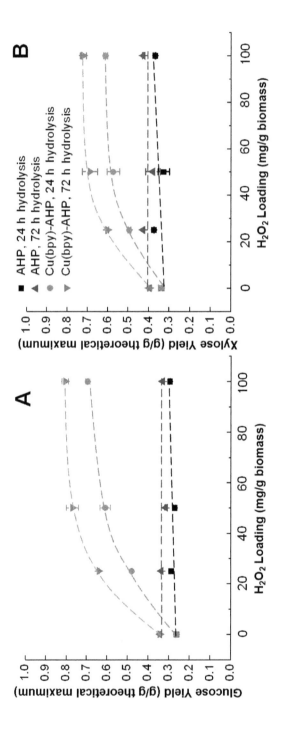

FIGURE 1: Effect of H$_2$O$_2$ loading during pretreatment on enzymatic hydrolysis yields. Results for (A) glucose and (B) xylose demonstrate that yields approach their saturation values near an H$_2$O$_2$ loading of 50 mg/g of biomass for Cu(bpy)-catalyzed AHP pretreatment. Pretreatment was performed for 24 h at 10% (w/v) solids with catalyzed pretreatment employing 2.0 mM catalyst concentration.

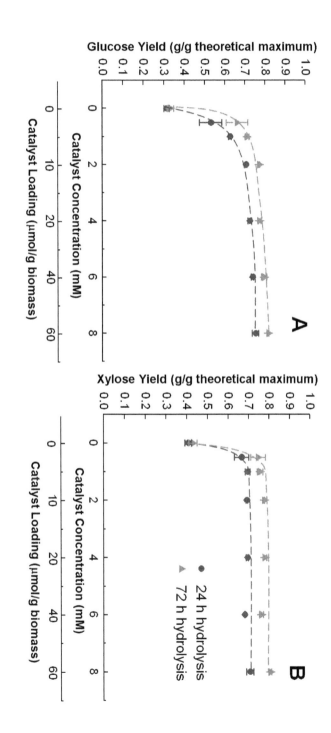

FIGURE 2: Effect of catalyst concentration during Cu(bpy)-AHP pretreatment on the enzymatic hydrolysis yields. Results for (A) glucose and (B) xylose, demonstrate that yields approach their saturation values near a Cu(bpy) concentration of 2 mM (or 10 μmol/g biomass). Pretreatment was performed for 24 h at a solids concentration of 20% (w/v) and a 100 mg/g H2O2 loading on the biomass.

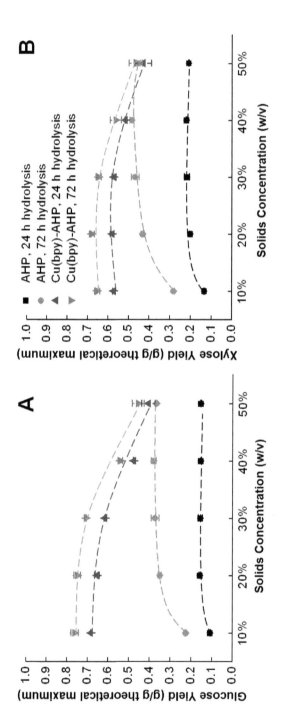

FIGURE 3: Effect of solids concentrations during pretreatment on enzymatic hydrolysis yields. The data for (A) glucose and (B) xylose, demonstrate inverse trends for Cu(bpy)-AHP versus uncatalyzed AHP. Pretreatment was performed for 24 h at 100 mg H_2O_2 per gram of biomass with catalyzed pretreatment employing 2.0 mM catalyst concentrations.

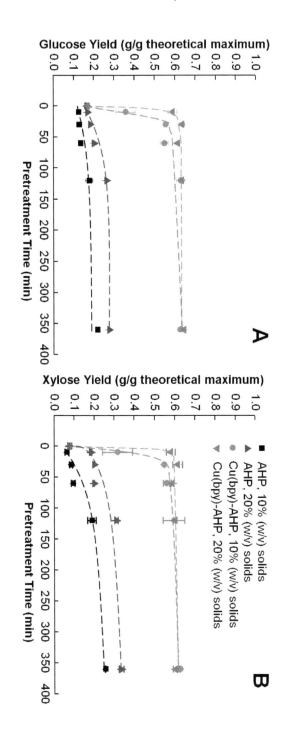

FIGURE 4: Effect of pretreatment time and solids concentration on enzymatic hydrolysis yields. The pretreatment kinetics showing enzymatic (A) glucose and (B) xylose release for catalyzed pretreatments were performed with 2.0 mM catalyst concentrations and 100 mg H2O2 per gram of biomass, and the hydrolysis was performed for 24 h.

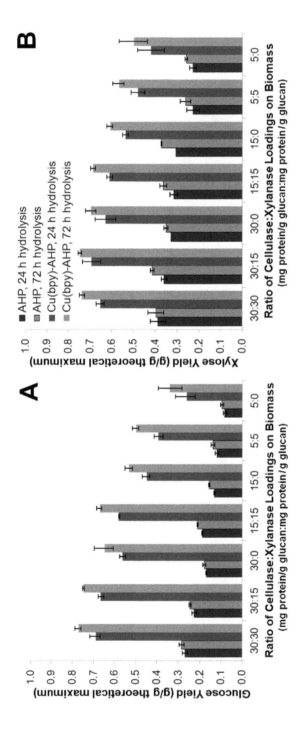

FIGURE 5: Effect of enzyme loading and xylanase supplementation on enzymatic hydrolysis yields. The results for (A) glucose and (B) xylose compare AHP and Cu(bpy)-catalyzed AHP pretreatment performed for 24 h with 100 mg H_2O_2 per gram of biomass, 10% (w/v) solids concentration, and a Cu(bpy) concentration of 2.0 mM for the catalyzed reaction.

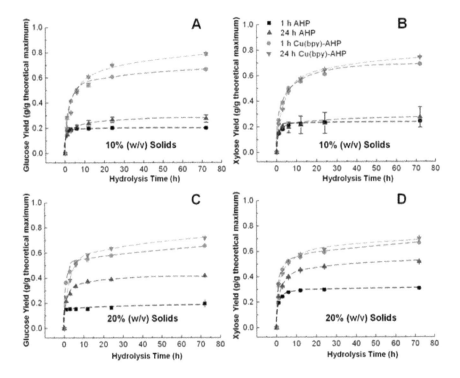

FIGURE 6: Enzymatic hydrolysis kinetics for uncatalyzed and Cu(bpy)-catalyzed AHP. The data show yields of glucose and (A and C) and xylose (B and D) and the solids concentrations (w/v) during pretreatment are 10% (A and B) and 20% (C and D). Pretreatment was performed for 24 h with 100 mg H2O2 per gram of biomass at a catalyst concentration of 2.0 mM for Cu(bpy)-catalyzed AHP pretreatment.

TABLE 1: Chemical inputs used for pretreatment of 500 mg biomass at different solids concentrations

Insoluble solids concentration (w/v)	Liquid volume (mL)	Cu catalyst	H_2O_2 loading on biomass (mg/g)	NaOH loading on biomass (mg/g)
		loading on biomass (µmol/g)[a]		
10%	5.00	20	100	108
20%	2.50	10	100	108
30%	1.67	6.67	100	108
40%	1.25	5	100	108
50%	1.00	4	100	108

[a]*In all cases the catalyst concentration is 2.0 mM.*

For both catalyzed and uncatalyzed AHP pretreatment, the xylose hydrolysis yields are highly correlated with the glucose hydrolysis yields (Figures 1, 2, 3, 4, 5 and 6). In fact, there is an almost perfect linear correlation (R-squared greater than 0.99) between glucose and xylose release for most conditions of Cu(bpy)-catalyzed AHP (Additional file 1: Figure S1A) with a slope of 1.5, meaning that every 15% increase in glucose yield corresponds to a 10% increase in xylose yield. This can be contrasted to uncatalyzed AHP pretreatment of hybrid poplar which results in lower yields of both glucose and xylose (less than 35% and 50% respectively), a weaker correlation (R-squared values ranging from 0.70 to greater than 0.99 for individual data sets), and slopes in the range of 0.60 to 0.85 (Additional file 1: Figure S1B). We speculate that these differences may have implications for both the mechanisms of pretreatment and its impact on the cell wall structure.

In general terms, improvements in enzymatic yields can be understood as cell wall alterations that provide hydrolytic enzymes greater accessibility to the cell wall matrix polysaccharides, and these differences in glucose versus xylose released may be interpreted as differences in how the pretreatments alter the polysaccharide accessibility to enzymes. Specifically, enzyme accessibility for cellulose hydrolysis is known to be dependent on xylan removal in conjunction with lignin removal or relocalization in diverse woody and herbaceous angiosperms. This relationship between xylan removal and cellulose hydrolysis has been shown for both dilute acid [46] and AHP [11] pretreatments, as well as during enzymatic hydrolysis whereby increased xylanase supplementation is well-known to increase the glucose yields for biomass where a significant fraction of the xylan is retained during pretreatment [44]. This is reasonable considering that xylan is believed to sheath the surface of cellulose microfibrils and modulate its interactions with other cellulose microfibrils as well as other cell wall matrix macromolecules [47]. We speculate that the uncatalyzed AHP pretreatment is only capable of removing easily accessible or easily extractable xylan and that only minor glucose hydrolysis yield improvements can be realized in this regime as the bulk of the cellulose may still be embedded within the lignified cell wall. The increase in the slope of the linear correlation in the high glucan digestibility regime (corresponding to pretreatment with Cu(bpy)-catalyzed AHP) can be interpreted as representing an enhanced total polysaccharide accessibility due to the action of pretreatment, whereby the increased accessibility of xylan and its hydrolysis results in an increase in cellulose hydrolysis. As shown in our previous work [35], this enhancement in polysaccharide accessibility during Cu(bpy)-catalyzed AHP pretreatment is possibly a consequence of improved lignin removal. Moreover, the depolymerization of lignocellulose macromolecules potentially disrupts cell wall structure, increases the area accessible to enzymes, and facilitates mass transfer during enzymatic hydrolysis.

10.3 CONCLUSIONS

In this work we characterized Cu(bpy)-catalyzed AHP pretreatment conditions that resulted in significantly increased enzymatic saccharification yields for hybrid poplar heartwood. Relative to our previously reported results from an initial screening of catalysts and biomass feedstocks [35], here we describe a number of important improvements in this catalyzed pretreatment of woody biomass that will enable more efficient use of process inputs during pretreatment and enzymatic hydrolysis. These include significant decreases in the loading of H_2O_2 on biomass, catalyst concentrations, enzyme loadings, and pretreatment duration as well as an increase in solids concentrations. Importantly, it was found that the H_2O_2 loadings on biomass could be decreased to as low as 35–50 mg/g without significant losses in sugar yield by enzymatic hydrolysis, bringing the oxidant loadings down into the range that may be employed in commercial pulp bleaching sequences. Future investigations on Cu(bpy)-catalzyed AHP pretreatment will focus on enhancing our mechanistic understanding of the catalyzed oxidation, characterizing the changes to cell wall polymers and solubilized organics, and improving process integration with fermentation.

10.4 METHODS

10.4.1 CHEMICAL COMPOSITION ANALYSIS OF HYBRID POPLAR HEARTWOOD

Heartwood from 18-year old hybrid poplar (*Populus nigra* var. *charkoviensis* x *caudina* cv. NE-19) grown at the University of Wisconsin Arlington Agricultural Research Station was hammermilled to pass through a 5 mm screen (Circ-U-Flow model 18-7-300, Schutte-Buffalo Hammermill, LLC). The initial composition of structural carbohydrates and acid-insoluble lignin (Klason lignin) were determined using the NREL two-stage acidolysis method [48] with modifications as described elsewhere [35].

10.4.2 CATALYTIC AHP PRETREATMENT

Copper(II) 2,2′-bipyridine complexes were prepared in situ in an aqueous stock solution containing 15.6 g/L CuSO4·5H₂O (EMD Chemicals, Billerica, MA) and 31.2 g/L 2,2′-bipyridine (Sigma-Aldrich, St. Louis, MO) to yield a molar ligand to metal (L:M) ratio of 5:1. Hybrid poplar (0.500 g dry basis; 3.0% moisture) was pretreated in a total of 5.0 mL aqueous solution containing 2.0 mM Cu catalyst, 20.0 g/L H_2O_2, and 21.6 g/L NaOH for pretreatment at 10% solids. For pretreatment at different solids concentrations, a different volume of aqueous solution with the 2.0 mM Cu catalyst was used, while the doses of H_2O_2 and NaOH on the basis of biomass were the same (Table 1). After vortex mixing the reactants with the biomass, the slurry was incubated with orbital shaking at 180 rpm at 30°C [35].

10.4.3 ENZYMATIC HYDROLYSIS

After pretreatment, 20 µL of 72% (w/w) H_2SO_4 and 0.5 mL of 1 M citric acid buffer (pH 4.8) were added to the pretreated slurry to adjust the pH to 5.0, a level suitable for enzymatic hydrolysis. Next, 40 µL of 10 mM tetracycline (Sigma-Aldrich) stock solution was added to inhibit microbial growth, followed by addition of the enzyme cocktail consisting of Cellic CTec2 and HTec2 (Novozymes A/S, Bagsværd, DK) at a loading of 30 mg protein/g glucan each on the untreated biomass unless otherwise noted. The total protein contents of enzyme cocktails used in determining enzyme loadings on biomass were quantified using the Bradford Assay (Sigma–Aldrich). The total volume was adjusted to 10 mL by the addition of deionized water, and the samples were incubated at 50°C with orbital shaking at 180 rpm. Following enzymatic hydrolysis, the solid and liquid phases were separated by centrifugation, and the amount of glucose and xylose released into the aqueous phase was quantified by HPLC (Agilent 1100 Series equipped with an Aminex HPX-87H column operating at 65°C, a mobile phase of 0.05 M H2SO4, a flow rate of 0.6 mL/min, and detection by refractive index) [35]. The yield of glucose and xylose released was defined as the amount of solubilized monosaccharide divided by the total sugar content of the biomass prior to pretreatment as determined by chemical composition analysis [48]. While xylan is solubilized during pretreatment, no monomeric sugars were detected in the pretreatment liquor. Error bars in figures represent the data range between biological replicates.

REFERENCES

1.   Scheffran J: The Global Demand for Biofuels: Technologies, Markets and Policies. In Biomass to Biofuels. Edited by Vertès AA, Qureshi N, Blaschek HP, Yukawa H. Chichester, UK: John Wiley & Sons Ltd; 2010:27-54.

2. Valentine J, Clifton-Brown J, Hastings A, Robson P, Allison G, Smith P: Food vs. fuel: the use of land for lignocellulosic 'next generation' energy crops that minimize competition with primary food production. GCB Bioenergy 2012, 4:1-19.

3. Walsh ME, Ugarte DGD, Shapouri H, Slinsky SP: Bioenergy crop production in the United States - Potential quantities, land use changes, and economic impacts on the agricultural sector. Environ Resour Econ 2003, 24:313-333.

4. Adler PR, Del Grosso SJ, Parton WJ: Life-cycle assessment of net greenhouse-gas flux for bioenergy cropping systems. Ecol Appl 2007, 17:675-691.

5. Kumar P, Barrett DM, Delwiche MJ, Stroeve P: Methods for pretreatment of ligno-cellulosic biomass for efficient hydrolysis and biofuel production. Ind Eng Chem Res 2009, 48:3713-3729.

6. Wei C-J, Cheng C-Y: Effect of hydrogen peroxide pretreatment on the structural features and the enzymatic hydrolysis of rice straw. Biotechnol Bioeng 1985, 27:1418-1426.

7. Gould JM: Alkaline peroxide delignification of agricultural residues to enhance enzymatic saccharification. Biotechnol Bioeng 1984, 26:46-52.

8. Gould JM, Jasberg BK, Fahey GC, Berger LL: Treatment of wheat straw with alkaline hydrogen peroxide in a modified extruder. Biotechnol Bioeng 1989, 33:233-236.

9. Qi B, Chen X, Shen F, Su Y, Wan Y: Optimization of enzymatic hydrolysis of wheat straw pretreated by alkaline peroxide using response surface methodology. Ind Eng Chem Res 2009, 48:7346-7353.

10. Banerjee G, Car S, Liu T, Williams DL, Meza SL, Walton JD, Hodge DB: Scale-up and integration of alkaline hydrogen peroxide pretreatment, enzymatic hydrolysis, and ethanolic fermentation. Biotechnol Bioeng 2012, 109:922-931.

11. Li M, Foster C, Kelkar S, Pu Y, Holmes D, Ragauskas A, Saffron C, Hodge D: Structural characterization of alkaline hydrogen peroxide pretreated grasses exhibiting diverse lignin phenotypes. Biotechnol Biofuels 2012, 5:38.

12. Doner LW, Hicks KB: Isolation of hemicellulose from corn fiber by alkaline hydrogen peroxide extraction. Cereal Chem J 1997, 74:176-181.

13. Nguyen XT, Simard L: On the delignification of OCC with hydrogen peroxide. 1996, 569-576. [TAPPI International Pulp Bleaching Conference]

14. Zhai HM, Lee ZZ: Ultrastructure and topochemistry of delignification in alkaline pulping of wheat straw. J Wood Chem Technol 1989, 9:387-406.

15. Sperry JS: Evolution of water transport and xylem structure. Int J Plant Sci 2003, 164:S115-S127.

16. Gupta R, Lee YY: Pretreatment of corn stover and hybrid poplar by sodium hydroxide and hydrogen peroxide. Biotechnol Prog 2010, 26:1180-1186.

17. Ayeni AO, Hymore FK, Mudliar SN, Deshmukh SC, Satpute DB, Omoleye JA, Pandey RA: Hydrogen peroxide and lime based oxidative pretreatment of wood waste to enhance enzymatic hydrolysis for a biorefinery: Process parameters optimization using response surface methodology. Fuel 2013, 106:187-194.

18. Nguyen Q, Tucker M, Keller F, Eddy F: Two-stage dilute-acid pretreatment of softwoods. Appl Biochem Biotechnol 2000, 84–86:561-576.

19. Zhao Y, Wang Y, Zhu JY, Ragauskas A, Deng Y: Enhanced enzymatic hydrolysis of spruce by alkaline pretreatment at low temperature. Biotechnol Bioeng 2008, 99:1320-1328.

20. Wang ZJ, Zhu JY, Zalesny RS, Chen KF: Ethanol production from poplar wood through enzymatic saccharification and fermentation by dilute acid and SPORL pretreatments. Fuel 2012, 95:606-614.

21. Pan X, Xie D, Kang K-Y, Yoon S-L, Saddler J: Effect of organosolv ethanol pretreatment variables on physical characteristics of hybrid poplar substrates. Appl Biochem Biotechnol 2007, 137–140:367-377.

22. Dale BE, Ong RG: Energy, wealth, and human development: Why and how biomass pretreatment research must improve. Biotechnol Prog 2012, 28:893-898.

23. Baldrian P, Valášková V: Degradation of cellulose by basidiomycetous fungi. FEMS Microbiol Rev 2008, 32:501-521.

24. Fujii K, Sugimura T, Nakatake K: Ascomycetes with cellulolytic, amylolytic, pectinolytic, and mannanolytic activities inhabiting dead beech (Fagus crenata) trees. Folia Microbiol 2010, 55:29-34.

25. Warshaw JE, Leschine SB, Canaleparola E: Anaerobic cellulolytic bacteria from wetwood of living trees. Appl Environ Microbiol 1985, 50:807-811.

26. Wenzel M, Schönig I, Berchtold M, Kämpfer P, König H: Aerobic and facultatively anaerobic cellulolytic bacteria from the gut of the termite Zootermopsis angusticollis. J Appl Microbiol 2002, 92:32-40.

27. Kerem Z, Jensen KA, Hammel KE: Biodegradative mechanism of the brown rot basidiomycete Gloeophyllum trabeum: evidence for an extracellular hydroquinone-driven Fenton reaction. FEBS Lett 1999, 446:49-54.

28. Leonowicz A, Matuszewska A, Luterek J, Ziegenhagen D, Wojtaś-Wasilewska M, Cho N-S, Hofrichter M, Rogalski J: Biodegradation of lignin by white rot fungi. Fungal Genet Biol 1999, 27:175-185.

29. Wong DWS: Structure and action mechanism of ligninolytic enzymes. Appl Biochem Biotechnol 2009, 157:174-209.

30. Baldrian P: Fungal laccases – occurrence and properties. FEMS Microbiol Rev 2006, 30:215-242.

31. Hatakka A: Lignin-modifying enzymes from selected white-rot fungi: Production and role from in lignin degradation. FEMS Microbiol Rev 1994, 13:125-135.

32. Béguin P, Aubert J-P: The biological degradation of cellulose. FEMS Microbiol Rev 1994, 13:25-58.

33. Rahmawati N, Ohashi Y, Honda Y, Kuwahara M, Fackler K, Messner K, Watanabe T: Pulp bleaching by hydrogen peroxide activated with copper 2,2'-dipyridylamine and 4-aminopyridine complexes. Chem Eng J 2005, 112:167-171.

34. Suchy M, Argyropoulos Dimitris S, American Chemical Society: Catalysis and Activation of Oxygen and Peroxide Delignification of Chemical Pulps: A Review. 2001, 2-43. [Oxidative Delignification Chemistry]

35. Li Z, Chen CH, Liu T, Mathrubootham V, Hegg EL, Hodge DB: Catalysis with CuII(bpy) improves alkaline hydrogen peroxide pretreatment. Biotechnol Bioeng 2013, 110:1078-1086.

36. Korpi H, Lahtinen P, Sippola V, Krause O, Leskelä M, Repo T: An efficient method to investigate metal–ligand combinations for oxygen bleaching. Appl Catal A-Gen 2004, 268:199-206.
37. Hakola M, Kallioinen A, Kemell M, Lahtinen P, Lankinen E, Leskelä M, Repo T, Riekkola T, Siika-aho M, Uusitalo J, Vuorela S, von Weymann N: Liberation of cellulose from the lignin cage: A catalytic pretreatment method for the production of cellulosic ethanol. Chem Sus Chem 2010, 3:1142-1145.
38. Rovio S, Kallioinen A, Tamminen T, Hakola M, Leskelä M, Siika-aho M: Catalysed alkaline oxidation as a wood fractionation technique. BioRes 2012, 7:756-776.
39. Lucas M, Hanson SK, Wagner GL, Kimball DB, Rector KD: Evidence for room temperature delignification of wood using hydrogen peroxide and manganese acetate as a catalyst. Bioresour Technol 2012, 119:174-180.
40. Bajpai P: ECF and TCF bleaching. In Environmentally Benign Approaches for Pulp Bleaching. Edited by Bajpai P. Amsterdam: Elsevier; 2005:177-192.
41. Xu E, Koefler H, Antensteiner P: Some latest developments in alkali peroxide mechanical pulping, Part 2: Low consistency secondary refining. Pulp Paper Canada 2003, 104:47-51.
42. Garribba E, Micera G, Sanna D, Strinna-Erre L: The Cu(II)-2,2 '-bipyridine system revisited. Inorg Chim Acta 2000, 299:253-261.
43. Modenbach AA, Nokes SE: The use of high-solids loadings in biomass pretreatment—a review. Biotechnol Bioeng 2012, 109:1430-1442.
44. Kumar R, Wyman CE: Effect of xylanase supplementation of cellulase on digestion of corn stover solids prepared by leading pretreatment technologies. Bioresour Technol 2009, 100:4203-4213.
45. McMillan J, Jennings E, Mohagheghi A, Zuccarello M: Comparative performance of precommercial cellulases hydrolyzing pretreated corn stover. Biotechnol Biofuels 2011, 4:29.
46. Yang B, Wyman CE: Effect of xylan and lignin removal by batch and flowthrough pretreatment on the enzymatic digestibility of corn stover cellulose. Biotechnol Bioeng 2004, 86:88-98.
47. Reis D, Vian B, Roland J-C: Cellulose-glucuronoxylans and plant cell wall structure. Micron 1994, 25:171-187.
48. Sluiter A, Hames B, Ruiz R, Scarlata C, Sluiter J, Templeton D, Crocker D: Determination of structural carbohydrates and lignin in biomass. 2011, 10-42618. [Technical Report NREL/TP]

PART III

METABOLIC ENGINEERING

CHAPTER 11

MOLECULAR CLONING AND EXPRESSION OF CELLULASE AND POLYGALACTURONASE GENES IN *E. COLI* AS A PROMISING APPLICATION FOR BIOFUEL PRODUCTION

EMAN IBRAHIM, KIM D. JONES, EBTESAM N. HOSSENY, AND JEAN ESCUDERO

11.1 INTRODUCTION

The massive usage of petroleum and petroleum products in the last decade, with the consequent reverse effect on minimizing consumption of these unsustainable resources, has increased the demand for the development of renewable sources [1,2]. Currently, based on the carbon neutrality concept, two sources of biofuels have entered the marketplace; ethanol from cellulosic materials and biodiesel from soybean or palm oil [3]. The bioconversion of lignocellulosic materials is a challenging process which requires two steps. During the bioconversion process, the lignin and the hemicellulosic parts are first degraded into simpler sugars and/or organic acids, followed by a deoxygenating step to produce a liquid fuel [4]. De-

Molecular Cloning and Expression of Cellulase and Polygalacturonase Genes in E. coli *as a Promising Application for Biofuel Production.* © *Ibrahim E, Jones KD, Hosseny EN and Escudero J.* Journal of Petroleum & Environmental Biotechnology *4,147 (2013), doi: 10.4172/2157-7463.1000147. Licensed under a Creative Commons Attribution License, http://creativecommons.org/licenses/by/2.0/.*

sign of a genetically modified microorganism for direct lignocellulosic biomass conversion purposes has recently been taken into consideration [5]. The production of several types of fuel through direct lignocellulosic biomass conversions has been demonstrated by various studies [6,7]. A genetically engineered *E. coli* capable of degrading pectin-rich lignocellulosic biomass by cellulolytic and pectinolytic activities has been developed [8]. *E. coli* has been considered a convenient biocatalyst in biofuel production for its fermentation of glucose into a wide range of short-chain alcohols [9,10], and production of highly deoxygenated hydrocarbon through fatty acid metabolism [11,12]. Moreover, the ability to ferment several pentoses and hexoses makes *E. coli* an ideal ethanologen for biofuel production [5,13].

The recalcitrant nature of lignocellulosic biomass was attributed to its complex constituents of cellulose and hemicellulose in addition to variable amounts of lignin [14,15]. Lignin has been reported to impede the enzymatic hydrolysis of plant cell wall polysaccharides [16,17]. However, a pectin-rich substance such as citrus waste residue has been proposed as an ideal substrate for ethanol production due to its low lignin content and high soluble sugar concentration in comparison to other lignocellulosic feedstocks [18,19]. Enzymatic hydrolysis of the insoluble carbohydrate parts of the citrus waste residue, such as cellulose, hemicellulose and pectin, can be performed using an enzyme cocktail of cellulase and pectinase [20].

Hydrolysis of the insoluble cellulosic biomass in citrus waste residue can be carried out by the action of a group of cellulase enzymes collectively known as the cellulosome. Typically, endoglucanases (EGs), and cellobiohydrolases (CBH) are the main cellulase enzymes that act synergistically in hydrolyzing a cellulosic substrate [21]. EGs are endo-acting enzymes that function in the hydrolysis of glycosidic bonds, making free ends available for the exo-action of CBH to produce cellobiose and some glucose molecules [22]. However, β-glucosidase is also necessary for reducing the end-product inhibitory effect of cellobiose on CBH via its hydrolytic conversion into glucose [23]. On the other hand, pectin is a homo-polysaccharide located in the middle lamella of the cell walls of the plant tissues and represents one-third of their dry weights [24]. Pectin typically consists of long chains of galacturonic acid with carboxyl groups and methyl ester residues [25]. Production of the polygalacturonase enzyme is an area of utmost concern for its role in releasing cell wall cellulosic fibrils, which are tightly cemented and embedded into the pectin matrix. The characterization of polygalacturonases has been reported by many researchers [26,27]. Polygalacturonases are a group of enzymes that collectively function in the hydrolysis of non-esterified polygalacturonide chains of pectin. Hydrolysis of these chains can be accomplished by either the random breakdown of the internal C1-1, C-glycosidic, linkages (endpolygalacturonases) or by the release of the digalacturonic acid (exo-poly-a-D-galacturonosidase) from the free ends of the pectin chains. Endo-polygalacturonases have been found to have variable forms which ranged in molecular weights from 30 to 80 kDa, with an acidic optimum pH range of 2.5-6.0 and temperature optima of 30-50°C [28,29]. Exopolygalacturonases are commonly found in Pectobacterium sp. and some other bacterial, fungal and plant species with molecular weight ranges of 30 and 50 kDa [30,31].

Pectobacterium carotovorum subsp. *carotovorum* (*P. carotovorum*), formerly referred to as *Erwinia carotovora* subsp. *carotovora*, is a phytopathogenic bacterium that causes soft rot. Plant cell wall maceration by *P. carotovorum* has been shown to occur through the action of cellulases, β-glucosidases and proteases. Moreover, a group of pectinolytic enzymes including polygalacturonase (PG), pectin lyase (PNL), and pectin methyl esterase (PME) have been recognized to have critical roles in plant cell wall penetration leading to maceration by *P. carotovorum* [32]. These enzymes have been found in various isozymes that are believed to bring about more effective hydrolysis of the plant cell wall polysaccharides [33]. Pectinases of *P. carotovorum*, along with a few other bacterial species, have the specific property of being alkaline tolerant, which is useful in many industrial applications [33]. celA, celB and celC encoded cellulases have been previously isolated and characterized from *P. carotovorum* LY43 [34]. The role of these cellulases in sugar uptake systems inside the cell was described in the cloned cellobiose phosphotransferase system operon from *Bacillus stearothermophilus* [35]. Sugar uptake is generally carried out through a phosphoenolpyruvate:phosphotransferase system (PEP:PTS) in which an enzyme I complex (EI), a histidine containing protein (HPr) as well as an enzyme II complex (EII) play important roles [36,37]. The polygalacturonase enzyme from *P. carotovorum* was also isolated and characterized years ago [38], which then led to the cloning of peh from different isolates of the bacterium.

The aim of this study was to clone and express the genes of *P. carotovorum* encoding cellulases and polygalacturinase into *E. coli* cells. Specifically, celB, celC and peh of *P. carotovorum* were amplified by polymerase chain reaction and cloned in an expression vector to be expressed in *E. coli*. The cloned genes were sequenced and catalytically active residues were identified in the deduced amino acid sequences based on homologies with known degradative enzymes. Characterization of the enzymes and evaluation of their ability to degrade recalcitrant products found in citrus waste are also part of this study. This work could lead to a low cost system using a genetically recombinant *E. coli* for the direct citrus waste bioconversions into bioethanol. A genetically recombinant *E. coli* with cellulase and pectinase activities would be used as an ideal candidate for biofuel production.

11.2 MATERIALS AND METHODS

11.2.1 BACTERIAL STRAINS, PLASMIDS AND MEDIA

Pectobacterium carotovorum subs *P. carotovorum* (*P. carotovorum*), ATCC™ no. 15359, was used as a source of DNA in this study. Typically, -20°C stored cells were revived by streaking on Luria Bertani (LB) agar, and incubated overnight at 26°C. A single colony was inoculated into 3 ml of LB broth (5 g/ml yeast extract, 10 g/l tryptone, 0.5g/l NaCl) and propagated overnight in an orbital shaking incubator (220 rpm) at 37°C. Approximately, 1.5 ml of the overnight culture was centrifuged at 4,000 x g for 10 min at room temperature and the genomic DNA was isolated from the pellet using the E.Z.N.A.™ bacterial DNA isolation kit protocol (Omega Biotek, cat. no. D3350-02, Norcross, GA). Isolated DNA was eluted in deionized, nuclease-free, distilled water and quantitated spectrophotometrically using an Eppendorf Biophotometer (AG. 22331, Hamburg, Germany). *E. coli* DH5α chemically competent cells (Lucigen, cat. no. 95040-456, Middleton, WI) were used for cloning of genes and expression of the enzymes. Plasmid-containing *E. coli* was grown in LB broth containing 100 µg/ml ampicillin or plated on antibiotic containing LB agar supplemented with

0.1 mM of isopropyl β-D-1-thiogalactopyranoside (IPTG) and 40µg/ml of 5-bromo-4- chloro-3-indolyl-β-D-galactopyranoside (X-gal) for the selection and differentiation between empty-plasmid transformed *E. coli* and clonetransformed *E. coli* (blue compared to white, respectively.

pGEM-Teasy vector with a molecular size of 3015bp (Promega, Madison, WI) facilitates cloning of PCR products without prior restriction digestion and was used for general DNA manipulation and DNA sequencing of cloned genes. The expression vector pTAC-MATTAG ®-2 with a 5178 bp molecular size (Sigma Aldrich, cat. no. E5405, St. Louis, MO) was used for heterologous expression of celB, celC and peh in *E. coli* DH5 α.

11.2.2 MOLECULAR BIOLOGICAL TECHNIQUES

Primer design: The sequences of celB, celC and peh were found on GenBank®, the NIH genetic sequence database (accession numbers AF025769.2, AY188753 and BAA74431.1, respectively). In order to amplify the 3 genes of interest, primers were designed corresponding to the open reading frames (ORFs) of the 3 genes. To facilitate cloning into pTACMAT, restriction enzyme sites were incorporated at the 5' end of each primer. The designed primers, as well as their corresponding restriction sites indicated by the underscore, are shown in Table 1.

PCR amplification: One hundred micrograms of genomic DNA was used as a template for PCR amplification using a PCR Sprint Thermocycler (Thermo Electron Corporation, Milford MA) under the following conditions: 1 cycle of 95°C for 5 min; 30 cycles of 95°C for 30s, 55°C for 30s, and 72°C for 1-2 min, according to the length of fragment. The PCR products with sizes of 795 bp (celB), 1105 bp (celC), and 1200 bp (peh) were purified following agarose gel electrophoresis, desalted and concentrated using QIAEX II® Gel extraction Kit (Qiagen, USA).

Cloning and sequencing: The cleaned products were ligated into the pGem-Teasy vector, transformed into *E. coli* DH5α chemically competent cells and plated on antibiotic containing LB agar supplemented with 0.1 mM IPTG and 40µg/ml X-gal. Following overnight incubation at 37°C, white colonies were picked and propagated in LB broth containing 100 µg/ml ampicillin. Plasmid DNA was isolated using the E.Z.N.A.® Plasmid Midi Kit (Omega Biotek, USA) and resulting DNA was digested with EcoRI to confirm the presence of the gene of interest and the DNA was sequenced by MCLAB (San Francisco, CA, USA). Nucleotide sequence translation, nucleotide alignments, and the deduced amino acid sequences were performed online using the bioinformatics tools available at http://www.justbio.com The alignment of the obtained sequences were also carried out using basic local alignment search tool (BLAST) available through National Center for Biotechnology information (NCBI: http://www.ncbi.nlm.nih.gov). Confirmed cloned sequences were digested with their respective restriction enzymes (Table 1) and ligated into pTAC-MAT vector and transformed into *E. coli* DH5α cells. Plasmid DNA was isolated as described above and inserts were confirmed by PCR with the primers used for the original amplification as well as with restriction digests.

TABLE 1: The designed oligonucleotide primers and their corresponding restriction enzyme sites.

Primer name	Restriction Enzyme sites	Sequence
celBF	XhoI	5′GCGCTCGAGATGCTTACAGTGAATA-AGAAG3′
celBR	SmaI	5′GCGCCCGGGTTATTTTACGTCTAC-GCTCC3′
celCF	EcoRI	5′GCGGAATTCATGCCACGCGTGCTG-CACTAC3′
celCR	Bgl II	5′GCGAGATCTTTACGGTGTTGTTATG-CATTGGC3′
pehF	EcoRI	5′GCGGAATTCATGGAATATCAATCAG-GCAAG3′
pehR	Bgl II	5′GCGAGATCTTTATTTCTTAACGTT-GACGTTCTTG3′

11.2.3 GENE EXPRESSION, ENZYME EXTRACTION AND PURIFICATION

Freshly inoculated *E. coli* cells harboring celB and celC and peh were grown in LB broth containing 100µg/ml ampicillin to an optical density of

0.5 at 595 nm. Gene expression was then induced by the addition of 0.1 mM IPTG and cells were harvested by centrifugation 4 hours later. The empty vector strain was propagated and induced in the same manner as a negative control. Over-expressed soluble proteins were extracted and partially purified using the B-PER bacterial protein extraction kit (Pierce Scientific, cat. no. 90078) with DNAse (1, 2,500 U/ml), lysozyme (50 mg/ml) and nonionic detergent in 20mM Tris- HCl buffer (pH 7.5). The extracted proteins were further purified by gel filtration chromatography using Sephadex G-100 (Sigma, St. Louis, MO) with a flow rate 0.75 ml/min. The gel filtration chromatography was performed in a CHROMAF-LEX™ column of 120 cm length and 2.5 cm diameter (KONTES®, cat. no. 4208301210), using 20 mM Tris- HCl, pH 7.5, containing 0.1 M NaCl and 1 mM EDTA. A total of 60 fractions were collected and were tested for their cellulolytic and pectinolytic activities using the 3,5-dinitrosalicylic acid (DNS) assay previously described [39] and the Nelson-Somogyi (NS) assay [40,41] with copper and arsenomolybdate reagents, respectively. The fractions with the highest cellulase and polygalacturonase activities were selected for further characterization and purity determination by SDS-PAGE.

11.2.4 SODIUM DODECYL SULFATE POLYACRYLAMIDE GEL ELECTROPHORESIS (SDS-PAGE)

SDS-PAGE was performed according to a published method [42]. An approximate equivalent amount of 100 µg of protein was denatured by incubation at 95°C for 5 min in loading buffer (0.05% bromophenol blue, 5% β-mercaptoethanol, 10% glycerol and 1% SDS in 0.25M Tris-HCl buffer, pH 6.8). Proteins were separated by electrophoresis through a 10% polyacrylamide gel at 50 mA for approximately 1 h and protein bands were visualized following staining with Gel Code blue stain reagent (Thermo Scientific, cat. no. 24590) for 1 h followed by destaining for another 1 h in deionized distilled water. The molecular weights of the detected protein bands were determined using a molecular weight standard protein marker kit (ProSieve®, cat. no. 50550, USA).

TABLE 2: Comparisons of celB, celC and peh translated amino acid sequences with various species using NCBI's BLAST search.

Tested clone	Aligned organism	Enzyme coded	% Identity	% positivity	Accession no.
celB	*P. carotovorum* subsp. *carotovorum*	beta(1,4)-glucan glucanohydrolase precursor	97%	98%	AAC02965.2
	P.carotovorum subsp.*carotovorum* PC1	Cellulase	96%	97%	YP_003017082.1
	P. wasabiae WPP163z	Cellulase	84%	90%	YP_003260319.1
	Bacillus licheniformis WX-02	glycoside hydrolase family protein	66%	81%	ZP_17658456.1
	Paenibacillus mucilaginosus 3016	glycoside hydrolase	53%	74%	YP_005314236.1
celC	*P. carotovorum* subsp. *carotovorum* WPP14	endo-1,4-D-glucanase	95%	95%	ZP_03832232.1
	P. carotovorum subsp. *carotovorum* PC1	Cellulase	92%	94%	ref-YP_003015672.1
	P. wasabiae WPP163	endo-1,4-D-glucanase	87%	91%	YP_003257546.1
	Serratia marcescens FGI94	endoglucanase Y	70%	79%	YP_007342699.1
	Yersinia intermedia ATCC 29909	Endoglucanase	63%	74%	ZP_04636877.1
peh	*P. carotovorum*	polygalacturonase	99%	99%	AAA03624.1
	P. carotovorum	polygalacturonase	99%	99%	BAA74431.1
	P. atrosepticum SCRI1043	endo-polygalacturonase	95%	97%	YP_049201.1
	Erwinia pyrifoliae DSM 12163	polygalacturonase	60%	73%	YP_005802329.1
	Erwinia amylovora ATCC BAA-2158	polygalacturonase	59%	73%	CBX81050.1

11.2.5 CMC-AGAR DIFFUSION METHOD FOR DETECTION OF CELLULASE ACTIVITIES

Carboxymethyl cellulose (CMC) agar was prepared using LB agar containing ampicillin (100 µg/ml), 0.1 mM IPTG, 40 µg/ml X-gal and 1% (w/v) of CMC. Expression of cellulase was confirmed by inoculating the celC or celB plasmid containing *E. coli* on the inducing medium for 3 days at 37°C and then staining with 0.5% Congo red for 30 min followed by 2 washings with 1M NaCl [43]. *E. coli* containing an empty pTAC-MAT vector was included as a negative control. Enzymatically active cellulases were indicated by a yellow halo against a red background where the cellulose had been digested and was no longer available to bind Congo red. Sterilized 1% CMC agar based medium was prepared for cellulolytic activity determination of the crude extracts. Twenty ml of the prepared medium was transferred to 100×15 mm Petri plates and 5 mm holes were made in the solidified medium. Thirty-five microliters of the crude *E. coli* expressed cellulases as well as the extract from the negative control were loaded separately in the corresponding well areas. The plates were incubated 24 h at 37°C and the cellulase activity was determined by staining with 0.5% Congo red as described.

11.2.6 DETECTION OF POLYGALACTURONASE ACTIVITY

The determination of polygalacturonase activity was done using a modified method described by [44]. Polygalacturonic acid-based substrate agar was prepared by dissolving 1% polygalacturonic acid (Sigma, cat. no., 9049-37-0) in 50 mM sodium acetate pH 4.6 containing 0.8% agarose (Sigma-Aldrich, cat. no., A9918). The medium was then heated to dissolve the polygalacturonic acid and agarose, and 0.2% of sodium azide was added after cooling to 60°C. Nine mm diameter wells were made in the agar and 100 µl the supernatant of the peh recombinant *E. coli* was loaded in the corresponding wells. Supernatant of empty vector transformed *E. coli* was used as a negative control. The plates were incubated 24 h and polygalacturonase activity was detected by appearance of a clear white halo after the addition of 6 M HCl.

11.3 RESULTS

11.3.1 CLONING AND RESTRICTION ANALYSIS OF THE RECOMBINANT CLONES

Primers were designed based on published DNA sequences for celB, celC and peh encoding two cellulases and a polygalacturonase enzyme from *P. cartovorum*. The genes were amplified by PCR and the products were separated by gel electrophoresis in 0.8% agarose and visualized following staining with ethidium bromide. The predicted 795 bp, 1105 bp and 1200 bp fragments were cloned into pGEM-Teasy, sequenced and then expressed in pTAC-MAT-TAG 2. DNA sequence analysis confirmed the identity of the genes and comparison with published sequences resulted in up to 95% similarities with the corresponding genes in *P. carotovorum*. Sequenced plasmids were digested using the restriction enzymes XhoI and SmaI for celB clones, and EcoRI, and Bgl II restriction enzymes for

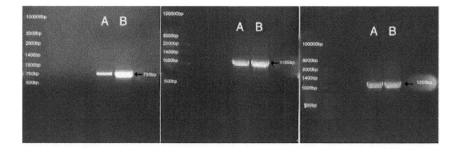

FIGURE 1: Gel electrophoresis showing a comparison of the PCR products of the amplified ORFs of celB, celC and peh of P. carotovorum (lanes B in each figure) and the amplified fragments of the celB, celC and peh clones (lanes A), respectively. Lane 1 of each figure contains Hi-lo™ DNA ladder, (Minnesota Molecular, Inc) with relevant size markers indicated. Bands with a molecular size of approximately 795bp, 1105bp, and 1200bp were detected in the lanes loaded with the amplified products from celB, celC and peh clones, which were the same size as the amplified products from genomic DNA of *P. carotovorum*.

celC and peh clones. The isolated fragments were cloned into pTAC-MAT-TAG®-2 expression vector and recombinant clones were picked based on blue/white selection, and propagated overnight at 37°C in LB broth containing ampicillin (100μg/ml) for DNA plasmid isolation. The isolated plasmids were digested and the products were analyzed on 0.8% agarose gel (data not shown). The inserts were confirmed by comparing PCR products amplified from the clones with products amplified using genomic DNA from *P. carotovorum* (Figure 1).

11.3.2 NUCLEOTIDE SEQUENCE ANALYSIS

The translated sequences of the cloned genes were compared with published sequences using the BLAST program available through NCBI. The celB, celC and peh translated ORFs were similar in sequences to those of several endoglucanases, glucan-glucanohydrolases and polygalacturonases, respectively. The nucleotide sequences along with their corresponding amino acids were found to exhibit a high degree of similarities with *P. carotovorum* and other organisms possessing cellulase and polygalacturonase activities (Table 2). Based on the available conserved domain sequences, the deduced amino acid sequences of the 3 tested ORFs were found to belong to glycosyl hydrolase families 12, 8 and 28 for celB, celC and peh, respectively [45,46]. The nucleotide length of the tested celB, celC and peh ORFs were determined to be 795 bp, 1105 bp and 1209 bp, with putatively encoding proteins of 266, 369 and 404 amino acids, and with estimated molecular weights of 29.5 kDa, 41.3 kDa and 42.5 kDa, respectively. The nucleotide sequences of the ORFs of the 3 cloned genes and their corresponding amino acids are shown in Figure 2.

11.3.3 CMC-AGAR DIFFUSION METHOD FOR DETECTION OF CELLULASE ACTIVITY

Three transformants each of celC and celB were inoculated on CMC-agar including the negative control, *E. coli* containing an empty pTAC-MAT vector. The agar plates were incubated 3 days at 37°C and then stained with 0.5% Congo red for 30 min followed by 2 washings with 1M NaCl [43]. Enzymatically active cellulases were indicated by a yellow halo against a red background where the cellulose had been digested and was no longer available to bind Congo red. All of the tested strains were positive for cellulases by visual inspection (data not shown) so one representative transformant of each of the cloned cellulases was used for gene expression and characterization. The transformed *E. coli* strains, including the empty-vector negative control strain, were grown and induced for gene expression for 4 hours. The cells were harvested and 35 µl of the supernatants were loaded separately in the corresponding wells in the CMC-agar. The plates were incubated 24 h at 37°C and the cellulase activity was determined by staining with 0.5% Congo red as described before. As shown in Figure 3, the crude supernatants of the celB and celC transformants contained cellulase activity as indicated by the appearance of the yellow halo against the red background. However, the CMC-cellulase activity of thecelC protein is much higher than was detected by celB encoded protein as indicated by the larger diameter of the yellow halo. Furthermore, no activity was detected in the well containing the negative control supernatant. These results confirm the expression and cellulase activity of the cloned celB and celC products.

11.3.4 DETECTION OF POLYGALACTURONASE ACTIVITY

The peh-pTAC-MAT transformed *E. coli* was grown in LB broth containing ampicillin (100µg/ml), 0.1 mM IPTG, 40µg/ml X-gal to induce the expression of the cloned peh product, polygalacturonase. After 4 hours of induction, the cells were harvested and the supernatant was collected and tested for enzymatic activity. The negative control strain of *E. coli* transformed with the empty pTAC-MAT vector was treated in the same manner. One hundred µl of the respective supernatants were incubated in the wells incorporated in the polygalacturonic acid agar plates and incubated for 24 h. Polygalacturonase activity was indicated by the appearance of a white halo around the wells after the addition of 6 M HCl. As shown in Figure

4, polygalacturonase activity was detected in the supernatant of the *E. coli* transformed with peh-pTACMAT but it was not detected in the well of the negative control (panels A and B, respectively). These results clearly indicate expression and activity of polygalacturonase from the cloned peh gene.

11.3.5 SDS-PAGE FOR MOLECULAR WEIGHT DETERMINATION

The molecular weights of the expressed celB, celC and peh products were determined using the SDS-PAGE method. The crude extracts of the expressed *E. coli* transformed with the cloned celB, celC and peh genes, as well as the corresponding semi-purified fractions with maximum cellulolytic and pectinolytic activities were used for molecular weight determination. The molecular weights of the protein products found in the supernatant of the negative control *E. coli* were also analyzed. As shown in Figure 5, protein bands of approximate molecular weights of 29.5, 40, 41.5 kDa were detected in the lanes loaded with the crude supernatants of the expressed celB, celC and peh along with their representative semi-purified fractions, respectively. Similar molecular weights were previously detected in the protein products of the cloned genes of *P. carotovorum* [34,47,48]. In contrast, the lane loaded with the crude supernatant of the negative control of *E. coli* had no bands in the aforementioned molecular weight ranges. These results indicate the putative expression of the 3 cloned genes.

11.4 DISCUSSION

Glycoside hydrolases (GHs) are comprised of a wide range of enzymes that are capable of hydrolyzing the glycosidic bonds in glycosides, glycans and glycoconjugates [49]. GHs are playing a critical role in biofuel production through their wide application in production of reducing sugars from pre-treated biomass materials. The reducing sugars formed are useful substrates for ethanol and butanol production which, indeed, can be used as renewable sources for gasoline [50].

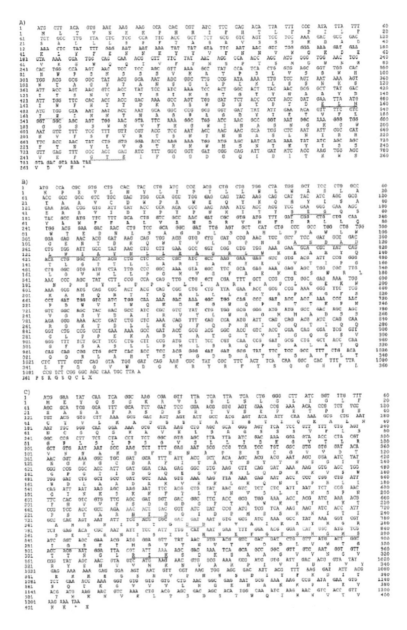

FIGURE 2: Nucleotide sequences of the ORFs of the cloned genes A) celB, B) celC and C) peh from P. carotovorum, ATCC no. 15359. The amino acid residues are placed below the nucleotides of the corresponding codon. The catalytic active site residues of each gene are underlined.

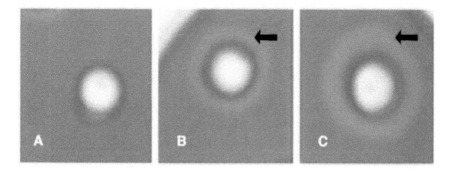

FIGURE 3: CMC agar for the detection of cellulase activity of the cloned celB and celC. A: negative control supernatant of E. coli with an empty vector. B: the supernatant of E. coli with celB gene. C: the supernatant of E. coli with celC gene. The detected halo against the background in B and C indicate the cellulase activity of the cloned gene products.

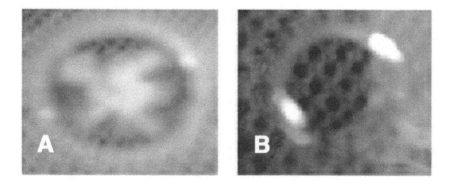

FIGURE 4: Cup plate method for determination of polygalacturonase activity. A: peh transformed *E. coli* supernatant. B: empty vector transformed *E. coli* supernatant. The white halo appearance after the addition of 6.0 M HCl indicates the presence of polygalacturonase activity from the expressed peh encoded protein.

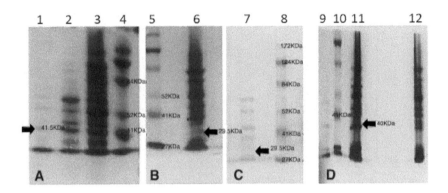

FIGURE 5: SDS-PAGE for molecular determination of expressed proteins from E. coli. Panel A – peh; Panels B and C – celB; Panel D – celC; Lanes 1, 2, 7, 9, semi-purified proteins; Lanes 4, 6, 8, 12, Prosieve protein marker mixture; Lanes 3, 6, 10 crude peh, celB and celC protein extract, respectively, and Lane 12, crude extract of the negative control. Bands with molecular weights of approximately 29.5, 40 and 41.5 kDa are indicated in the representative lanes of peh, celB and celC, respectively. No bands in these ranges were detected in Lane 12 confirming the expression of the tested cloned genes.

Based on the available conserved domain sequences in the NCBI web-site, the deduced amino acid sequences of the 3 cloned ORFs were found to belong to GHs-12, 8, 28 for the cloned celB, celC and peh, respectively [45,46]. The enzymes belonging to GH-28 family are classified into several categories in accordance with their catalytic hydrolysis mechanism and are given specific E.C. numbers based on the Nomenclature Committee of the International Union of Biochemistry and Molecular Biology (http://www.brenda-enzymes. org/). Comparison of the deduced amino acid sequences encoded by peh with those GH amino acid sequences of various species using NCBI′s BLAST search revealed that the cloned gene belongs to the polygalacturonase category of the GH-28 enzymes. Polygalacturonases (PG) are a group of enzymes that function in the hydrolysis of the α-linkage of the galacturonic acid (GalA) monomer residues in pectin (E.C.'s 3.2.1.15 [endo-PG] and 3.2.1.67 [exo-PG]) [51]. Genes encoding GH-28 enzymes have been identified in a number of plant pathogens and fungal species [52]. These degradative enzymes found in pathogenic microorganisms may act as virulence factors and play a critical role in plant cell wall maceration [53,54]. *P. carotovorum* is a known plant pathogen and the presence of the GH-28 polygalacturonase gene in the genome supports its pathogenic role in the hydrolysis of plant cell wall pectin. The role of the peh product in plant tissue maceration along with the other expressed enzymes has been investigated by many researchers [55,56]. The GH-28 polygalacturonases are characterized by the presence of 4 conserved amino acid groups (NTD, DD, HG, and RIK) which were thought to be implicated in the catalytic mechanism [57]. As indicated in Figure 2, the aforementioned conserved amino acid regions were found to be present in the deduced amino acid sequence of the peh encoded protein. Comparison of these results with those of others [58] reveals replacement of a histidine residue with an arginine residue in their identified pehN gene of E. chrysanthemi 3937. The possible replacement of the pehN gene was inferred to suggest a mechanism adapted to different substrate specificities [58]. Histidine (H) was also recognized as a conserved active site residue among fungal and bacterial GH-28 polygalacturonases with an identified motif of (G/S/D/E/N/K/R/H*)-x(2) (V/M/F/C)-x(2)-(G/S)-H*-G- (L/I/V/M/A/G)- x(1,2)-(L/I/V/M)-G-S [59]. The deduced amino acid sequence of the cloned peh was found to include the histidine active site residue in the following motif [(H*)- N-E- (F) -G-T- (G)- H*-G- (M)- S- (I)- G -S]. Rye et al. noted that the cleavage of glycosidic bonds by several glycoside hydrolases can be achieved by either a single- or double-displacement mechanism, which gives rise to inversion or retention of anomeric configuration, respectively [60]. Hydrolysis of the homogalacturonan and the rhamnogalacturonan components of the pectin chain by GH-28 polygalacturonases has been revealed to be a single-displacement inverting mechanism [59,61]. Pickersgill et al have mapped the active site of the pehA encoded GH-28 polygalacturonase of *P. carotovorum*. Three aspartate residues: D202, D223 and D224 have been found to be involved in the catalytic mechanism of this polygalacturonase [62]. The 3 catalytic aspartate residues have also been recognized to be functionally conserved between exo-and endoacting polygalacturonases [59]. Site-directed mutagenesis studies of endopolygalacturonase II from Aspergillus niger revealed that these residues have a role in protonation and deprotonation actions throughout the hydrolysis process [63]. Mutation in any of these residues was found to have a significant negative influence in the catalytic activity of the enzyme. As shown in Figure 2, the 3 putative catalytically active aspartate residues D228, D249 and D250, are present and have the same spatial alignment in the amino acid sequences of the cloned peh protein as in the pehA protein. These results along with those obtained from NCBI`S BLAST-x alignments of the other bacterial and fungal species are evidence for identifying the cloned peh gene as an encoding gene for a GH-28 polygalacturonase enzyme.

The NCBI BLAST results of the deduced amino acid sequences of cloned celB and celC with the other available conserved sequences showed a high degree of homology with the cellulases of *P. carotovorum* and other bacterial species. These results indicate that the cloned aforementioned genes are a part of a cellusome cluster encoding cellulases that may be transcribed by a promoter placed directly upstream of the genes in the chromosome of *P. carotovorum*. A similar explanation has been proposed regarding the celB encoding endo-1,4-p-glucanase and constituting a putative cellulase gene cluster in *Clostridium josui* [64]. The tertiary structure of celB enzyme belongs to the GH-12 grouping of glycoside hydrolases as identified by [65,66]. The cellulase encoded by celB has a tertiary structure which includes a N-terminal signal peptide and a catalytic domain connected to a C-terminal cellulose-binding domain by a remarkably flexible conjunction. Analysis of the three-dimensional structures of the GHfamilies' proteins is considered the best approach for identifying the enzymatic nucleophilic residues. These residues are generally located in the active site clefts with notable characteristics of being conserved, polar and hydrogen bonded [67]. The GH-12 family was among many other families whose members were found to have a glycosidase mechanism with net retention of anomeric configuration [68,69]. In the retention mechanism, hydrolyses are generally achieved through a double-displacement mechanism in which a glycosyl-enzyme intermediate is initially formed followed by its hydrolysis through oxocarbenium-ion transition states [70]. On the other hand, the GH-8 family includes enzymes that catalyze the glycosidic hydrolysis through a single-displacement inverting mechanism with α-configuration products [71,72]. In the inverting mechanism, a carboxyl group (acting as the general acid) and a carboxylate group (acting as the general base) are considered as the two main functional groups. These two functional groups were found to have an important role in the hydrolysis process and are generally conserved among a particular family. GH-12 family members were found to have a highly conserved glutamic acid (E) residue with the following motif: [E*-(I/L)-M-(I/V)-W], [73], which was believed to play a key catalytic role as a nucleophile [74]. Another conserved glutamic acid residue identified in a conserved motif of [G-(T/F)-E*] was also recognized as acid-base residue catalysts in the BLAST search comparing twelve known GH-12 cellulase sequences performed [73]. Comparisons with other retaining glycosidases revealed that a general acid-base catalytic mechanism is less consistent compared with a nucleophile catalytic mechanism [75,76]. Both the nucleophile conserved sequence (E*-(L)-M-(I)-W) and the acid-base residues sequence (G-(T)-E*) have been detected in cloned celB (amino acid residues 178-182 and 244-246, respectively). This investigation along with the data obtained from NCBI's BLAST-x search strongly confirm that the cloned celB is an encoding gene for a GH-12 group member.

Concerning the GH-8 group, amino acid residues with the following pattern [A-(S/T)-D-(A/G)-D-X(2)-(I/M)-A-X-(S/A)- (L/I/V/M)-(L/I/V/M/G)-X-A-X(3)-(F/W)] have been identified as a conserved region (77). The deduced amino acid sequences of the cloned celC have been shown to involve similar GH-8 conserved residues with the following pattern: [A-(S)-D-(A)-D-(L-W)-(I)-A-(Y)-(N)-(L)-(L)- (E)-A-(G-R-L)-(W)], (Figure 2b). Glutamic acid and aspartic acid have been recognized as conserved catalytically active amino acid residues in GH-8 [77,78] with the following motif [(T/V)-S-E*- (G/A)-(Q/H/ LM)] and [D-(G/A)- D*-(L/M/E)], respectively [79]. These residues have been observed in the amino acid sequence of the cloned celC with similar motif: [(T)-S-E*- (G)-(Q)] and [D-(A) - D*-(L)]. The conserved glutamic acid active site residue was suggested to act as a proton donor in the catalytic reaction, while the aspartic acid residue was inferred to have a catalytic role as a nucleophile [79]. The significant ablation that has been observed in the enzymatic activity upon the replacement of either active-site residue by site-directed mutagenesis emphasizes their main catalytic role [77].

The basal expression of the 3 tested clones has been examined using agar diffusion methods. Polygalacturonase activity was detected in the supernatant of the cloned peh as indicated by the appearance of a white halo surrounding the well in the polygalacturonic acid based medium (Figure

4). This observation was confirmed by the absence of the white precipitate around the negative control well. Similar findings have been described of the appearance of the white halo surrounding the well loaded with poly-galacturonase of *Fusarium moniliforme* and no halo was detected with the same sample mixed with polygalacturonaseinhibitor protein of *Phaseolus vulgaris* (common bean) [44]. The cellulase activities of the cloned celB and celC products were also confirmed by the appearance of a yellow halo against a red background in CMC-based medium. As indicated in Figure 3, cellulase activity of the celC product is relatively high in comparison with that of the celB product. It has also been reported a marked reduction in the cellulase activity of celB of *P. carotovorum* - LY34 relative to that of celA product [80]. The relative dissimilarities in cellulase activity may be attributed to the heterogeneous mode of their catalytic activities. As shown in Table 2, the deduced amino acid sequences of celB and celC encoded proteins showed a high degree of similarities with that of β-(1,4)-glucan glucanohydrolase precursor and endo-1,4-D-glucanase of *P. carotovorum*, respectively. The endogluconases exhibit a general hydrolytic activity for internal glycosidic linkages with a consequential increase in the reducing sugar concentration and decrease in the polymer length [81,82]. Further-more, the action of glucan glucanohydrolases was reported on cello-oligo-mers with glucose production as a primary product [83,84]. However, the action of the GH-12 cellulase of *Thermotoga neapolitana* on cellotriose and cellotretrose resulted in the production of small quantities of glucose after prolonged periods of incubation with cellobiose as the main product [85]. On the other hand, the activity of the celB enzyme on CMC sub-strate, reported by [86,87], suggests that this enzyme has both endo- and exo-cellulase activities [85]. This suggestion has been supported by other reports of cellulases with both endo- and exo-activities [88,89]. Analysis of β-glucan hydrolysis products of a GH-12 member β-(1,4)-glucanase from *Aspergillus japonicas* by gel-permeation chromatography revealed the formation of glucose in the initial part of the reaction [90]. This report also suggested the exo-acting activity of that enzyme. The suggestion of exo-acting activity of the celB enzyme is consistent with its original role inside the cell as a transmembrane channel for cellobiose [91,92]. CelC encodes the enzyme that is reported to be responsible for the hydrolysis of cellobiose-phosphate into hydrophilic glucose products [35].

The successful cloning of genes encoding cellulases and polygalacturonase enzymes will lead to the development of a low-cost effective biorefinery strategy to achieve significant reduction in the consumption of unsustainable resources for fuel. Understanding the kinetic action of the expressed enzymes needs further investigations. The role of such enzymes in lignocellulosic waste bioconversions will also be investigated in the ongoing research. Further studies are anticipated involving the optimization of the experimental conditions necessary for achieving maximum biomass conversions into fermentable sugars. An improved gas chromatography-mass spectrophotometry method that has been developed by researchers at TAMUK will be a helpful tool in understanding the catalytic action of the expressed enzymes in the bioconversion process.

REFERENCES

1. Kerr RA (2007) Climate change. Global warming is changing the world. Science 316: 188-190.
2. Stephanopoulos G (2007) Challenges in engineering microbes for biofuels production. Science 315: 801-804.
3. Hill J, Nelson E, Tilman D, Polasky S, Tiffany D (2006) Environmental, economic, and energetic costs and benefits of biodiesel and ethanol biofuels. Proc Natl Acad Sci U S A 103: 11206-11210.
4. Schmidt LD, Dauenhauer PJ (2007) Chemical engineering: hybrid routes to biofuels. Nature 447: 914-915.
5. Huffer S, Roche CM, Blanch HW, Clark DS (2012) Escherichia coli for biofuel production: bridging the gap from promise to practice. Trends Biotechnol 30: 538-545.
6. Bokinsky G, Peralta-Yahya PP, George A, Holmes BM, Steen EJ, et al. (2011) Synthesis of three advanced biofuels from ionic liquid-pretreated switchgrass using engineered Escherichia coli. Proc Natl Acad Sci U S A 108: 19949-19954.
7. Steen EJ, Kang Y, Bokinsky G, Hu Z, Schirmer A, et al. (2010) Microbial production of fatty-acid-derived fuels and chemicals from plant biomass. Nature 463: 559-562.
8. Edwards MC, Henriksen ED, Yomano LP, Gardner BC, Sharma LN, et al. (2011) Addition of genes for cellobiase and pectinolytic activity in Escherichia coli for fuel ethanol production from pectin-rich lignocellulosic biomass. Appl Environ Microbiol 77: 5184-5191.
9. Hanai T, Atsumi S, Liao JC (2007) Engineered synthetic pathway for isopropanol production in Escherichia coli. Appl Environ Microbiol 73: 7814-7818.
10. Atsumi S, Cann AF, Connor MR, Shen CR, Smith KM, et al. (2008) Metabolic engineering of Escherichia coli for 1-butanol production. Metab Eng 10: 305-311.
11. Lee SK, Chou H, Ham TS, Lee TS, Keasling JD (2008) Metabolic engineering of microorganisms for biofuels production: from bugs to synthetic biology to fuels. Curr Opin Biotechnol 19: 556-563.

12. Rude MA, Schirmer A (2009) New microbial fuels: a biotech perspective. Curr Opin Microbiol 12: 274-281.
13. Edwards MC, Doran-Peterson J (2012) Pectin-rich biomass as feedstock for fuel ethanol production. Appl Microbiol Biotechnol 95: 565-575.
14. Dale BE, Leong CK, Pham TK, Esquivel VM, Rios I, et al. (1996) Hydrolysis of lignocellulosics at low enzyme levels: Application of the AFEX process. Bioresource Technol 56:111-116.
15. Sun Y, Cheng J (2002) Hydrolysis of lignocellulosic materials for ethanol production: a review. Bioresour Technol 83: 1-11.
16. Berlin A, Gilkes N, Kurabi A, Bura R, Tu M, et al. (2005) Weak lignin-binding enzymes: a novel approach to improve activity of cellulases for hydrolysis of lignocellulosics. Appl Biochem Biotechnol 121-124: 163-70.
17. Guo G-L, Hsu D-C, Chen W-H, Chen W-H, Hwang W-S (2009) Characterization of enzymatic saccharification for acid-pretreated lignocellulosic materials with different lignin composition. Enzyme Microb Tech 45: 80-87.
18. Rivas B, Torrado A, Torre P, Converti A, Domínguez JM (2008) Submerged citric acid fermentation on orange peel autohydrolysate. J Agric Food Chem 56: 2380-2387.
19. Boluda-Aguilar M, García-Vidal L, González-Castañeda Fdel P, López-Gómez A (2010) Mandarin peel wastes pretreatment with steam explosion for bioethanol production. Bioresour Technol 101: 3506-3513.
20. Wilkins MR, Widmer WW, Grohmann K, Cameron RG (2007) Hydrolysis of grapefruit peel waste with cellulase and pectinase enzymes. Bioresour Technol 98: 1596-1601.
21. Dashtban M, Schraft H, Qin W (2009) Fungal bioconversion of lignocellulosic residues; opportunities & perspectives. Int J Biol Sci 5: 578-595.
22. Enari TM, Niku-Paavola ML (1987) Enzymatic hydrolysis of cellulose: is the current theory of the mechanisms of hydrolysis valid? Crit Rev Biotechnol 5: 67-87.
23. Maki M, Leung KT, Qin W (2009) The prospects of cellulase-producing bacteria for the bioconversion of lignocellulosic biomass. Int J Biol Sci 5: 500-516.
24. Gupta S, Kapoor M, Sharma KK, Nair LM, Kuhad RC (2008) Production and recovery of an alkaline exo-polygalacturonase from Bacillus subtilis RCK under solid-state fermentation using statistical approach. Bioresour Technol 99: 937-945.
25. Voragen AGJ, Coenen GJ, Verhoef RP, Schols HA (2009) Pectin, a versatile polysaccharide present in plant cell walls. Struct Chem 20: 263-275.
26. Martin N. DSRS, De Silva R., Gomes E (2004) Pectinases production by fungal strain in solid state fermentation agroindustrial by-product. Braz Arch Biol Technol 47: 813-819.
27. Sharma N, Rathore M, Sharma M (2013) Microbial pectinase: sources, characterization and applications. Reviews in Environmental Science and Bio/Technology 12: 45-60.
28. Singh SA, Appu Rao AG (2002) A simple fractionation protocol for, and a comprehensive study of the molecular properties of, two major endopolygalacturonases from Aspergillus niger. Biotechnol Appl Biochem 35: 115-123.
29. Takao M, Nakaniwa T, Yoshikawa K, Terashita T, Sakai T (2001) Molecular cloning, DNA sequence, and expression of the gene encoding for thermostable pectate lyase of thermophilic Bacillus sp. TS 47. Biosci Biotechnol Biochem 65: 322-329.

30. Alonso J, Canet W, Howell N, Alique R (2003) Purification and characterisation of carrot (Daucus carota L) pectinesterase. J Sci Food Agr 83: 1600-1606.
31. Pathak N, Sanwal GG (1998) Multiple forms of polygalacturonase from banana fruits. Phytochemistry 48: 249-255.
32. Collmer A, Keen NT (1986) The Role of Pectic Enzymes in Plant Pathogenesis. Annu Rev Phytopathol 24: 383-409.
33. Béguin P (1990) Molecular biology of cellulose degradation. Annu Rev Microbiol 44: 219-248.
34. Woo Park Y, Tech Lim S, Dae Yun H (1998) Cloning and characterization of a CMCase gene, celB, of Erwinia carotovora subsp. carotovora LY34 and its comparison to celA. Mol Cells 8: 280-285.
35. Lai X, Ingram LO (1993) Cloning and sequencing of a cellobiose phosphotransferase system operon from Bacillus stearothermophilus XL-65-6 and functional expression in Escherichia coli. J Bacteriol 175: 6441-6450.
36. Cote CK, Honeyman AL (2003) The LicT protein acts as both a positive and a negative regulator of loci within the bgl regulon of Streptococcus mutans. Microbiology 149: 1333-1340.
37. Warner JB, Lolkema JS (2003) A Crh-specific function in carbon catabolite repression in Bacillus subtilis. FEMS Microbiol Lett 220: 277-280.
38. Nasuno S, Starr MP (1966) Polygalacturonase of Erwinia carotovora. J Biol Chem 241: 5298-5306.
39. Miller GL (1959) Use of Dinitrosalicylic Acid Reagent for Determination of Reducing Sugar. Anal Chem 31: 426-428.
40. Nelson N (1944) A Photometric Adaptation of the Somogyi Method for the Determination of Glucose. J Biol Chem 153: 375-380.
41. SMOGYI M (1952) Notes on sugar determination. J Biol Chem 195: 19-23.
42. Laemmli UK (1970) Cleavage of structural proteins during the assembly of the head of bacteriophage T4. Nature 227: 680-685.
43. Teather RM, Wood PJ (1982) Use of Congo red-polysaccharide interactions in enumeration and characterization of cellulolytic bacteria from the bovine rumen. Appl Environ Microbiol 43: 777-780.
44. Sella L, Castiglioni C, Roberti S, D'Ovidio R, Favaron F (2004) An endo-polygalacturonase (PG) of Fusarium moniliforme escaping inhibition by plant polygalacturonase-inhibiting proteins (PGIPs) provides new insights into the PG-PGIP interaction. FEMS Microbiol Lett 240: 117-124.
45. Marchler-Bauer A, Bryant SH (2004) CD-Search: protein domain annotations on the fly. Nucleic Acids Res 32: W327-331.
46. Marchler-Bauer A, Anderson JB, Chitsaz F, Derbyshire MK, DeWeese-Scott C, et al. (2009) CDD: specific functional annotation with the Conserved Domain Database. Nucleic Acids Res 37: D205-210.
47. Hinton JC, Gill DR, Lalo D, Plastow GS, Salmond GP (1990) Sequence of the peh gene of Erwinia carotovora: homology between Erwinia and plant enzymes. Mol Microbiol 4: 1029-1036.
48. Lei SP, Lin HC, Wang SS, Higaki P, Wilcox G (1992) Characterization of the Erwinia carotovora peh gene and its product polygalacturonase. Gene 117: 119-124.

49. Vuong TV, Wilson DB (2010) Glycoside hydrolases: catalytic base/nucleophile diversity. Biotechnol Bioeng 107: 195-205.
50. Wilson DB (2009) Cellulases and biofuels. Curr Opin Biotechnol 20: 295-299.
51. Markovic O, Janecek S (2001) Pectin degrading glycoside hydrolases of family 28: sequence-structural features, specificities and evolution. Protein Eng 14: 615-631.
52. Sprockett DD, Piontkivska H, Blackwood CB (2011) Evolutionary analysis of glycosyl hydrolase family 28 (GH28) suggests lineage-specific expansions in necrotrophic fungal pathogens. Gene 479: 29-36.
53. Scott-Craig JS, Panaccione DG, Cervone F, Walton JD (1990) Endopolygalacturonase is not required for pathogenicity of Cochliobolus carbonum on maize. Plant Cell 2: 1191-1200.
54. Gao S, Choi GH, Shain L, Nuss DL (1996) Cloning and targeted disruption of enpg-1, encoding the major in vitro extracellular endopolygalacturonase of the chestnut blight fungus, Cryphonectria parasitica. Appl Environ Microbiol 62: 1984-1990.
55. Saarilahti HT, Heino P, Pakkanen R, Kalkkinen N, Palva I, et al. (1990) Structural analysis of the pehA gene and characterization of its protein product, endopolygalacturonase, of Erwinia carotovora subspecies carotovora. Mol Microbiol 4: 1037-1044.
56. Pirhonen M, Saarilahti HT, Karlsson MB, Palva ET (1991) Identification of pathogenicity determinants of Erwinia carotovora subsp. carotovora by transposon mutagenesis. Plant-Microbe Interact 4: 276-283.
57. Bussink HJ, Buxton FP, Visser J (1991) Expression and sequence comparison of the Aspergillus niger and Aspergillus tubigensis genes encoding polygalacturonase II. Curr Genet 19: 467-474.
58. Hugouvieux-Cotte-Pattat N, Shevchik VE, Nasser W (2002) PehN, a polygalacturonase homologue with a low hydrolase activity, is coregulated with the other Erwinia chrysanthemi polygalacturonases. J Bacteriol 184: 2664-2673.
59. Abbott DW, Boraston AB (2007) The structural basis for exopolygalacturonase activity in a family 28 glycoside hydrolase. J Mol Biol 368: 1215-1222.
60. Rye CS, Withers SG (2000) Glycosidase mechanisms. Curr Opin Chem Biol 4: 573-580.
61. Pitson SM, Mutter M, van den Broek LA, Voragen AG, Beldman G (1998) Stereochemical course of hydrolysis catalysed by alpha-L-rhamnosyl and alpha-D-galacturonosyl hydrolases from Aspergillus aculeatus. Biochem Biophys Res Commun 242: 552-559.
62. Pickersgill R, Smith D, Worboys K, Jenkins J (1998) Crystal structure of polygalacturonase from Erwinia carotovora ssp. carotovora. J Biol Chem 273: 24660-24664.
63. van Santen Y, Benen JA, Schröter KH, Kalk KH, Armand S, et al. (1999) 1.68-A crystal structure of endopolygalacturonase II from Aspergillus niger and identification of active site residues by site-directed mutagenesis. J Biol Chem 274: 30474-30480.
64. Fujino T, Karita S, Ohmiya K (1993) Nucleotide sequences of the celB gene encoding endo-1,4-ß-ß-glucanase-2, ORF1 and ORF2 forming a putative cellulase gene cluster of Clostridium josui. J Ferment Bioeng 76: 243-250.

65. Wittmann S, Shareck F, Kluepfel D, Morosoli R (1994) Purification and characterization of the CelB endoglucanase from Streptomyces lividans 66 and DNA sequence of the encoding gene. Appl Environ Microbiol 60: 1701-1703.

66. Davies G, Henrissat B (1995) Structures and mechanisms of glycosyl hydrolases. Structure 3: 853-859.

67. Bartlett GJ, Porter CT, Borkakoti N, Thornton JM (2002) Analysis of catalytic residues in enzyme active sites. J Mol Biol 324: 105-121.

68. Schou C, Rasmussen G, Kaltoft MB, Henrissat B, Schülein M (1993) Stereochemistry, specificity and kinetics of the hydrolysis of reduced cellodextrins by nine cellulases. Eur J Biochem 217: 947-953.

69. Schülein M (1997) Enzymatic properties of cellulases from Humicola insolens. J Biotechnol 57: 71-81.

70. Davies G, Sinnott, ML, and Withers, SG (1997) Comprehensive Biological Catalysis. Academic Press, London, UK.

71. Fierobe HP, Bagnara-Tardif C, Gaudin C, Guerlesquin F, Sauve P, et al. (1993) Purification and characterization of endoglucanase C from Clostridium cellulolyticum. Catalytic comparison with endoglucanase A. Eur J Biochem 217: 557-565.

72. Collins T, Meuwis MA, Stals I, Claeyssens M, Feller G, et al. (2002) A novel family 8 xylanase, functional and physicochemical characterization. J Biol Chem 277: 35133-35139.

73. Altschul SF, Gish W, Miller W, Myers EW, Lipman DJ (1990) Basic local alignment search tool. J Mol Biol 215: 403-410.

74. Zechel DL, He S, Dupont C, Withers SG (1998) Identification of Glu-120 as the catalytic nucleophile in Streptomyces lividans endoglucanase celB. Biochem J 336 : 139-145.

75. Wang Q, Graham RW, Trimbur D, Warren RAJ, Withers SG (1994) Changing Enzymic Reaction Mechanisms by Mutagenesis: Conversion of a Retaining Glucosidase to an Inverting Enzyme. J Am Chem Soc 116: 11594-11595.

76. Lawson SL, Wakarchuk WW, Withers SG (1997) Positioning the acid/base catalyst in a glycosidase: studies with Bacillus circulans xylanase. Biochemistry 36: 2257-2265.

77. Kimoto H, Kusaoke H, Yamamoto I, Fujii Y, Onodera T, et al. (2002) Biochemical and genetic properties of Paenibacillus glycosyl hydrolase having chitosanase activity and discoidin domain. J Biol Chem 277: 14695-14702.

78. Alzari PM, Souchon H, Dominguez R (1996) The crystal structure of endoglucanase CelA, a family 8 glycosyl hydrolase from Clostridium thermocellum. Structure 4: 265-275.

79. Choi YJ, Kim EJ, Piao Z, Yun YC, Shin YC (2004) Purification and characterization of chitosanase from Bacillus sp. strain KCTC 0377BP and its application for the production of chitosan oligosaccharides. Appl Environ Microbiol 70: 4522-4531.

80. Park YW, Lim ST, Cho SJ, Yun HD (1997) Characterization of Erwinia carotovora subsp. carotovora LY34 endo-1,4-beta-glucanase genes and rapid identification of their gene products. Biochem Biophys Res Commun 241: 636-641.

81. Wood TM, Bhat KM (1988) Methods for measuring cellulase activities. Methods in Enzymology. (Vol. 160)Academic Press, London, UK.

82. Béguin P, Aubert JP (1994) The biological degradation of cellulose. FEMS Microbiol Rev 13: 25-58.

83. Rixon JE, Ferreira LM, Durrant AJ, Laurie JI, Hazlewood GP, et al. (1992) Characterization of the gene celD and its encoded product 1,4-beta-D-glucan glucohydrolase D from Pseudomonas fluorescens subsp. cellulosa. Biochem J 285: 947-955.

84. Goyal AK, Eveleigh DE (1996) Cloning, sequencing and analysis of the ggh-A gene encoding a 1,4-beta-D-glucan glucohydrolase from Microbispora bispora. Gene 172: 93-98.

85. Bok JD, Yernool DA, Eveleigh DE (1998) Purification, characterization, and molecular analysis of thermostable cellulases CelA and CelB from Thermotoga neapolitana. Appl Environ Microbiol 64: 4774-4781.

86. Bronnenmeier K, Kern A, Liebl W, Staudenbauer WL (1995) Purification of Thermotoga maritima enzymes for the degradation of cellulosic materials. Appl Environ Microbiol 61: 1399-1407.

87. Ruttersmith LD, Daniel RM (1991) Thermostable cellobiohydrolase from the thermophilic eubacterium Thermotoga sp. strain FjSS3-B.1. Purification and properties. Biochem J 277 : 887-890.

88. Barr BK, Hsieh YL, Ganem B, Wilson DB (1996) Identification of two functionally different classes of exocellulases. Biochemistry 35: 586-592.

89. Tomme P, Kwan E, Gilkes NR, Kilburn DG, Warren RA (1996) Characterization of CenC, an enzyme from Cellulomonas fimi with both endo- and exoglucanase activities. J Bacteriol 178: 4216-4223.

90. Grishutin SG, Gusakov AV, Dzedzyulya EI, Sinitsyn AP (2006) A lichenase-like family 12 endo-(1-->4)-beta-glucanase from Aspergillus japonicus: study of the substrate specificity and mode of action on beta-glucans in comparison with other glycoside hydrolases. Carbohydr Res 341: 218-229.

91. Hu KY, Saier MH Jr (2002) Phylogeny of phosphoryl transfer proteins of the phosphoenolpyruvate-dependent sugar-transporting phosphotransferase system. Res Microbiol 153: 405-415.

92. Kotrba P, Inui M, Yukawa H (2003) A single V317A or V317M substitution in Enzyme II of a newly identified beta-glucoside phosphotransferase and utilization system of Corynebacterium glutamicum R extends its specificity towards cellobiose. Microbiology 149: 1569-1580.

CHAPTER 12

DIRECTED EVOLUTION OF AN *E. COLI* INNER MEMBRANE TRANSPORTER FOR IMPROVED EFFLUX OF BIOFUEL MOLECULES

JEE LOON FOO AND SUSANNA SU JAN LEONG

12.1 BACKGROUND

Worldwide industrialization has resulted in an increasing global demand for energy, particularly fossil fuel. To meet the growing energy needs and tackle escalating environmental concerns, much effort has been directed towards biosynthesis of biofuels from renewable sources to substitute petroleum-based fuels [1-6]. Although bioethanol synthesis from biorenewable feedstocks formed the earliest success story in the biofuel arena, its low energy density and hygroscopicity led to efforts to produce longer chain hydrocarbons with higher energy density and lower hygroscopicity.

Many hydrocarbons are, however, toxic to microorganisms and endogenous biosynthesis of these compounds is expected to have adverse effects on cell growth and production yield [7]. One of the ways microorganisms mitigate the deleterious effect of toxic compounds is to employ efflux pumps to expel noxious molecules, hence enhancing cell survival

[8]. While cell viability may not be a major concern for production of less toxic biofuels such as medium and long chain alkanes [8], these efflux pumps can help to accelerate release of the hydrocarbons into the growth media for harvest. Thus, utilizing efflux pumps to complement biosynthesis of biofuel candidates can be extremely beneficial for improving production yield.

Dunlop et al. [9] recently discovered novel bacteria efflux pumps that could be expressed heterologously in *E. coli* to improve hydrocarbon tolerance. Although it may seem straightforward to increase extrusion of biofuel molecules by over-expressing the efflux pumps, this could compromise cell viability, which is undesirable [10]. A more favorable approach is to improve the pump efficiency for secreting hydrocarbons. Alternatively, efflux pump systems native to *E. coli* can be optimized for enhanced efflux efficiency by protein engineering. The latter strategy minimizes problems with protein aggregation and inactivity that may be associated with heterologous expression of foreign proteins [11,12], which forms the basis of this study.

We chose to study the well-characterized AcrAB-TolC efflux pump system, which consists of an inner membrane transporter AcrB and an outer membrane channel TolC held together by a periplasmic membrane fusion protein AcrA [13]. It has a remarkably wide substrate specificity, including antibiotics, dyes, detergents, as well as many solvent molecules [13]. Since AcrB plays a critical role in substrate recognition and transport in the tripartite system [14], we endeavored to engineer this protein to improve the efflux efficiency of the pump system, particularly to accelerate the exportation of biofuel molecules into the growth medium for easy recovery. The target molecule chosen for this study is n-octane, a potential fuel substitute [5,15]. To this end, we employed directed evolution to screen for AcrB variants with improved n-octane efflux rate. Unlike fluorescent substrates like Nile Red [16], n-octane has no chromophore, thus efflux rate of n-octane from *E. coli* cannot be conveniently determined spectroscopically in real time. n-Octane is also chemically inert and difficult to derivatize for colorimetric analysis. In the absence of a high-throughput analytical method, we sought to increase the screening throughput by competitive growth selection of AcrB mutant pools. Based on the hypothesis that solvent resistance correlates to increased growth and efflux rate, subjecting *E. coli* cells expressing a mixture of AcrB variants to the toxicity of n-octane would enable the superior mutants to dominate the population, thus enriching the culture with AcrB mutants with improved efflux efficiency. Owing to the mild toxicity of n-octane to *E. coli*[8], growth selection using 1-octanol as a more toxic substrate surrogate [17] to exert stronger selection pressure was also subsequently investigated. Efflux rates of n-octane were determined to select for variants with enhancement in efflux efficiency. Finally, we studied the locations of the mutations found in the selected variants based on the reported crystal structure of AcrB (PDB ID: 2DHH) to rationalize the possible roles that the amino acid substitutions play in improving efflux of small biofuel molecules.

12.2 RESULTS AND DISCUSSION

12.2.1 COMPETITIVE GROWTH ASSAY FOR ESTABLISHING CONDITIONS FOR LIBRARY ENRICHMENT

To identify AcrB variants with improved n-octane efflux efficiency, we first set out to establish suitable conditions for growth selection by determining the hydrocarbon tolerance of *E. coli* K-12. To achieve this aim, the *E. coli* K-12 JA300, which has been used in several studies in solvent resistance [8,18-20], was manipulated to generate the acrB-inactivated derivative JA300A. The acrB gene was cloned into the low copy number plasmid pMW119 to create pAcrB. JA300A was transformed with (i) pAcrB (for plasmid complementation) and (ii) pMW119 (for a non-AcrB-expressing negative control strain) and examined for the inhibitory effects of n-octane and 1-octanol on the growth rates (Figure 1). In the absence of n-octane, JA300A/pMW119 and JA300A/pAcrB grew at very similar rates (Figure 1a). Addition of n-octane almost fully inhibited the growth of JA300A/pMW119 in the biphasic culture system saturated with n-octane and significantly retarded the proliferation of JA300A/pAcrB. Supplementation of the growth media with 0.1 mM isopropyl-β-D-thiogalactopyranoside (IPTG) to induce AcrB expression markedly increased the cell growth of JA300A/pAcrB to nearly the rate of the strains when n-octane was absent, but had no effect on JA300A/pMW119 (Figure 1a). The addition of

IPTG neither enhanced nor inhibited growth rates of JA300A/pMW119 and JA300A/pAcrB in the absence of n-octane (Additional file 1: Figure S1). We therefore established that competitive growth enrichment of the mutant libraries with n-octane would be performed at basal expression level. Under this condition, the growth rate of cells with wild-type AcrB n-octane efflux efficiency is sufficiently lower than the maximum growth rate achievable by the cells. Thus, this condition applied adequate selection pressure to enrich the culture with cells possessing higher n-octane tolerance.

Competitive growth assays with 1-octanol were performed with varying concentrations of the substrate in growth media supplemented with 0.1 mM IPTG (Figure 1b). The difference in growth rates of JA300A/pMW119 and JA300A/pAcrB was insignificant in the presence of 0.025% 1-octanol, although growth was markedly inhibited for both compared to that without 1-octanol. When the 1-octanol concentration was further increased to 0.050%, the growth of JA300A/pMW119 was more significantly inhibited than JA300A/pAcrB, with a distinct difference in growth rate apparent between the two strains. Thus, competitive growth enrichment of the mutant libraries with 1-octanol was performed using induced cells with 0.050% of the substrate.

The ability of the chosen selection condition to allow cells with higher hydrocarbon tolerance to dominate the population in the culture was tested. JA300A/pMW119 and JA300A/pAcrB grown separately were mixed in equal amounts and the diluted mixture was subjected to the selection conditions for growth. Cultures with no hydrocarbons were grown in parallel as controls. Plasmids isolated from the control cultures consisted of both pMW119 and pAcrB while those extracted from cultures grown under the selection pressure of the hydrocarbons contained only pAcrB (Figure 2). The elimination of non-AcrB-expressing JA300A/pMW119 under the selection conditions illustrates the successful enrichment of the cultures with JA300A/pAcrB, which possess higher tolerance to the hydrocarbons. These competitive growth assay results demonstrate the potential of selecting variants with improved hydrocarbon tolerance and possibly increased efflux efficiency from an AcrB mutant library.

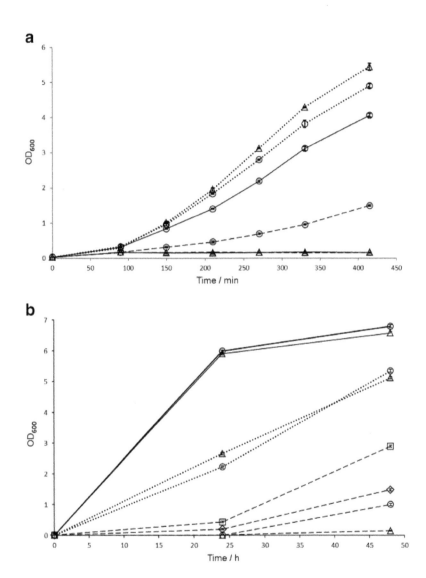

FIGURE 1: Effects of (a)n-octane and (b) 1-octanol on the growth rates of JA300A/ pMW119 (Δ) and JA300A/pAcrB (○). In (a), cell growth in the absence of n-octane are represented by dotted lines. Dashed lines depict cell growth in the presence of n-octane without IPTG induction. Solid lines illustrate the growth of IPTG-induced cells in the presence of n-octane. In (b), all cells were induced with IPTG. Cell growth in the presence of 0, 0.025 and 0.050% 1-octanol are shown in solid, dotted and dashed lines, respectively. Growth of mutants 1G1 (□) and 1G2 (◊) in 0.050% 1-octanol are overlaid.

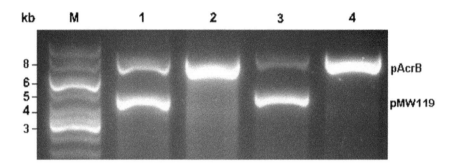

FIGURE 2: Plasmids isolated from a mixture of JA300A/pMW119 and JA300A/pAcrB that have been subjected to selection conditions. AcrB was expressed at basal level in lanes 1 and 2 while IPTG was used to induce protein expression in lanes 3 and 4. Lane 2 was subjected to growth selection with n-octane and lane 4 with 1-octanol. M denotes marker. The isolated plasmids were linearized with BamHI and the sizes correspond to those of pAcrB (7.4 kb) and pMW119 (4.2 kb).

12.2.2 EFFLUX RATE DETERMINATION OF N-OCTANE

Owing to the chemical inertness of n-octane and a lack of chromophore, a high-throughput method for quantifying n-octane was not available. Therefore, in this study, the n-octane efflux rate was determined by a discontinuous assay using gas chromatography to quantify the intracellular n-octane over time by adaptation of reported protocols [8,16] (Figure 3a). JA300A/pMW119 and JA300A/pAcrB were cultivated and induced with IPTG for protein expression. AcrB, a proton-motive efflux pump, was inactivated by the ionophore carbonyl cyanide 3-chlorophenylhydrazone (CCCP) before n-octane was added to the cells to form a biphasic mixture saturated with n-octane. After incubating the cells with n-octane for uptake, the cells were resuspended in a buffer with glucose to reactivate AcrB. The cells were pelleted at regular time intervals and subjected to chloroform extraction to recover intracellular n-octane, which was subsequently quantified by gas chromatography analyses. The change in intracellular n-octane content over time was curve-fitted as a first-order process to estimate the rate constants for n-octane release (Figure 4) [8].

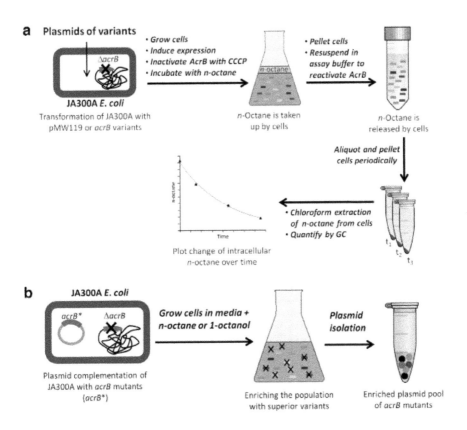

FIGURE 3: Schematic diagram demonstrating the protocols for efflux assay and competitive growth enrichment. (a) Efflux assay was performed in the acrB-inactivated *E. coli* strain JA300A transformed with the plasmids of interest. Expressed AcrB were inactivated to allow substantial uptake of n-octane. Resuspension of the cells in assay buffer containing glucose reactivated the cells and the change in intracellular n-octane content was quantified by GC-MS to determine the efflux rate. Discrete clones were isolated from libraries with improved AcrB mutants. These mutants were assayed using the same method except the cells were cultivated from single colonies. (b) JA300A was transformed with plasmids harboring acrB mutants. The cultures containing cells expressing a mixture of AcrB variants were grown in the presence of n-octane or 1-octanol. As inferior mutants were killed by the hydrocarbons, the population became enriched with superior AcrB that conferred increased tolerance. The plasmid mixtures were isolated after enrichment for efflux assay.

The efflux rate of n-octane by AcrB was calculated by taking the difference between the n-octane release rate constants of JA300A/pMW119
(expressing no AcrB) and JA300A/pAcrB (expressing wild-type AcrB)
(Table 1).

12.2.3 GROWTH ENRICHMENT OF MUTANT LIBRARIES AND SELECTION BASED ON DETERMINATION OF EFFLUX RATES

Eight mutant libraries of acrB were generated by mutagenic PCR and
cloned into pMW119. Each library consisted of approximately 20,000
variants and had a mean mutation rate of 3.0 per kilobase (standard deviation 1.9). Library enrichment was initially performed using n-octane. Each
library was transformed into JA300A and grown. The pre-culture contain-

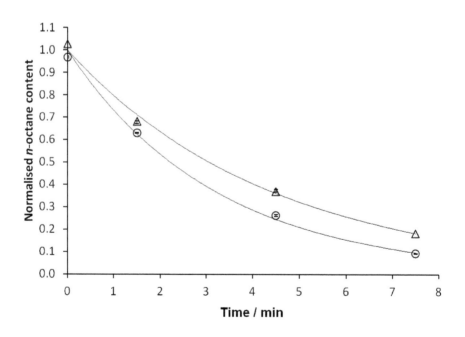

FIGURE 4: Change of intracellularn-octane with time in JA300A/pMW119 (Δ) and
JA300A/pAcrB (○). Intracellular n-octane content has been normalized to the respective
initial mass in mg.

ing a pool of acrB variants being expressed at basal level was subjected to the growth enrichment condition as described for n-octane. After two additional rounds of sub-cultivation for enrichment, the plasmid mixture for each library was isolated (Figure 3b).

TABLE 1: Rates of n-octane and α-pinene efflux from JA300A expressing AcrB variants

Hydrocarbons	AcrB variants	Hydrocarbon release rate constant/min^{-1}	AcrB hydrocarbon efflux rate constant[a]/min^{-1}	Relative AcrB efflux rate
n-octane	none	0.227 ± 0.007	-	-
	Wild-type	0.312 ± 0.010	0.085	1.00
	1G1	0.352 ± 0.011	0.125	1.47
	1G2	0.349 ± 0.009	0.122	1.44
	N189H	0.318 ± 0.018	0.091	1.07
	L357Q	0.290 ± 0.011	0.063	0.74
	T678S	0.318 ± 0.025	0.091	1.07
	T678A	0.334 ± 0.015	0.107	1.26
	Q737L	0.355 ± 0.023	0.128	1.51
	M844L	0.325 ± 0.010	0.098	1.15
	M844A	0.339 ± 0.015	0.112	1.31
	M987T	0.292 ± 0.019	0.065	0.77
	V1028M	0.304 ± 0.008	0.077	0.91
α-pinene	none	0.282 ± 0.018	-	-
	Wild-type	0.297 ± 0.014	0.015	1.00
	1G1	0.342 ± 0.015	0.060	4.00
	1G2	0.328 ± 0.018	0.046	3.07
	N189H	0.322 ± 0.012	0.040	2.69
	L357Q	0.296 ± 0.009	0.014	0.96
	T678S	0.319 ± 0.008	0.037	2.44
	T678A	0.330 ± 0.012	0.048	3.20
	Q737L	0.340 ± 0.008	0.058	3.89
	M844L	0.318 ± 0.008	0.036	2.41
	M844A	0.327 ± 0.019	0.045	3.00
	M987T	0.293 ± 0.005	0.011	0.75
	V1028M	0.297 ± 0.005	0.015	0.97

[a] Calculated by taking the difference between the hydrocarbon release rate constants of JA300A expressing no AcrB and JA300A expressi

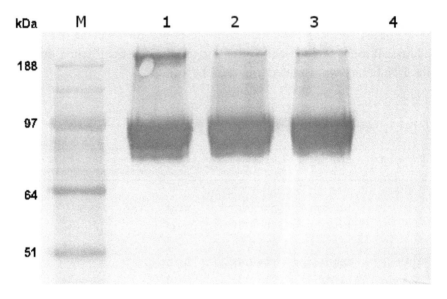

FIGURE 5: Expression levels of AcrB variants. M, marker; lane 1, wild-type AcrB; lane 2, 1G1; lane 3, 1G2; lane 4, negative control (pMW119).

To facilitate rapid selection of improved AcrB from the enriched libraries in the absence of a high-throughput method for measuring efflux rate of n-octane, the enriched libraries were preliminarily assayed as variant mixtures to identify the pools with AcrB mutants that improved extrusion of n-octane. Discrete clones could then be isolated from these pools for further assay and final selection. Thus, the plasmids of the eight enriched libraries were transformed into JA300A to determine the average rate of n-octane efflux for each pool of AcrB variant mixture.

However, the rates of n-octane release for the libraries were observed to be similar to that of wild-type AcrB. Sequencing of five discrete clones isolated from a library revealed mutations in the lac operator in four clones, even though this region was not subjected to random PCR mutagenesis. This could have arisen because AcrB variants had to be expressed only at basal level for growth selection to maintain sensitivity to n-octane such that growth improvement of cells with superior AcrB mutants would be apparent. Consequently, clones with spontaneous mutations in the lac operator would have conferred constitutive expression of AcrB, thus providing increased tolerance of n-octane and growth advantage over cells expressing AcrB at basal level under the regulation of wild-type lac operator. As a result, the libraries were dominated by plasmids with mutant lac operator expressing AcrB variants with average performance rather than AcrB mutants with improved efflux efficiency expressed from wild-type lac operator. Clearly, n-octane could not apply sufficient selection pressure to enable differentiation of variants with good performance from the average ones. To overcome this limitation, we directed our effort towards growth enrichment using 1-octanol.

In contrast to selection conditions with n-octane, AcrB expression was induced with IPTG before and during growth enrichment with 1-octanol to overcome the higher toxicity to *E. coli* compared to n-octane. After two rounds of sub-cultivation, the enriched plasmid mixture was purified and the variants were assayed for n-octane release rate. Two of the eight enriched libraries contained variants with enhanced n-octane efflux efficiency. Five discrete clones were isolated from each of the two libraries. Sequencing of the acrB mutants identified only one variant from each library and no mutation was found in the lac operator for both variants. This result indicates that the selection pressure from the growth enrichment conditions effectively eliminated less tolerant cells harboring less efficient AcrB mutants. These two AcrB variants are hereafter designated 1G1 and 1G2. The plasmids containing the genes are named p1G1 and p1G2, respectively. The mutations in 1G1 are N189H and T678S, and those in 1G2 are L357Q, Q737L, M844L, M987T, V1028M.

12.2.4 CHARACTERIZATION OF ACRB VARIANTS AND STRUCTURAL ANALYSIS

JA300A harboring p1G1 and p1G2 were cultivated from single colonies and induced with IPTG to express the AcrB mutants. The growth rates of the cells in the presence of 1-octanol and the efficiency of the AcrB variants in effluxing n-octane were investigated. Both 1G1- and 1G2-expressing JA300A showed significantly faster growth rates in the presence of 1-octanol, reaching OD600 of 2.90 and 1.48, respectively, compared to

1.05 of wild-type AcrB after 48 h of growth (Figure 1b). The variants also exhibited over 40% increase in n-octane efflux rate relative to wild-type AcrB (Table 1). To examine the possibility that improvements in performance were due to increased AcrB expression of the mutants, western blotting was performed to compare the AcrB expression levels between the variants. Constructs of pAcrB, p1G1 and p1G2 with hexahistidine-tags inserted at the C-termini were synthesized, and the proteins expressed with IPTG induction were probed using anti-hexahistidine antibodies. There was negligible difference in the expression levels of 1G1 and 1G2 compared to wild-type AcrB (Figure 5), indicating that the increased efflux of n-octane and enhanced tolerance of the cells expressing 1G1 and 1G2 were due to improved performance of the mutant proteins. Thus, we have demonstrated successful isolation of AcrB mutants with improved n-octane efflux efficiency from a competitive growth selection based on 1-octanol tolerance.

Single-site mutants of the seven mutations identified in 1G1 and 1G2 were generated to investigate their individual effects on n-octane efflux (Table 1). L357Q, M987T, V1028M showed reduced rate of n-octane extrusion. Q737L and M844L increased the rate of n-octane efflux by 51 and 15%, respectively. Interestingly, N189H and T678S improved efflux of n-octane only slightly (less than 10%) although 1G1, which possess these two mutations, exhibited 47% faster efflux of n-octane. These results suggest that N189H and T678S may work synergistically.

To gain structural-functional insights of the seven mutations in 1G1 and 1G2, the positions of the amino acid substitutions were located on a known asymmetric structure of AcrB (PDB ID: 2DHH) [21]. Analysis of the structure shows that L357Q, M987T, V1028M were in the trans-membrane domain at locations unlikely to contribute to substrate efflux, consistent with the lowered n-octane efflux rates of the single-site mutants relative to wild-type AcrB. An overview of the positions of N189, T678, Q737 and M844 on the structure of AcrB is illustrated in Figure 6 using the protomer with the "Access" conformation. N189H and Q737L are located near the hairpins in the TolC docking domains (Figure 6a), which are responsible for interaction with TolC to form an AcrB-TolC complex [22]. This interaction is important to facilitate transfer of substrates from AcrB to TolC for exportation out of the cell via the TolC channel. The exact mechanism how N189H and Q737L promotes efflux of n-octane is unclear. However, the proximity of the two mutations to the hairpins in the TolC docking domains suggests they might assist the assembly of the AcrB-TolC complex. Particularly, Q737 is in the vicinity of a hairpin on the adjacent protomer, thus the shortening of the side chain due to Q737L might stabilize or improve the assembly or interactions between AcrB and TolC for efficient export of n-octane.

T678S and M844L occur close to each other along two proposed transport paths i.e. the vestibule between protomers and the cleft in the periplasmic domain (Figure 6b) [23,24]. Analysis of the structure of AcrB, which shows functional rotation of the protomers, revealed that T678 is near the entrance at the lower cleft in the periplasmic domain leading to the binding pocket of AcrB and exhibits large displacement when transitioning between the "Access", "Binding" and "Extrusion" conformations (Additional file 1: Figure S2) [21]. In the "Extrusion" conformation, T678, which is in the PC2 subdomain, is extremely close to N667 (2.6 Å) in the PC1 subdomain as the entrance is closed off. The distance between T678 and N667 increases as the entrance opens to allow substrate access into AcrB when the conformation transitions to "Access" and "Binding", before closing off the entrance again to assume the "Extrusion" conformation. Thus, we propose that shortening the side chain of threonine in the T678S mutant would likely widen the entrance at the lower cleft in the periplasmic domain and permit hydrocarbons more access time into the binding pocket for extrusion. The role of M844L is less obvious. M844 is on the same α-helix as E842 that has been identified as one of the residues lining the vestibule leading to the binding pocket [23], thus mutation to a leucine might cause a conformational change that facilitates hydrocarbon entry into the transport channel through the vestibule. Additionally, the side chain of M844 is pointing in the opposite direction of the vestibule towards adjacent loops bearing D681, R717, L828 that line another channel to the binding pocket in the cleft [23-25]. In fact, it is in close proximity to F682, V716 and I827 (3.9-4.3Å) that flank D681, R717 and L828, respectively (Figure 6b). Incidentally, M844 is located in the PC2 subdomain of AcrB in a region that exhibits large conformational changes during the displacement of the drug doxorubicin [26]. It is possible that the shorter side chain of leucine causes distal conformational changes that re-position

D681, R717 and L828 such that the channel leading to the binding pocket is enlarged or the dynamics of substrate efflux is altered. This phenomenon has been observed in other proteins [27,28]. Crystal structures would be required to validate these hypotheses. Nevertheless, these propositions on the roles of the T678S and M844L mutations are supported by the increased n-octane efflux by the T678A and M844A variants (Table 1), which have the side chains shortened to a methyl group. The effect of T678A is also consistent with the observation that increasing the bulk of residue A677, which is adjacent to T678, reduced drug efflux [25].

To investigate the ability of 1G1, 1G2 and the single-site mutants in effluxing biofuels that are structurally dissimilar to 1-octanol and n-octane, which are conformationally flexible linear compounds, we chose a cyclic and conformationally more rigid biofuel candidate, α-pinene, as the test substrate. The rate of α-pinene release was assayed (Table 1) and interestingly, the mutants that improved efflux of n-octane also enhanced the efflux rate of α-pinene (Table 1) despite the structural dissimilarity of the substrates (Figure 7). This suggests that the beneficial effects of the mutations are not specific to 1-octanol and n-octane only, but may be generally important for improving efflux of small hydrophobic hydrocarbon molecules.

Three known antibiotic substrates of AcrB [29] of varying molecular weight—nalidixic acid (232.24 g mol^{-1}), chloramphenicol (323.13 g mol-1) and tetracycline (444.44 g mol^{-1})—were chosen to investigate the antibiotic resistance of the mutants that improved n-octane efflux (i.e. 1G1, 1G2, N189H, T678S, T678A, Q737L, M844L and M844A). Interestingly, none of the mutants showed increased antibiotic resistance towards the antibiotics. JA300A expressing wild-type and AcrB mutants have minimum inhibitory concentrations (MICs) of 5.0, 5.0 and 1.0 mg/L for chloramphenicol, nalidixic acid and tetracycline, respectively. The MICs for the control strain (JA300A with pMW119) towards chloramphenicol, nalidixic acid and tetracycline are 1.5, 1.5 and 0.1 mg/L, respectively. The neutral effect of the mutations towards antibiotic resistance could be because the mechanisms for extrusion of antibiotics and small hydrophobic molecules are different. For example, substrate recognition through interactions between amino acids in the transport pathway of AcrB and the polar functional groups in antibiotics is required for antibiotic efflux [21] whereas such interactions are absent with small hydrophobic molecules. Nevertheless, it is advantageous that the mutations beneficial for n-octane efflux do not increase antibiotic resistance because that would eliminate the possibility of generating pathogenic bacteria with high antibiotic resistance. This would be particularly important if strains with the mutant AcrBs were used in large volumes for mass production of hydrocarbons.

12.3 CONCLUSIONS

A 'competitive growth'-based selection strategy was developed in this study for rapid isolation of AcrB mutants with improved efflux efficiency of n-octane. Although the crystal structure and substrate path of AcrB are known, the mechanism of AcrB involves a highly dynamic peristaltic action and functional rotation between the protomers [21]. Therefore, we chose directed evolution over rational design to engineer AcrB because it could enable discovery of variants with mutations that would not have been obvious enough to be identified. Indeed, with the exception of T678S which lies in the substrate path, the other amino acid changes that benefit n-octane efflux, i.e. N189H, Q737L and M844L, are distant from the substrate path. We reasoned that these mutations may improve substrate efflux by enlarging the substrate path, altering the substrate efflux dynamics and facilitating the assembly of the AcrB-TolC complex. These mutations are also beneficial for effluxing other small hydrophobic molecules, as illustrated by the increased extrusion rate of α-pinene.

We successfully engineered AcrB for enhanced efflux efficiency of n-octane by competitive growth selection for tolerance towards the more toxic substrate surrogate, 1-octanol. This strategy of employing a toxic substrate surrogate for competitive growth enrichment is particularly beneficial for engineering efflux pumps to accelerate extrusion of biofuel with low toxicity, such as n-octane and higher alkanes [8], since mildly toxic compounds could not select for AcrB mutants effectively for enhanced efflux. These improved AcrB can now be used to improve the yield of biofuel production and/or increase cell tolerance to the synthesized hydrocarbon [9]. The next step forward would be to validate the use of these mutant pumps in engineered bacterial systems that biosynthesize small biofuel molecules such as n-octane and α-pinene in vivo. Much efforts are currently focused on metabolic engineering activities for biofuel production in microbes [1-6] and we expect the outcome of this work to underpin the development of scalable microbial platforms for biofuels production.

12.4 METHODS

12.4.1 STRAINS, PLASMIDS, OLIGONUCLEOTIDES, CHEMICALS AND CULTURE MEDIA

E. coli XL10-Gold {(δ(mcrA)183 δ(mcrCB-hsdSMR-mrr)173 endA1 supE44 thi-1 recA1 gyrA96 relA1 lac Hte [F' proAB lacIqZ δ15 Tn10 (Tetr) Tn5 (Kanr) Amy]} (Stratagene) was the host strain used for plasmid construction. JW0451-2 [F-, δ(araD-araB)567, δlacZ4787(::rrnB-3), δacrB747::kan, λ-, rph-1, δ(rhaD-rhaB)568, hsdR514] strain and P1vir phage were purchased from The Coli Genetic Stock Center (Yale University, USA). JA300 (F-thr leuB6 trpC1117 thi rpsL20 hsdS) strain was obtained from ATCC. Its acrB-inactivated derivative, JA300A, was generated by P1vir phage transduction [30] with JW0451-2. JA300A was used for library enrichment, hydrocarbon resistance studies and efflux assays. acrB gene was cloned from the *E. coli* strain W3110 [F- IN(rrnD-rrnE)1]. The low-copy-number vector pMW119 (WAKO Pure Chemical Industries, Osaka, Japan) was used for cloning and expression under the control of the lac promoter. Oligonucleotides were synthesized by 1st Base (Singapore) and the sequences are shown in Additional file 1: Table S1. All chemicals were purchased from Sigma-Aldrich (Singapore) and of purity higher than 98%.

Lysogeny broth (LB) consisting of 1% tryptone, 0.5% yeast extract and 1% NaCl was used for cell cultivation. Where appropriate, ampicillin (100 μg/mL) and/or IPTG (0.1 mM) were added to the medium. Hereafter, LBA and LBI refers to LB medium supplemented with ampicillin and IPTG, respectively, and LBAI refers to LB medium with both ampicillin and IPTG added.

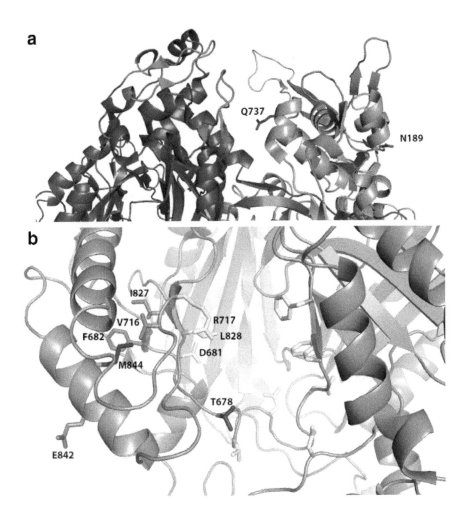

FIGURE 6: Structural analysis of the mutations at N189, T678, Q737 and M844 (PDB ID: 2DHH). The "Access" conformation of the structure was used. In (a), N189 and Q737 (in blue) are near the hairpins (in red) that interact with TolC to form the AcrB-TolC complex. Q737 is in the vicinity of a hairpin on the adjacent protomer (in dark gray). In (b), T678 and M844 are shown in blue. Residues that were reported to be in the substrate path leading to the binding pocket are colored yellow [24,25]. The amino acids F682, V716 and I827 (shown in pink) are in close proximity to M844 and flank residues in the substrate path (D681, R717 and L828). M844 is also on the same α-helix as E842 (shown in green) which is part of the vestibule leading to the binding pocket.

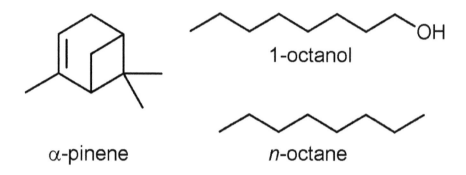

FIGURE 7: Structures of hydrocarbons used in this work.

12.4.2 DNA MANIPULATION

Standard techniques were used for DNA manipulation, including isolation and purification of plasmids and DNA fragments, agarose gel electrophoresis, restriction enzyme digestion, ligation and transformation of *E. coli*[31]. All restriction enzymes and DNA-modifying enzymes were purchased from Fermentas (Singapore).

12.4.3 CLONING OF ACRB

acrB was amplified from W3110 chromosomal DNA by PCR using the primers acrB-F and acrBCln-R, which contain the BamHI and SacI restriction sites, respectively. The 3.2-kb PCR fragment was digested with BamHI and SacI and the restricted DNA was cloned into the 4.2-kb vector pMW119 to generate the 7.4-kb plasmid pAcrB. The sequence of the cloned acrB was verified by DNA sequencing.

12.4.4 STUDY OF GROWTH INHIBITION BY N-OCTANE AND 1-OCTANOL

pMW119 and plasmids of acrB variants were transformed into electrocompetent JA300A. Overnight cultures of the strains were diluted in 3 mL LB medium to OD_{600}~0.03. To study the growth inhibition of cells by n-octane, the diluted cells were shaken at 37°C with 1 mL n-octane, both in the presence and absence of 0.1 mM IPTG, and the OD_{600} was measured hourly. The growth inhibition of the strains by 1-octanol was performed similarly with 0.025% and 0.050% (v/v) 1-octanol in LBI medium and OD_{600} absorbance was measured every 24 h. Cultures without n-octane and 1-octanol were grown in parallel as controls.

12.4.5 COMPETITIVE GROWTH STUDIES

JA300A was transformed with either pMW119 or pAcrB. Overnight cultures of the two strains were grown separately and mixed in a 1:1 ratio. A 400 μL aliquot of the mixed cells was inoculated into 20 mL of LB medium and split into two 10 mL portions. To one of the portions, 3 mL of n-octane was added. Both cultures were grown overnight at 37°C with shaking and subcultivated by 50-fold dilution in the respective growth media. After two more rounds of subcultivating, 200 μL of the final cultures served as inoculum for 10 mL of LBA medium. Plasmids were isolated from the overnight cultures for agarose gel electrophoresis after BamHI digestion.

For competitive growth study with 1-octanol, two 10 mL portions of mixed cells were similarly prepared except LBI medium was used. 1-Octanol was added to one of the portions to a final concentration of 0.05% and both cultures were grown for 2 days at 37°C with shaking. After another round of subcultivating, the cells were diluted 50-fold in LBA medium and grown for plasmid isolation. The plasmids were subsequently digested with BamHI and analyzed by agarose gel electrophoresis.

12.4.6 MUTANT LIBRARY CONSTRUCTION

Mutagenic PCR was performed to introduce random mutations into the acrB gene. Eight 50 μL PCR reaction mixtures were prepared and handled individually. Each reaction consisted of 10 ng of pAcrB as template, 0.2 μM of each primer acrB-F and acrB-R, 0.2 mM dATP, 0.2 mM dGTP, 1 mM dCTP, 1 mM dTTP, 10 mM Tris–HCl (pH 8.8), 50 mM KCl, 0.08%

(v/v) Nonidet P40, 7 mM MgCl2, 1 μL DMSO, 5 U Taq DNA polymerase (Fermentas). The thermocycling program used consisted of an initial denaturation at 94°C for 2 min, followed by 20 cycles of denaturation at 94°C for 10 s and concurrent annealing and extension at 72°C for 210 s. The PCR products were restricted with BamHI and SacI and cloned into pMW119, effectively generating eight mutant libraries. The ligated products were recovered by transformation into electrocompetent XL-10 Gold and isolation of the plasmid mixtures from cells cultured in LBA medium. Mutation rates of the libraries were determined by plating 10 μL of the initial transformants on LBA-agar plates and sequencing plasmids isolated from 10 random single colonies picked from the plates.

12.4.7 COMPETITIVE GROWTH ENRICHMENT OF MUTANT LIBRARIES WITH N-OCTANE AND 1-OCTANOL

The purified mutant library plasmid mixtures were transformed into electrocompetent JA300A. After 1 h of outgrowth, a pre-culture was prepared by growing the transformants overnight in 10 mL of LBA medium at 37°C with continuous shaking.

For growth enrichment with n-octane, a 200 μL aliquot of the pre-culture was inoculated into 10 mL of LB medium saturated with a 3-mL overlay of n-octane, and shaken overnight at 37°C. The cultures were enriched by two more rounds of sub-culturing.

For growth enrichment with 1-octanol, a 200 μL aliquot of culture was inoculated into 10 mL of LBAI medium. The cells were grown overnight with shaking at 37°C to induce expression of AcrB. The overnight culture served as inoculum for 10 mL of LBI medium with 0.05% of 1-octanol. This culture was grown at 37°C with shaking for 2 days. The cultures were subcultivated once more for enrichment.

To isolate plasmid DNA from the enriched libraries, a 200 μL aliquot of the final enrichment culture was inoculated into 10 mL of LBA medium and the cells were grown overnight at 37°C for miniprep. The enriched library DNA was transformed into XL-10 Gold and plated on LBA agar. Plasmids of discrete clones were isolated from cultures grown from single colonies picked from the plate of cells and sequenced (Figure 3b).

12.4.8 ASSAY FOR RATE OF HYDROCARBON RELEASE FROM E. COLI

The rate of hydrocarbon release from *E. coli* cells were determined by adapting published methods [8,16]. Pre-cultures of JA300A cells transformed with various plasmids were prepared. For preliminary assays of enriched acrB libraries, the pre-culture was prepared by transformation of the mixture of plasmids into electrocompetent JA300A cells and growing the cells overnight in 10 mL LBA medium after 1 h of outgrowth. For assays of individual variants, cells for the pre-cultures were grown from a single colony of JA300A transformed with the plasmid picked from an LBA agar plate.

The pre-cultures were diluted 50-fold with 10 mL LBAI medium and grown for 16 h at 37°C. The cells were washed thrice by centrifugation (2000g, 3 min) and resuspended in 10 mL PPB (20 mM sodium phosphate pH 7 and 1mM $MgCl_2$). After another centrifugation cycle to pellet the cells, the cells were resuspended in 10 mL PPB at OD_{600}~1.5. Carbonyl cyanide 3-chlorophenylhydrazone (10 mM stock solution in DMSO) was added to the cells to a final concentration of 10 µM and the cell suspensions were shaken at 160 rpm for 15 min at 37°C. n-Octane or α-pinene (0.5 mL) was added to each sample and the biphasic mixture was further shaken at 200 rpm for 1 h at 37°C. The cells were then pelleted at 2000g for 3 min. The supernatant and hydrocarbon were carefully and thoroughly aspirated. The cells were resuspended in 10 mL PPB with 50 mM glucose, then incubated at 37°C and shaken at 160 rpm. One milliliter aliquots were transferred into microcentrifuge tubes in duplicate periodically (0-7min). The cell aliquots were immediately centrifuged (21000 g, 1 min) and the supernatant was removed. The cell pellets were resuspended in 500 µL dH_2O before extraction with 750 µL $CHCl_3$ for 3 h. The organic extracts were removed and analyzed by gas chromatography-mass selective detector (GC-MS). (Figure 3a).

12.4.9 SITE-DIRECTED AND INSERTIONAL MUTAGENESIS

Single-site N189H, T678A, Q737L and M844A acrB mutants were generated from pAcrB using the QuikChange (Stratagene) protocol for site-directed mutagenesis. The primers used were as follow: N189H-F, N189H-R for N189H acrB; T678A-F, T678A-R for T678A acrB; Q737L-F, Q737L-R for Q737L acrB; M844A-F, M844A-R for M844A acrB. The PCR reaction mixtures consisted of 20 ng pAcrB, 200 µM dNTP, 1.75 mM MgSO$_4$, 1 µL DMSO, 1 U KOD Hot Start DNA polymerase (Novagen), 0.3 µM mutagenic primers and 5 µL 10x KOD Hot Start DNA polymerase PCR buffer made up to 50 µL with deionized water. The reaction mixtures were subjected to an initial denaturation at 95°C for 2 min, followed by 20 cycles of denaturation at 95°C for 20 s, annealing at 60°C for 20 s and extension at 70°C for 240 s.

Hexahistidine-tags were added to the C-termini of acrB genes by appending four histidines to the two that are naturally present in protein via insertional mutagenesis [32]. PCR mixtures consisting of 20 ng plasmid templates, 200 µM dNTP, 1.75 mM MgSO$_4$, 1 µL DMSO, 1 U KOD Hot Start DNA polymerase (Novagen), 0.2 µM of mutagenic primers acrBHis-F and acrBHis-R each and 5 µL 10x KOD Hot Start DNA polymerase PCR buffer made up to 50 µL with deionized water were prepared. The reaction mixtures were subjected to an initial denaturation at 95°C for 2 min, followed by 15 cycles of denaturation at 95°C for 20 s, annealing at 50°C for 20 s and extension at 70°C for 210 s.

To isolate plasmids of the mutants, the PCR mixtures were digested with DpnI at 37°C for 3 h and transformed into electrocompetent XL-10 Gold cells. The cells were plated onto LBA agar and incubated at 37°C overnight. Cells were grown from single colonies and plasmids were isolated by miniprep kit. Mutations were verified by sequencing.

12.4.10 IMMUNOBLOT ANALYSIS

Cells expressing hexahistidine-tagged AcrB were cultivated as described for assaying the rate of hydrocarbon release. The OD$_{600}$ of the cultures were adjusted to 1.0, and 1-mL aliquots of the cells were centrifuged (21000g, 2 min). The cell pellets were resuspended in a buffer containing 20 mM Tris–HCl (pH 8.0), 5 mM EDTA, 8 M urea, 10 mM phenylmethanesulfonyl fluoride and 2% sodium dodecyl sulfate, and shaken gently at room temperature for 1 h to lyse the cells and solubilize AcrB. The lysates were mixed with an equal volume of 2x loading dye (Sigma-Aldrich) and incubated at 37°C for 10 min before being analyzed by SDS-PAGE. The proteins were transferred to polyvinylidene fluoride membranes (Bio-Rad) electrophoretically and probed using anti-hexahistidine antibody conjugated with horse radish peroxidase. The hexahisitidine-tagged AcrBs were detected colorimetrically on the membrane using 3,3',5,5'-tetramethylbenzidine.

12.4.11 ANTIBIOTIC RESISTANCE ASSAY

JA300A harboring pMW119 and plasmids with genes for wild-type, 1G1, 1G2, N189H, T678S, T678A, Q737L, M844L and M844A acrB were grown overnight in LBA at 37°C with shaking at 200 rpm. Approximately 10^3 cells were added to LB with various concentrations of chloramphenicol (0.5-10 mg/L), nalidixic acid (0.5-10 mg/L) and tetracycline (0.05 - 3mg/L) in 96-well microtitre plates and grown at 37°C with shaking at 200 rpm for 20 h. The MIC is the lowest antibiotic concentration that inhibited growth. Triplicated experiments gave consistent MIC.

12.4.12 STRUCTURAL ANALYSIS

The locations of the mutations were studied on a published structure of AcrB (PDB ID: 2DHH). Rendered images of protein structures were generated with Pymol [33].

REFERENCES

1. Fortman JL, Chhabra S, Mukhopadhyay A, Chou H, Lee TS, Steen E, Keasling JD: Biofuel alternatives to ethanol: pumping the microbial well. Trends Biotechnol 2008, 26:375-381.
2. Atsumi S, Liao JC: Directed evolution of Methanococcus jannaschii citramalate synthase for biosynthesis of 1-propanol and 1-butanol by Escherichia coli. Appl Environ Microbiol 2008, 74:7802-7808.
3. Zhang K, Sawaya MR, Eisenberg DS, Liao JC: Expanding metabolism for biosynthesis of nonnatural alcohols. Proc Natl Acad Sci USA 2008, 105:20653-20658.

4. Steen EJ, Kang Y, Bokinsky G, Hu Z, Schirmer A, McClure A, Del Cardayre SB, Keasling JD: Microbial production of fatty-acid-derived fuels and chemicals from plant biomass. Nature 2010, 463:559-562.

5. Lee SK, Chou H, Ham TS, Lee TS, Keasling JD: Metabolic engineering of microorganisms for biofuels production: from bugs to synthetic biology to fuels. Curr Opin Biotechnol 2008, 19:556-563.

6. Schirmer A, Rude MA, Li X, Popova E, del Cardayre SB: Microbial biosynthesis of alkanes. Science 2010, 329:559-562.

7. Connor MR, Cann AF, Liao JC: 3-Methyl-1-butanol production in Escherichia coli: random mutagenesis and two-phase fermentation. Appl Microbiol Biotechnol 2010, 86:1155-1164.

8. Tsukagoshi N, Aono R: Entry into and release of solvents by Escherichia coli in an organic-aqueous two-liquid-phase system and substrate specificity of the AcrAB-TolC solvent-extruding pump. J Bacteriol 2000, 182:4803-4810.

9. Dunlop MJ, Dossani ZY, Szmidt HL, Chu HC, Lee TS, Keasling JD, Hadi MZ, Mukhopadhyay A: Engineering microbial biofuel tolerance and export using efflux pumps. Mol Syst Biol 2011, 7:487.

10. Dunlop M, Keasling J, Mukhopadhyay A: A model for improving microbial biofuel production using a synthetic feedback loop. Syst Synth Biol 2010, 4:95-104.

11. Bokma E, Koronakis E, Lobedanz S, Hughes C, Koronakis V: Directed evolution of a bacterial efflux pump: adaptation of the E. coli TolC exit duct to the Pseudomonas MexAB translocase. FEBS Lett 2006, 580:5339-5343.

12. Niwa T, Ying BW, Saito K, Jin W, Takada S, Ueda T, Taguchi H: Bimodal protein solubility distribution revealed by an aggregation analysis of the entire ensemble of Escherichia coli proteins. Proc Natl Acad Sci USA 2009, 106:4201-4206.

13. Pos KM: Drug transport mechanism of the AcrB efflux pump. Biochim Biophys Acta 2009, 1794:782-793.

14. Yu EW, Aires JR, Nikaido H: AcrB multidrug efflux pump of Escherichia coli: composite substrate-binding cavity of exceptional flexibility generates its extremely wide substrate specificity. J Bacteriol 2003, 185:5657-5664.

15. Violi A, Yan S, Eddings EG, Sarofim F, Granata S, Faravelli T, Ranzi E: Experimental formulation and kinetic model for JP-8 surrogate mixtures. Combust Sci Technol 2002, 174:399-417.

16. Bohnert JA, Karamian B, Nikaido H: Optimized Nile Red efflux assay of AcrAB-TolC multidrug efflux system shows competition between substrates. Antimicrob Agents Chemother 2010, 54:3770-3775.

17. Vermue M, Sikkema J, Verheul A, Bakker R, Tramper J: Toxicity of homologous series of organic solvents for the gram-positive bacteria Arthrobacter and Nocardia Sp. and the gram-negative bacteria Acinetobacter and Pseudomonas Sp. Biotechnol Bioeng 1993, 42:747-758.

18. Aono R, Kobayashi M, Nakajima H, Kobayashi H: A close correlation between improvement of organic solvent tolerance levels and alteration of resistance toward low levels of multiple antibiotics in Escherichia coli. Biosci Biotechnol Biochem 1995, 59:213-218.

19. Asako H, Nakajima H, Kobayashi K, Kobayashi M, Aono R: Organic solvent tolerance and antibiotic resistance increased by overexpression of marA in Escherichia coli. Appl Environ Microbiol 1997, 63:1428-1433.

20. Hayashi S, Aono R, Hanai T, Mori H, Kobayashi T, Honda H: Analysis of organic solvent tolerance in Escherichia coli using gene expression profiles from DNA microarrays. J Biosci Bioeng 2003, 95:379-383.

21. Murakami S, Nakashima R, Yamashita E, Matsumoto T, Yamaguchi A: Crystal structures of a multidrug transporter reveal a functionally rotating mechanism. Nature 2006, 443:173-179.

22. Tamura N, Murakami S, Oyama Y, Ishiguro M, Yamaguchi A: Direct interaction of multidrug efflux transporter AcrB and outer membrane channel TolC detected via site-directed disulfide cross-linking. Biochemistry 2005, 44:11115-11121.

23. Husain F, Bikhchandani M, Nikaido H: Vestibules are part of the substrate path in the multidrug efflux transporter AcrB of Escherichia coli. J Bacteriol 2011, 193:5847-5849.

24. Husain F, Nikaido H: Substrate path in the AcrB multidrug efflux pump of Escherichia coli. Mol Microbiol 2010, 78:320-330.

25. Nakashima R, Sakurai K, Yamasaki S, Nishino K, Yamaguchi A: Structures of the multidrug exporter AcrB reveal a proximal multisite drug-binding pocket. Nature 2011, 480:565-569.

26. Schulz R, Vargiu AV, Collu F, Kleinekathofer U, Ruggerone P: Functional rotation of the transporter AcrB: insights into drug extrusion from simulations. PLoS Comput Biol 2010, 6:e1000806.

27. Foo JL, Jackson CJ, Carr PD, Kim HK, Schenk G, Gahan LR, Ollis DL: Mutation of outer-shell residues modulates metal ion co-ordination strength in a metalloenzyme. Biochem J 2010, 429:313-321.

28. Jackson CJ, Foo JL, Tokuriki N, Afriat L, Carr PD, Kim HK, Schenk G, Tawfik DS, Ollis DL: Conformational sampling, catalysis, and evolution of the bacterial phosphotriesterase. Proc Natl Acad Sci USA 2009, 106:21631-21636.

29. Okusu H, Ma D, Nikaido H: AcrAB efflux pump plays a major role in the antibiotic resistance phenotype of Escherichia coli multiple-antibiotic-resistance (Mar) mutants. J Bacteriol 1996, 178:306-308.

30. Miller JH: A short course in bacterial genetics: a laboratory manual and handbook for Escherichia coli and related bacteria. Plainview, N.Y.: Cold Spring Harbor Laboratory Press; 1992.

31. Sambrook J, Russell DW: Molecular cloning: a laboratory manual. 3rd edition. Cold Spring Harbor, N.Y.: Cold Spring Harbor Laboratory Press; 2001.

32. Rabhi I, Guedel N, Chouk I, Zerria K, Barbouche MR, Dellagi K, Fathallah DM: A novel simple and rapid PCR-based site-directed mutagenesis method. Mol Biotechnol 2004, 26:27-34.

33. Schrodinger LLC: The PyMOL Molecular Graphics System, Version 1.3r1. 2010.

GENERATING PHENOTYPIC DIVERSITY IN A FUNGAL BIOCATALYST TO INVESTIGATE ALCOHOL STRESS TOLERANCE ENCOUNTERED DURING MICROBIAL CELLULOSIC BIOFUEL PRODUCTION

ROSANNA C. HENNESSY, FIONA DOOHAN, AND EWEN MULLINS

13.1 INTRODUCTION

Lignocellulosic biomass is an abundant feedstock and attractive source of sugars for biofuel production. Large-scale utilisation is however challenged by the general lack of low-cost technologies that can overcome cellulose recalcitrance [1]. One potential route to eco-friendly sustainable energy production is the consolidated bioprocessing (CBP) of biomass into biofuels [2].

Several yeasts (e.g. *Saccharomyces cerevisiae, Pichia stipitis, Candida shehatae, Pachysolen tannophilus*) and bacteria (e.g. *Escherichia coli, Klebsiella oxytoca, Zymomomas mobilis*) have been engineered for CBP ethanol production however their capacity to secrete saccharification and

Generating Phenotypic Diversity in a Fungal Biocatalyst to Investigate Alcohol Stress Tolerance Encountered during Microbial Cellulosic Biofuel Production. © Hennessy RC, Doohan F, and Mullins E. PLoS ONE *8,10 (2013), doi:10.1371/journal.pone.0077501. Licensed under a Creative Commons Attribution License, creativecommons.org/licenses/by/3.0/.*

fermentation enzymes at sufficient yield remains an obstacle [3-7]. In contrast, filamentous fungi including *Trichoderma sp., Neurospora sp., Aspergillus sp., Monilinia sp., Rhizopus sp., Mucor sp., Paecilomyces sp.* and *Fusarium sp.*, possess a large repertoire of lignocellulolytic enzymes due to their co-evolution with plants, and can convert released plant-derived sugars into ethanol [7–10]. In particular, the broad host range phytopathogen *Fusarium oxysporum* [11] can degrade and produce ethanol from various cellulosic substrates (e.g. untreated and pre-treated straw [12,13], brewer's spent grain [14], potato waste [15]). Previous work [13] identified *F. oxysporum* strain 11C as a promising microbial biocatalyst capable of producing high bioethanol yields from delignified wheat straw.

At present no organism can ferment cellulosic materials to ethanol at rates and titres necessary to achieve economic feasibility [5]. CBP requires robust microbes able to: (i) degrade lignocellulosic materials in the absence of supplementary exogenous enzymes, (ii) utilise at high efficiency both hexose and pentose sugars including their conversion into high-titer ethanol and (iii) tolerate toxic compounds notably the primary alcohol product formed during fermentation [16]. Whereas the endogenous ability of *F. oxysporum* to tolerate inhibitory compounds encountered during CBP including lignocellulosic hydrosylates (carboxylic acids, phenolic compounds, furan derivatives) [17] and the fermentation by-product acetic acid [18] has been reported, a knowledge deficit exists in regards to *F. oxysporum*'s capability to tolerate ethanol.

Ethanol stress affects cell growth and viability in addition to productivity [19,20]. While other compounds compatible as biofuels such as n-butanol offer advantages over ethanol (e.g. high energy content and miscibility with gasoline), production is limited by the sensitivity of native producers such as *Clostridium sp.* to the alcohol [21]. This has led to the investigation of alternative butanol production hosts [22-24], prompting the investigation of n-butanol (hereafter referred to as butanol) tolerance in addition to ethanol in this study. For the cost-effective and commercially viable production of biofuels, product yields must exceed native microbial tolerance levels [25] therefore engineering of stress tolerant strains is critical to achieve high survival and production rates [25,26].

Several strategies have been investigated to improve alcohol tolerance in yeast and bacteria, including gene over-expression or random knockout libraries, genome shuffling and transcriptional or translational engineering [27]. For filamentous fungi ATMT has established itself as a valuable and powerful tool for genetic studies, greatly facilitating the identification and analysis of genes involved in complex biological processes such as host pathogenicity [28] and, as we hypothesise here, in regards to alcohol tolerance. As the gram-negative soil bacterium *A. tumefaciens* facilitates the stable transfer and integration of the tumour inducing (Ti) plasmid segment (known as T- DNA) into the targeted host cell [29], ATMT holds significant advantages over more traditional methods of gene disruption. For example, *A. tumefaciens'* broad host range, its potential to transform a range of starting materials and its propensity for low copy number insertion events [30]. ATMT also provides an ability to 'tag' the disrupted gene(s) thereby facilitating their identification via genome walking. In the case of *F. oxysporum* this is further assisted by the recent availability of the organism's genome sequence [31] (http://www.broadinstitute.org).

In this study, we hypothesised that ATMT could be exploited as a tool to generate significant degrees of phenotypic diversity in *F. oxysporum* strain 11C in response to alcohol stress. To this end, a modified ATMT protocol was established to generate a library of random gene disruption transformants, which underwent a three tier physiological evaluation. As a result, the initial population of *F. oxysporum* transformants was reduced from 1,563 to 29 individuals, which were subjected to ethanol tolerance screening and cross-resistance to butanol. The level of recorded phenotypic variation confirmed the occurrence of ATMT-derived genotypes with several *F. oxysporum* transformants identified for future analysis. To the best of our knowledge this is the first report detailing the use of random insertional mutagenesis to investigate alcohol tolerance in a fungal CBP agent.

13.2 MATERIALS AND METHODS

13.2.1 ORIGIN AND MAINTENANCE OF FUNGI

Fusarium oxysporum strain 11C originated from Irish soils as previously described [13]. For ATMT, *F. oxysporum* was sub-cultured onto potato dextrose agar (PDA; Difco, UK) plates, incubated at 25°C for 5 days prior to producing fungal conidial inoculum in Mung bean broth [32].

13.2.2 PLASMID AND ATMT TRANSFORMATION

The binary vector used for fungal gene disruption was pSK1019, which is equipped with the hph antibiotic resistance marker gene under the control of an *Aspergillus nidulans* TrpC promoter and was donated by Professor Seogchan Kang (The Pennsylvania State University, USA). Agrobacterium tumefaciens strain AGL-1 was transformed with the binary vector pSK1019 [33] and ATMT of *F. oxysporum* strain 11C was optimised based on a previously published protocol [34] (see Methods S1). Three independent ATMTs of *F. oxysporum* strain 11C were conducted.

13.2.3 MOLECULAR ANALYSIS OF TRANSFORMANTS

For PCR analysis, fungal genomic DNA was extracted from mycelia using a modified method [35] (see Methods S1). DNA was quantified using a QubitTM Quant-iT assay (Invitrogen, U.S) and stored at -80°C. PCR was used to detect the presence/absence of the hph transgene in DNA extracts from putative transformants (see Methods S1). For Southern blot analysis, genomic DNA was extracted [36] and quantified using a NanoDrop® ND-1000 Spectrophotometer (NanoDrop Technologies, USA). Southern blot analysis was used to determine plasmid copy number in transformants (see Methods S1).

13.2.4 PRIMARY ALCOHOL TOLERANCE SCREEN

A total of 1,563 putatively transformed hygromycin-resistant colonies were isolated into microplates (96-well) containing 200 μl minimal medium (supplemented with hygryomcinB, 60 μg ml^{-1}, Sigma, UK) per well [37] which were then sealed and incubated at 25°C for 3 days and subsequently used as inoculum for primary screening of alcohol tolerance. For the primary screen (first of three), 96-well plates were prepared containing 140µl minimal medium [37] per well supplemented with six treatments either hygromycinB (60 µg ml^{-1}), no ethanol or 0.5, 6.0 or 10% vv^{-1} ethyl alcohol (Sigma, UK) plus or minus hygromycinB (60 µg ml^{-1}; Sigma, UK). To target different response levels, low (0.5% vv^{-1} ethanol), medium (6% vv^{-1} ethanol) and high (10% vv^{-1}) ethanol concentrations were then selected. Wells were inoculated with 10 µl of fungal material (one well per treatment per transformant/wild type fungus; plate layout is shown in Figure S1A). Plates were sealed with a lid and parafilm and incubated at 25°C with fungal growth (OD$_{600}$nm) measured after both 48 and 96 hours using a spectrophotometer (Safire2, Tecan, Austria). Putative transformants with a $\geq$ 2-fold higher OD$_{600}$nm in 10% (vv^{-1}) ethanol relative to wild type fungus 11C were purified prior to secondary screening. This primary screening of the transformant collection was conducted once followed by the selection of putative transformants for purification and graduation to the secondary screening (see Methods S1).

13.2.5 SECONDARY ALCOHOL TOLERANCE SCREEN

dConidial suspensions of the *F. oxysporum* 11C wild type strain and transformants of interest (20 µl from the from monoconidial 20% (vv-1) -80°C glycerol stock) were inoculated into 220 µl minimal medium in microtiter (96-well) plate wells and incubated at 25°C for 24 hours. This served as inoculum for the secondary screen. For secondary screening; 100 µl of fungal culture was transferred into microtiter plate wells containing 100 µl minimal medium [37] supplemented with either 0, 6, 7, 8, 9 or 10% vv^{-1} ethanol (single well per treatment per transformant/wild type; see Figure S1B for plate layout). Plates were sealed with a lid and parafilm and incubated at 25°C for 96 hours after which fungal growth (OD$_{600}$nm) was recorded. Data was normalised relative to the control (wild-type strain ^{11}C). All datasets (0-10 vv^{-1} ethanol) were used for selecting transformants of interest. From the secondary screen, *F. oxysporum* strain ^{11}C putative transformants demonstrating either ethanol sensitive or ethanol tolerant phenotypes relative to the wild-type strain ^{11}C control, at concentrations ranging between 6-10% (vv^{-1}) were selected for tertiary screening. This experiment was conducted once.

13.2.6 TERTIARY ALCOHOL TOLERANCE SCREEN

In the tertiary screen, tolerance to both ethanol and butanol was assessed, with the goal of examining the potential for cross-correlation between ethanol and butanol tolerance. Conidial inoculum for the tertiary assessment was generated by individually inoculating each transformant into Mung bean broth [32] (100 ml conical flask) with three fungal plugs and incubating at 25°C and 150 rpm for 5 days. Inoculum was harvested by centrifugation and washed twice with sterile distilled water and resuspended in minimal medium [37] to a concentration of 106 ml^{-1} conidia. Microtiter (96-well) plates were prepared with wells containing 100 μl minimal medium supplemented with either no alcohol, ethanol [2, 4, 6, 8 or 10% (vv-1)] or butanol (0.25, 0.5, 0.75, 1 and 1.25% (vv^{-1})] with a single treatment tested per 96-well plate (see Figure S1C for plate layout). Each plate was inoculated with 100 μl fungal conidia (106 ml^{-1}) (eight wells per treatment per wild type/transformant strain). Plates were sealed with a lid and parafilm and incubated at 25°C for 7 days. Fungal growth was measured using a spectrophotometer OD_{600nm} (Spectra Max 340 PC 96-well Plate Reader, Molecular Devices, USA), and percentage increase or decrease in growth relative to wild-type control was determined. For data analysis see Methods S1. This experimental screening was repeated three times.

13.2.7 PHENOTYPIC ANALYSIS OF F. OXYSPORUM ^{11}C AND TR. 259

A series of phenotypic assays of *F. oxysporum* [11]C and Tr. 259 were conducted to investigate the effect of increasing alcohol concentration on growth, temporal analysis of alcohol tolerance, the effect of alcohol tolerance stress on spore germination and biomass production (see Methods S1). All experiments were conducted either twice or three times. For data analysis see Methods S1.

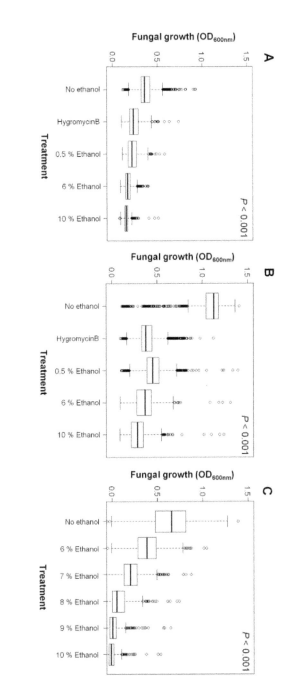

FIGURE 1: Primary and second-tier screening of putative Fusarium oxysporum strain ^{11}C transformants. Box and Whisker plot depicting the distribution of putative *Fusarium oxysporum* strain ^{11}C transformants (n =1,563) screened for altered ethanol tolerance during primary screening at 48 hours (A) and 96 hours (B) and transformants (n =402) screened for altered ethanol tolerance during secondary screening (C). Fungi were grown at 25°C in minimal medium [37] with five different treatments: no ethanol control; hygromycinB 60 μg ml⁻¹ control; 0.5% (vv⁻1) ethanol; 6% (vv⁻1) ethanol; 10% (vv⁻1) ethanol) (Primary-tier) (A, B) or six treatments: no alcohol control; 6% (vv⁻1) ethanol; 7% (vv⁻1) ethanol; 8% (vv⁻1) ethanol; 9% (vv⁻1) ethanol; 10% (vv⁻1) ethanol (Second-tier) (C). Fungal growth (OD$_{600nm}$) was measured after 96 hours using a spectrophotometer (Safire2, Tecan, Austria).

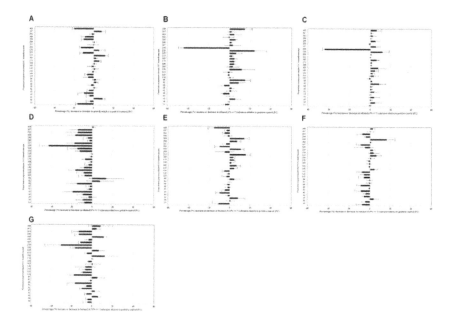

FIGURE 2: Tertiary screening of *Fusarium oxysporum* strain ^{11}C transformants (n=29). Fungi were grown in minimal medium [37] supplemented with; no alcohol (A); 2% (vv⁻¹) ethanol (ethyl alcohol; Sigma, UK) (B), 4% (vv⁻¹) ethanol (C), 6% (vv⁻¹) ethanol (D) or 0.25% (vv⁻¹) butanol (n-butanol; Sigma, UK) (E), 0.5% (vv-1) butanol (F) or 0.75% (vv⁻¹) butanol (G) and incubated at 25°C for 7 days. Percentage increase or decrease in fungal growth relative to the positive control (wild-type strain ^{11}C) was determined. Results represent the mean of three independent experiments and bars indicate SEM. Transformants significantly different from the positive control (wild-type strain ^{11}C) are highlighted with an asterisk (level of significance: <0.050 *, <0.01**, <0.001***).

13.2.8 GENOME WALKING AND SEQUENCE ANALYSIS

Genome walking was conducted on Tr. 259 to characterise the degree of genetic disruption underpinning the transformant's observed phenotype (see Methods S1 & Table S1).

13.2.9 RT-PCR OF PUTATIVE HEXOSE TRANSPORTER (FOXG_09625) UNDER ALCOHOL STRESS

An experiment was designed to investigate the temporal accumulation of a putative hexose transporter under alcohol stress (Figure S2) (see Methods S1& Table S1). Each experiment was conducted either three times (no alcohol, ethanol) or twice (butanol) with three replicates per strain per treatment per time point. Real-time quantification of the target and housekeeping gene respectively were performed in separate reactions. The threshold cycle (C_T) values obtained from RT-PCR were used to calculate the accumulation of the target gene (relative mRNA accumulation) relative to β-tubulin transcript by the $2^{-\Delta\Delta CT}$ method, where $\Delta\Delta C_T = (C_T, \text{ target gene - } C_T, \beta\text{-tubulin})$ [38]. Results were based on the average obtained for two replica RT-PCR reactions per sample. For data analysis see Methods S1.

13.2.10 PHYLOGENETIC ANALYSIS OF PUTATIVE SUGAR TRANSPORTERS IN F. OXYSPORUM

A phylogram of putative *F. oxysporum* strain 4287 sugar transporters was constructed using amino acid sequences from transcripts showing $\geq 30\%$ homology to FOXG_09625 (See Methods S1).

13.3 DATA ANALYSIS

The primary/secondary screen dataset was analysed using the Boxplot function in R (R v2.15.2 R Development Core Team, 2012) and a Box and Whisker plot was used to depict the distribution of transformant response across various treatments. The primary/secondary screen dataset was non-normally distributed as determined within Minitab (Minitab release © 16, 2011 Minitab Inc.). The significance between treatments at each time point (48 and 96 hours respectively) or ethanol treatments was analysed within the Statistical Package for the Social Sciences (SPSS 18, SPSS Inc.) using Kruskal-Wallis H-test for non-parametrical data. The significance between treatments and transformants amongst the 29 individuals selected from secondary screening was determined using one-way ANOVA as the data could be transformed to fit a normal distribution using Johnson transformation within Minitab. For tertiary screen analysis see Methods S1.

13.4 RESULTS

13.4.1 ATMT LIBRARY CONSTRUCTION FOR PRIMARY ALCOHOL SCREENING

A transformation platform was established for *F. oxysporum* strain 11C to generate a library of gene disruption transformants (n=1,563) (Figure S3). The library was screened using a high throughput alcohol screen based on microtiter (96-well) plates to discriminate against low to high ethanol (0.5-10% (vv^{-1})) response levels measured across 48 and 96 hours (Figure 1A & 1B & Figure S3). Transformant selection for second – tier screening, as previously described [Methods], was based on recordings at 96 hours for which greater phenotypic variation was observed compared to 48 hours. An established threshold [Methods] led to the isolation of 182 putative transformants, which when purified (4 monoconidial cultures produced per putative transformant) yielded 443 viable monoconidial-derived cultures for PCR analysis and secondary alcohol screening.

13.4.2 PCR ANALYSIS AND SECONDARY ALCOHOL SCREENING

Over 90% (402/443) of the monoconidial lines generated from primary transformants were PCR-positive for the hph transgene (Figure S4A). On re-screening the ethanol tolerance of this population, a significant differ-ence between treatments was observed between the derived monoconidial lines (P<0.001) (Figure 1C). A total of 29 PCR-verified transformants in-dicating hypo-or hyper-ethanol growth phenotypes were selected for ter-tiary screening.

TABLE 1: Summary of tertiary screening of Fusarium oxysporum strain 11C transformants to investigate the level of altered alcohol tolerance across a selected sub-population of individuals (n=29).

Treat (vv⁻¹)a	Population (%)b	Individuals (N)c	Hyper (N)d	Hypo (N)e
Control 0	66	19	7	12
Ethanol 2	17	5	4	1
Ethanol 4	31	9	8	1
Ethanol 6	31	9	0	9
Butanol 0.25	62	18	9	9
Butanol 0.5	38	11	3	8
Butanol 0.75	45	13	0	13

ᵃPositive control (no alcohol) or alcohol (ethanol or butanol) tested ᵇPercentage (%) population tested (n=29) showing hypo-hyper- phenotype relative to wild-type strain ¹¹C (P<0.05) ᶜTotal number of individuals showing hypo-hyper- phenotype relative to wild-type strain ¹¹C (P<0.05) ᵈTotal number of hyper-altered phenotypes (P<0.05) ᵉTotal number of hypo-altered phenotypes (P<0.05)

13.4.3 TERTIARY ALCOHOL SCREENING AND TRANSFORMANT SELECTION

Tertiary screening was used to assess tolerance of the 29 selected transfor-mants to ethanol and cross-tolerance to butanol (Figure 2 & Table 1). A sig-nificant level of phenotypic diversity was recorded between strains across all treatments (P<0.001) (Figure 2). At 2 and 4% (vv⁻¹) ethanol the highest level of hyper-or hypo-tolerance relative to parent strain ¹¹C was observed for Tr. 230 and Tr. 259 (Figure 2B & 2C). At the highest ethanol concen-tration tested (6% vv⁻¹), Tr. 259 continued to demonstrate a substantial hypo-tolerance to ethanol relative to the parental strain (-43.01%±6.40)

(Figure 2D). A strong positive correlation was noted between ethanol and butanol treatments respectively (r=0.414; n=6, P=0.01). Similar to results recorded for ethanol, at the highest butanol concentration (0.75% vv⁻¹) Tr. 259 (-30.61%±13.49) showed the highest hypo-butanol phenotype and Tr. 185 (12.57%±13.16) the most hyper-butanol phenotype relative to strain ¹¹C (Figure 2G).

Southern blot analysis of a random sub-set of nine transformants using an hph probe indicated an average transgene copy number of 1.55 (Figure S4B). PCA analysis (Figure 3) highlighted the degree of phenotypic diversity among transformants and isolated Tr. 259 as a candidate strain of interest for further analysis. Southern analysis of this transformant suggests a single insertion event (data not shown). From the presented evidence, nine transformants were identified as candidate strains for future studies showing significantly (P<0.05) altered tolerance relative to the positive control (wild type strain ¹¹C) to ethanol (respectively Tr. 14, 51,133,144 and 364), butanol (respectively Tr. 92, 230 and 408) or significantly(P<0.05) for both alcohols, Tr. 259.

13.4.4 PHENOTYPIC ANALYSIS OF TR. 259

The effect of alcohol stress on temporal growth, spore germination and biomass production was investigated to confirm the hypo-tolerant phenotype of Tr. 259 observed during tertiary screening. The effect of increasing alcohol concentration on strain growth was firstly assessed (Figure S5). Growth of both Tr. 259 and ¹¹C decreased with increasing ethanol (r≥-0.967, n=5, P≤0.02) and butanol (r≥-0.929, n=8, P≤0.01) concentration whereas ¹¹C showed significantly better growth than Tr. 259 up to 6% (vv⁻¹) ethanol and 1% (vv⁻¹) butanol (P<0.05) (Figure S5). At 4% (vv⁻¹) ethanol and 0.75% (vv⁻¹) butanol a substantial difference in growth (≥42%) was recorded between Tr. 259 and ¹¹C prompting the selection of these concentrations for further studies (Figure S5).

In the absence of alcohol, both Tr. 259 and ¹¹C exhibited similar mycelial growth and spore germination rates (Figure 4A & Figure S6D). In contrast, ¹¹C achieved better mycelial growth with either 4% ethanol or 0.75% butanol present in the medium, as compared to Tr. 259 (Figure 4B & 4C).

Both 4% ethanol and 0.75% butanol severely impacted spore germination, however ^{11}C achieved significantly greater spore germination than Tr. 259 (P≤0.01) (Figure S6B-C & S6E-F).

13.4.5 GENOMIC ANALYSIS OF TR. 259

Genome walking was conducted to determine the potential genomic region associated with Tr. 259's hypo-tolerant phenotype (Figure S7). This resulted in the cloning of a 1031nt sequence external to the T-DNA right border (RB) (sequence analysis confirmed the amplicon was flanked by the universal and RB specific primer as expected; Figure S8A). Further analysis did not result in the recovery of full-length T-DNA in the identified coding region, indicating partial T-DNA integration. In silico analysis of the recovered coding sequence showed the highest homology in the ORF of a putative sugar transporter gene FOXG_09625 in *F. oxysporum* f.sp.lycospersici (strain 4287) at both the protein (68% identity) and nucleotide (83% identity) level respectively (Figure S8B & S8C).

FOXG_09625 showed homology (percent identity ≥ 30%) with 30 different proteins all containing MFS and sugar transporter domains respectively, encoded within the genome of *F. oxysporum* f.sp.lycospersici (strain 4287) (FGCD). Five of these proteins were annotated as hexose transporters (FOXG_02491, FOXG_09722, FOXG_05876, FOXG_13263, FOXG_014666) including FOXG_09625, and the remaining as hypothetical proteins. Comparison of these proteins showed considerable sequence diversity indicating potentially diverse roles of these proteins in *F. oxysporum* (Figure S9A). In total, the genome of *F. oxysporum* f.sp. lycospersici (strain 4287) encodes 13 proteins annotated as hexose transporters (Figure S9B).

RT-PCR analysis was used to analyse FOXG_09625 transcript levels in strain ^{11}C compared to Tr. 259, following short-term (0.5-12 hours) cultivation of germinated spores (pre-cultured for 12 hours) in the absence or presence of alcohol (Figure 5). In the absence of alcohol, basal expression was low in both ^{11}C and Tr. 259 (P>0.05). (Figure 5A). Reduced transcript levels were recorded for both strains under ethanol stress compared to normal conditions from 0.5 to 12 hours with a significant difference between strains at 2 hours (P<0.01) (Figure 5B). In the presence of butanol, a significant difference between Tr. 259 and ^{11}C was recorded for all time points tested (P<0.001) with transcript levels highest at 0.5 hours (Figure 5C).

13.5 DISCUSSION

This study has shown that ATMT is an effective tool to generate phenotypic diversity in response to alcohol stress within *F. oxysporum* strain [11]C. Although the generation of fungal insertion libraries is well documented, the integration of a three-tier alcohol screen into an ATMT platform is novel, and led to the generation and isolation of phenotypically diverse transformants from the library, indicating genotype-dependent variation and a functional ATMT system.

Ethanol tolerance of strain [11]C (>6% (vv-1) ethanol) was high compared to *Candida albicans*, which like *F. oxysporum* can ferment glucose to ethanol but can only tolerate 1% (vv[-1]) ethanol stress [39]. In contrast, yeast can tolerate high ethanol concentrations (>15% (vv[-1])), however at lower concentrations (4-6% (vv[-1])) a 50% reduction in yeast's growth has been reported [40]. Unsurprisingly, strain ^{11}C could not tolerate equivalent butanol concentrations as a barrier between 1-2% (vv[-1]) exists for most microbes with the exception of some *Lactobacillus* strains (3% (vv-1)) [41] and *Bacillus subtilis* (5% (vv-1)) [42].

Large-scale screening of filamentous fungi can be challenging as manipulating fungal material (e.g. mycelium) can interfere with pipette-aided transfer between plates [43]. Fungal growth rates or hyphal growth patterns can vary greatly among isolates with some strains growing poorly in low volume space-limited microplates. Regardless, microplate-based assays facilitate the screening of large populations: a significant advantage over the labour-intensive regime of individual shake flasks or tubes [43]. Screening of *F. oxysporum* ATMT-derived transformants for alcohol tolerance has not yet been described. In other ascomycetous fungi notably the rice blast fungus Magnaporthe grisea, large-scale ATMT mutagenesis coupled with phenotypic plate-based screens has been used in pathogenicity related studies [44,45].

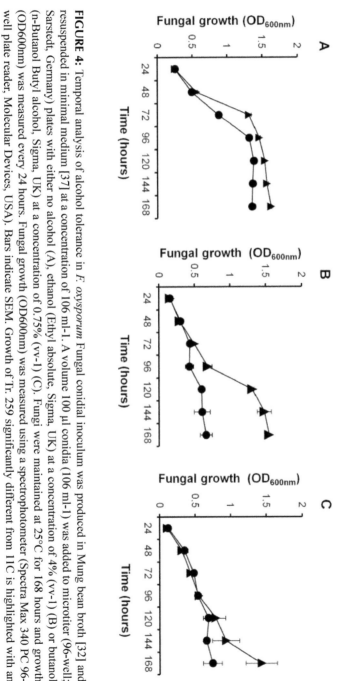

FIGURE 4: Temporal analysis of alcohol tolerance in *F. oxysporum* Fungal conidial inoculum was produced in Mung bean broth [32] and resuspended in minimal medium [37] at a concentration of 106 ml-1. A volume 100 μl conidia (106 ml-1) was added to microtiter (96-well; Sarstedt, Germany) plates with either no alcohol (A), ethanol (Ethyl absolute, Sigma, UK) at a concentration of 4% (vv-1) (B) or butanol (n-Butanol Butyl alcohol, Sigma, UK) at a concentration of 0.75% (vv-1) (C). Fungi were maintained at 25°C for 168 hours and growth (OD600nm) was measured every 24 hours. Fungal growth (OD600nm) was measured using a spectrophotometer (Spectra Max 340 PC 96-well plate reader, Molecular Devices, USA). Bars indicate SEM. Growth of Tr. 259 significantly different from 11C is highlighted with an asterisk (level of significance: *<0.05, ** < 0.01).

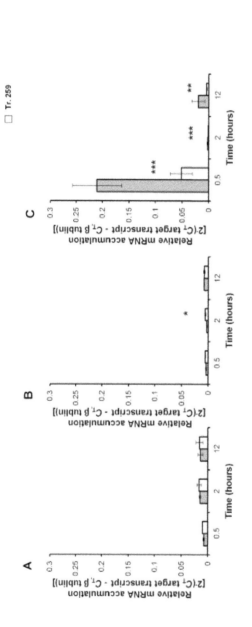

FIGURE 5: RT-PCR analysis of putative hexose transporter under alcohol stress. Temporal analysis of putative hexose transporter transcript (FOXG_09625) during shake flask growth of Fusarium oxysporum strain 11C and transformant Tr. 259 under normal conditions (A), ethanol stress (B) and butanol stress (C). *F. oxysporum* (2ml 106 ml-1) was aerobically cultured in Erlenmeyer flasks (100 ml) in 48 ml of minimal medium [37] for 12 hours shaking at 150 rpm at 25°C. At 12 hours, flasks were amended with 2ml of water, ethanol or butanol to a final concentration of 0%, 4% and 0.75% (vv-1), respectively. RT-PCR was conducted on RNA samples harvested at 0.5, 2, 12 hours respectively post – inoculation. Putative HXT accumulation was quantified relative to the housekeeping gene β-tubulin (FOXG_06228) [38] Results are based on three experiments (no alcohol, ethanol) or two experiments (butanol), each with three replicates per time point per strain tested. RT-PCR was conducted twice on each experiment. Bars indicate SEM. Tr. 259 FOXG_09625 mRNA accumulation significantly different from strain 11C is highlighted with an asterisk (level of significance: *<0.05, **≤0.01, ***<0.001).

The three-tier screening system was designed to specifically discriminate either alcohol sensitive or tolerant phenotypes. Integration of the primary screen into the ATMT protocol provided a course but effective method to instigate a logistically feasible sorting of the library, thereby avoiding the laborious and time-consuming task of purifying thousands of putative transformants prior to screening. Non-purified, putative transformants underwent primary screening in contrast to the secondary screen, which used single-spored hph-PCR positive transformants. It cannot be ruled out that false negatives were lost to the analysis during these two stages. These primary rounds of screening were based on single reads to accelerate the phenotyping process and isolate transformants of interest for further analysis using an intensive confirmatory tertiary screen. This latter screen highlighted transformant Tr. 259 which underwent phenotypic and genomic analysis to confirm its hypo-alcohol tolerant phenotype and the ATMT-based disruption of a coding region showing homology to a putative sugar transporter (FOXG_09625) in *F. oxysporum* f.sp lycospersici (strain 4287). Southern hybridisation suggested a single insertion site within Tr. 259 with genome walking indicating partial T-DNA integration but it cannot be ruled out that additional T-DNA insertion events have also occurred through the genome of Tr. 259.

Eukaryotic transporters are best characterised in the yeast 20-member family of major hexose transporter genes (HXT1-4, HXT6-7, HXT8-17) which includes two sensor-coding genes (SNF3, RGT2) and the GAL2 gene [46]. In contrast to yeasts (*S. cerevisiae, Pichia stipitis, Schizosaccharomyces pombe, Kluveromyces lactis*), few sugar transporters have been identified and characterised in filamentous fungi [47,48]. The genome of *F. oxysporum* f.sp lycospersici (strain 4287) has at least 13 putative hexose transporters including FOXG_09625, similar to *Aspergillus nidulans* reported to have 17 [47]. RT-PCR analysis indicated FOXG_09625 was differentially expressed in Tr. 259 compared to WT during alcohol-induced stress and it would be interesting to investigate by Western blotting whether FOXG_09625 is also differentially expressed at the protein level. In yeast, the closest protein homologue of FOXG_09625 is HXT13, which is induced in the presence of non-fermentable carbon sources (e.g. alcohols), low sugar levels and repressed under high sugar concentrations [49]. It was hypothesised that FOXG_09625 would be poorly expressed under normal conditions and highly expressed in the presence of ethanol or butanol. Whilst this hypothesis was not observed under ethanol stress, results recorded for butanol were as expected. For yeast, the association of hexose transport and glycolysis associated genes with higher expression levels under 4% (vv^{-1}) ethanol stress led to the proposal that the cell enters a pseudo-starvation state during ethanol stress, whereby sugar-based nutrients are no longer accessible to the cell despite being present in the culture medium [50,51]. In this study, the highest FOXG_09625 transcript levels were recorded under butanol stress (0.5 hours) suggesting that cells may have immediately entered a pseudo-starvation state in response to the increased toxicity of butanol, possibly inhibiting sugar up-take. In yeast, increased expression of hexose transporters and glycolysis genes improve both rapid production and consumption of ethanol [52].

A previous study [53] identified two sugar transporters overexpressed in strain ^{11}C during CBP. In the analysis of additional transformants we would expect to identify an overlap with the SSH library for genes holding dual function i.e. relating to both production and tolerance. However, ethanol stress studies to date have identified genes belonging to diverse functional categories hence the identification of novel genes could also be anticipated. Interestingly, FOXG_09625 showed 29% protein homology to the SSH-identified hexose transporter and no homology to the high affinity glucose transporter indicating potentially different roles for all three genes in either tolerance or production. Yet, a recent study [54] has indicated FOXG_09625 is associated (directly/indirectly) with ethanol production since mutants overexpressing a hexose transporter resulted in increased ethanol yield coupled with compensatory changes in the expression of other transporters notably the up-regulation of FOXG_09625. Investigations, including overexpression/knockout studies of FOXG_09625, while not possible due to a limitation in resources for this study will be required to fully elucidate the complexity of this gene interaction with alcohol and determine the degree to which over-expression/silencing of FOXG_09625 could improve consolidated bioprocessing yield.

Filamentous fungal transporters appear to hold diverse functional roles notably in plant-fungi symbiosis [55-57] pathogenicity [58,59], and ethanol production and tolerance [48,50,60-62]. In regards to tolerance, transcriptional studies have identified alcohol responsive genes associated with sugar transport [50,63,64]. This is not surprising since the rate of sugar transport is limited by accumulation of alcohol in the fermentation broth, which can become a significant stress during fermentation resulting in a 'survival versus production' conundrum for the producing microbe [65]. While major advances have been made investigating alcohol tolerance in model systems (yeast, *E. coli*), a knowledge deficit exists for filamentous fungi [66-68]. A better understanding of the molecular basis of alcohol stress and tolerance in such fungi is important if enhanced tolerance and production is to be achieved. Previous work [54] coupled with this study point towards sugar transporters as key targets for improving fungal-enabled CBP.

In conclusion, the work completed in this study represents a first step in the investigation of alcohol tolerance in *F. oxysporum* via ATMT, which has led to the generation of a collection of transformants, which are now available to the research community for future studies. This is all the more relevant in light of the recent advances and application of RNA sequencing to elucidate complex biological processes at the transcript level in an unlimited array of organisms [69]. On the basis of the results presented here, ATMT can be exploited as a tool to generate diverse alcohol tolerant phenotypes in the CBP agent *F. oxysporum*.

REFERENCES

1. Brodeur G, Yau E, Badal K, Collier J, Ramachandran KB et al. (2011) Chemical and Physicochemical Pretreatment of Lignocellulosic Biomass: A Review. Enzyme Res 2011: 787532. PubMed: 21687609.
2. Bhatia L, Johri S, Ahmad R (2012) A economic and ecological perspective of ethanol production from renewable agro waste: a review. AMB Express 2: 65. doi:10.1186/2191-0855-2-65. PubMed: 23217124.
3. Elkins JG, Raman B, Keller M (2010) Engineered microbial systems for enhanced conversion of lignocellulosic biomass. Curr Opin Biotechnol 21(5): 657-662. doi:10.1016/j.copbio.2010.05.008. PubMed: 20579868.
4. La Grange DC, den Haan R, van Zyl WH (2010) Engineering cellulolytic ability into bioprocessing organisms. Appl Microbiol Biotechnol 87(4): 1195-1208. doi:10.1007/s00253-010-2660-x. PubMed: 20508932.
5. Sakuragi H, Kuroda K, Ueda M (2011) Molecular Breeding of Advanced Microorganisms for Biofuel Production. J Biomed Biotechnol.Volume 2011: 416931. PubMed: 21318120.

6. Amore A, Faraco V (2012) Potential of fungi as category I Consolidated BioProcessing organisms for cellulosic ethanol production. Renewable Sustain Energ Rev 16 (5): 3286-3301. doi:10.1016/j.rser.2012.02.050.
7. Olson DG, McBride JE, Shaw AJ, Lynd LR (2012). ecent Progress Consolidated Bioprocessing 12 (3): 396–405.
8. Lübbehüsen TL, Nielsen J, McIntyre M (2004) Aerobic and anaerobic ethanol production by Mucor circinelloides during submerged growth. Appl Microbiol Biotechnol 63: 543–548. doi:10.1007/s00253-003-1394-4. PubMed: 12879305.
9. Dashtban M, Schraft H, Qin W (2009) Fungal Bioconversion of Lignocellulosic Residues; Opportunities & Perspectives. Int J Bios Sci. 5(6): 578-595. PubMed: 19774110.
10. Fan Z, Wu W, Hildebrand A, Kasuga T, Zhang R et al. (2012) A Novel Biochemical Route for Fuels and Chemicals Production from Cellulosic Biomass. PLOS ONE 7 (2). doi: 10.1371/journal.pone.0031693
11. Lakshman DK, Pandey R, Kamo K, Bauchan G, Mitra A (2012) Genetic transformation of Fusarium oxysporum f.sp. gladioli with Agrobacterium to study pathogenesis in Gladiolus. Eur J Plant Pathol 133 (3): 729-738. doi:10.1007/s10658-012-9953-0.
12. Christakopoulos PK, Koullas DP, Kekos D, Koukios EG, Macris BJ (1991) Direct conversion of straw to ethanol by Fusarium oxysporum: effect of cellulose crystallinity. Enzyme Microb Technol. 13(3): 272-274. doi:10.1016/0141-0229(91)90141-V.
13. Ali SS, Khan M, Fagan B, Mullins E, Doohan FM (2012) Exploiting the inter-strain divergence of Fusarium oxysporum for microbial bioprocessing of lignocellulose to bioethanol. AMB Express. 2(1): 1-9. doi:10.1186/2191-0855-2-1. PubMed: 22214346.
14. Xiros C, Christakopoulos P (2009) Enhanced ethanol production from brewer's spent grain by a Fusarium oxysporum consolidated system. Biotechnol Biofuels 2(1): 4. doi:10.1186/1754-6834-2-4. PubMed: 19208239.
15. Hossain SM, Anantharaman N, Das M (2012) Bioethanol fermentation from untreated and pretreated lignocellulosic wheat straw using fungi Fusarium oxysporum. Indian J Chem Technol 19: 63-70. doi: 10.1080/00194506.2011.659540
16. Favaro L, Jooste T, Basaglia M, Rose SH, Saayman M et al. (2013) Designing industrial yeasts for the consolidated bioprocessing of starchy biomass to ethanol. Bioengineered 4(2): 97-10. doi:10.4161/bioe.22268. PubMed: 22989992.
17. Xiros C, Vafiadi M, Paschos T, Christakopoulos P (2011) Toxicity tolerance of Fusarium oxysporum towards inhibitory compounds formed during pretreatment of lignocellulosic materials. J Chem Technol Biotechnol 86(2): 223-230. doi:10.1002/jctb.2499.
18. Panagiotou G, Pachidou F, Petroutsos D, Olsson L, Christakopoulos P (2008) Fermentation characteristics of Fusarium oxysporum grown on acetate. Bioresour Technol 99(15): 7397-7401. doi:10.1016/j.biortech.2008.01.017. PubMed: 18304808.
19. Singh A, Kumar PKR (1991) Fusarium oxysporum: Status in Bioethanol Production. Crit Rev Biotechnol 11(2): 129-147. doi:10.3109/07388559109040619. PubMed: 1913845.
20. Brown SD, Guss AM, Karpinets TV, Parks JM, Smolin N et al. (2011) Mutant alcohol dehydrogenase leads to improved ethanol tolerance in Clostridium thermocellum. Proc Natl Acad Sci USA, 108: 2011. PubMed: 21825121.

21. Pfromm PH, Amanor-Boadu V, Nelson R, Vadlani P, Madl P (2010) Bio-butanol vs. bio-ethanol: A technical and economic assessment for corn and switchgrass fermented by yeast or Clostridium acetobutylicum. Biomass Bioenerg 34: 515-524. doi:10.1016/j.biombioe.2009.12.017.

22. Atsumi S, Cann AF, Connor MR, Shen CR, Smith KM et al. (2008) Metabolic engineering of Escherichia coli for 1-butanol production. Metab Eng 10(6): 305-311. doi:10.1016/j.ymben.2007.08.003. PubMed: 17942358.

23. Knoshaug E, Zhang M (2009) Butanol Tolerance in a Selection of Microorganisms. Appl Biochem Biotechnol. 153(1): 13-20. doi: 10.1007/s12010-008-8460-4

24. Rühl J, Schmid A, Blank LM (2009) Selected Pseudomonas putida Strains Able To Grow in the Presence of High Butanol Concentrations. Appl Environ Microbiol 75(13): 4653-4656. doi:10.1128/AEM.00225-09. PubMed: 19411419.

25. Dunlop MJ, Dossani ZY, Szmidt HL, Chu HC, Lee TS et al. (2011) Engineering microbial biofuel tolerance and export using efflux pumps. Mol Syst Biol 10(7): 487. PubMed: 21556065.

26. Zingaro KA, Papoutsakis ET (2012) Toward a Semisynthetic Stress Response System To Engineer Microbial Solvent Tolerance. mBio 3(5): ([MedlinePgn:]) PubMed: 23033472.

27. Jia K, Zhang Y, Li Y (2010) Systematic engineering of microorganisms to improve alcohol tolerance. Eng Life Sci 10(5): 422-429. doi:10.1002/elsc.201000076.

28. Maruthachalam K, Klosterman SJ, Kang S, Hayes RJ, Subbarao KV (2011) Identification of pathogenicity-related genes in the vascular wilt fungus Verticillium dahliae by Agrobacterium tumefaciens-mediated T-DNA insertional mutagenesis. Mol Biotechnol 49(3): 209-221. doi:10.1007/s12033-011-9392-8. PubMed: 21424547.

29. de Groot MJ, Bundock P, Hooykaas PJ, Beijersbergen AG (1998) Agrobacterium-mediated transformation of filamentous fungi. Nat Biotechnol 16: 839-842. doi:10.1038/nbt0998-839. PubMed: 9743116.

30. Michielse CB, Hooykaas PJJ, van den Hondel Camjj, Ram AFJ (2005) Agrobacterium -mediated transformation as a tool for functional genomics in fungi. Curr Genet 48(1): 1-17. doi:10.1007/s00294-005-0578-0. PubMed: 15889258.

31. Ma LJ, van der Does C, Borkovich KA, Coleman JJ, Daboussi MJ et al. (2010) Comparative genomics reveals mobile pathogenicity chromosomes in Fusarium. Nature 464(7287): 367-373. doi:10.1038/nature08850. PubMed: 20237561.

32. Brennan J, Leonard G, Cooke B, Doohan F (2005) Effect of temperature on head blight of wheat caused by Fusarium culmorum and F. graminearum. Plant Pathol. 54: 156-160. doi:10.1111/j.1365-3059.2005.01157.x.

33. Sambrook J, Russell DW (2011) Molecular Cloning: A Laboratory Manual. Cold Spring Harbor Laboratory Press.

34. Mullins ED, Chen X, Romaine P, Raina R, Geiser DM et al. (2001) Agrobacterium-Mediated Transformation of Fusarium oxysporum: An Efficient Tool for Insertional Mutagenesis and Gene Transfer. Phytopathology. 91(2): 173-180. doi:10.1094/PHYTO.2001.91.2.173. PubMed: 18944391.

35. Liu D, Coloe S, Baird R, John P (2000) Rapid Mini-Preparation of Fungal DNA for PCR. J Clin Microbiol 38(1): 471pp. PubMed: 10681211.

36. Edel V, Steinberg C, Gautheron N, Alabouvette C (2000) Ribosomal DNA-targeted oligonucleotide probe and PCR assay specific for Fusarium oxysporum. Mycol Res. 104: 518-526. doi:10.1017/S0953756299001896.
37. Mishra C, Keskar S, Rao M (1984) Production and Properties of Extracellular Endoxylanase from Neurospora crassa. Appl Environ Microbiol 48(1): 224-228. PubMed: 16346591.
38. Livak KJ, Schmittgen TD (2001) Analysis of relative gene expression data using real-time quantitative PCR and the 2-[Delta][Delta] CT method. Methods 25(4): 402-408. doi:10.1006/meth.2001.1262. PubMed: 11846609.
39. Zeuthen ML, Aniebo CM, Howard DH (1988) Ethanol Tolerance and the Induction of Stress Proteins by Ethanol in Candida albicans. J Gen Microbiol 134(5): 1375-1384. PubMed: 3058867.
40. Snowdon C, Schierholtz R, Poliszczuk P, Hughes S, van der Merwe G (2009) ETP1/YHL010c is a novel gene needed for the adaptation of Saccharomyces cerevisiae to ethanol. FEMS_Yeast_Res 9: 372–380. PubMed: 19416103.
41. Liu S, Qureshi N (2009) How microbes tolerate ethanol and butanol. New Biotechnol 26(3–4): 117-121. PubMed: 19577017.
42. Kataoka N, Tajima T, Kato J, Rachadech W, Vangnai AS (2011) Development of butanol-tolerant Bacillus subtilis strain GRSW2-B1 as a potential bioproduction host. AMB Expr 1(1): 10. doi:10.1186/2191-0855-1-10. PubMed: 21906347.
43. Bills GF, Platas G, Fillola A, Jiménez MR, Collado J et al. (2007) Enhancement of antibiotic and secondary metabolite detection from filamentous fungi by growth on nutritional arrays. J Appl Microbiol. 10(4): 1644-1658. PubMed: 18298532.
44. Betts MF, Tucker SL, Galadima N, Meng Y, Patel G et al. (2007) Development of a high throughput transformation system for insertional mutagenesis in Magnaporthe oryzae. Fungal Genet Biol 44(10): 1035-1049. doi:10.1016/j.fgb.2007.05.001. PubMed: 17600737.
45. Jeon J, Park SY, Chi MH, Choi J, Park J et al. (2007) Genome-wide functional analysis of pathogenicity genes in the rice blast fungus. Nat Genet 39(4): 561-565. doi:10.1038/ng2002. PubMed: 17353894.
46. Horák J (2013) Regulations of sugar transporters: insights from yeast. Curr Genet 59(1-2): 1-31. doi:10.1007/s00294-013-0388-8. PubMed: 23455612.
47. Wei H, Vienken K, Weber R, Bunting S, Requena N et al. (2003) A putative high affinity hexose transporter, hxtA, of Aspergillus nidulans is induced in vegetative hyphae upon starvation and in ascogenous hyphae during cleistothecium formation. Fungal Genet Biol 41: 148-156. doi: 10.1016/j.fgb.2003.10.006
48. Fernandes S, Murray P (2010) Metabolic engineering for improved microbial pentose fermentation. Bioeng Bugs. 1(6): 424–428. doi:10.4161/bbug.1.6.12724. PubMed: 21468211.
49. Greatrix BW, van Vuuren HJ (2006) Expression of the HXT13, HXT15 and HXT17 genes in Saccharomyces cerevisiae and stabilization of the HXT1 gene transcript by sugar-induced osmotic stress. Curr Genet 49(4): 205-217. doi:10.1007/s00294-005-0046-x. PubMed: 16397765.
50. Chandler M, Stanley GA, Rogers P, Chambers P (2004) A genomic approach to defining the ethanol stress response in the yeast Saccharomyces cerevisiae. Ann Microbiol 54(4): 427-454.

51. Stanley D, Bandara A, Fraser S, Chambers PJ, Stanley GA (2010) The ethanol stress response and ethanol tolerance of Saccharomyces cerevisiae. J Appl Microbiol 109(1): 13-24. PubMed: 20070446.

52. Anderson MJ, Barker SL, Boone C, Measday V (2012) Identification of RCN1 and RSA3 as ethanol-tolerant genes in Saccharomyces cerevisiae using a high copy barcoded library. FEMS Yeast Res 12(1): 48-60. doi:10.1111/j.1567-1364.2011.00762.x. PubMed: 22093065.

53. Ali SS, Khan M, Mullins E, Doohan F (2013) Identification of Fusarium oxysporum Genes Associated with Lignocellulose Bioconversion Competency. Bioenergy Research.

54. Ali SS, Nugent B, Mullins E, Doohan FM (2013) Insights from the Fungus Fusarium oxysporum Point to High Affinity Glucose Transporters as Targets for Enhancing Ethanol Production from Lignocellulose. PLOS ONE 8(1): e54701. PubMed: 23382943.

55. López MF, Dietz S, Grunze N, Bloschies J, Weiß M et al. (2008) The sugar porter gene family of Laccaria bicolor: function in ectomycorrhizal symbiosis and soil-growing hyphae. New Phytol 180: 365–378. doi:10.1111/j.1469-8137.2008.02539.x. PubMed: 18627493.

56. Eckhardt N (2011) A Symbiotic Sugar Transporter in the Arbuscular Mycorrhizal Fungus Glomus sp. Plant Cell 23(10): 3561. doi:10.1105/tpc.111.231010. PubMed: 22003078.

57. Helber N, Wippel K, Sauer N, Schaarschmidt S, Hause B et al. (2011) A Versatile Monosaccharide Transporter That Operates in the Arbuscular Mycorrhizal Fungus Glomus sp Is Crucial for the Symbiotic Relationship with Plants. Plant Cell 23: 3812-3823. doi:10.1105/tpc.111.089813. PubMed: 21972259.

58. Talbot NJ (2010) Living the Sweet Life: How Does a Plant Pathogenic Fungus Acquire Sugar from Plants? PLOS Biol 8(2): e1000308. PubMed: 20161721.

59. Wahl R, Wippel K, Goos S, Kamper J, Sauer N (2010) A Novel High-Affinity Sucrose Transporter Is Required for Virulence of the Plant Pathogen Ustilago maydis. PLOS Biol 8(2). doi: 10.1371/journal.pbio.1000303

60. Olofsson K, Bertilsson M, Lidén G (2008) A short review on SSF – an interesting process option for ethanol production from lignocellulosic feedstocks. Biotechnol Biofuels 1: 7. doi:10.1186/1754-6834-1-7. PubMed: 18471273.

61. Patrascu E, Rapeanu G, Hopulele T (2009) Current approaches to efficient biotechnological production of ethanol. Innovative Romanian. Food Biotechnol Vol. 4.

62. Xu Q, Arjun Singh A, Himmel ME (2009) Perspectives and new directions for the production of bioethanol using consolidated bioprocessing of lignocellulose. Curr Opin Biotechnol 20: 364-371. doi:10.1016/j.copbio.2009.05.006. PubMed: 19520566.

63. Alexandre H, Ansanay-Galeote V, Dequin S, Blondin B (2001) Global gene expression during short-term ethanol stress in Saccharomyces cerevisiae. FEBS Lett 498(1): 98-103. doi:10.1016/S0014-5793(01)02503-0. PubMed: 11389906.

64. Reyes LH, Almario MP, Kao KC (2011) Genomic Library Screens for Genes Involved in n-Butanol Tolerance in Escherichia coli. PLOS ONE 6(3): e17678. PubMed: 21408113.

65. Yarris L (2011) Striking the Right Balance: JBEI Researchers Counteract Biofuel Toxicity in Microbes (Press Release). Available: http://newscenter.lbl.gov/news-releases/2011/05/11/efflux-pumps-for-biofuels/.
66. Takó M, Farkas E, Lung S, Krisch J, Vágvölgyi C et al. (2010) Identification of acid- and thermotolerant extracellular β-glucosidase activities in Zygomycetes fungi. Acta Biol Hung 61(1): 101-110. doi:10.1556/ABiol.61.2010.1.10. PubMed: 20194103.
67. Asiimwe T, Krause K, Schlunk I, Kothe E (2012) Modulation of ethanol stress tolerance by aldehyde dehydrogenase in the mycorrhizal fungus Tricholoma vaccinum. Mycorrhiza 22(6): 471-484. doi:10.1007/s00572-011-0424-9. PubMed: 22159964.
68. Vinche HM, Asachi R, Zamani A, Karimi K (2012) Ethanol and chitosan production from wheat hydrolysate by Mucor hiemalis. J Chem Technol Biotechnol, 87: 1222-1228. doi:10.1002/jctb.3822. PubMed: 23329859.
69. Ozsolak F, Milos PM (2011) RNA sequencing: advances, challenges and opportunities. Nat Rev Genet 12(2): 87-98. doi:10.1038/nrg2934. PubMed: 21191423.

There are several supplemental files that are not available in this version of the article. To view this additional information, please use the citation on the first page of this chapter.

CHAPTER 14

GENOME REPLICATION ENGINEERING ASSISTED CONTINUOUS EVOLUTION (GREACE) TO IMPROVE MICROBIAL TOLERANCE FOR BIOFUELS PRODUCTION

GUODONG LUAN, ZHEN CAI, YIN LI, AND YANHE MA

14.1 BACKGROUND

Efficient microbial production of biofuels from renewable resources requires robust cell growth and stable metabolism under tough industrial conditions, represented by inhibitory components in substrates and toxic products [1,2]. Microbial tolerance to these inhibitory environmental factors is a complex phenotype usually controlled by multiple genes [3,4], and thus is difficult to be engineered by targeted metabolic engineering approaches [5]. Instead, such complex phenotypes can be more effectively improved by evolutionary engineering approaches [6]. Examples of evolutionary engineering include successive passage for metabolic evolution [7-9], physical and chemical mutagenesis [10], global transcription

Genome Replication Engineering Assisted Continuous Evolution (GREACE) to Improve Microbial Tolerance for Biofuels Production. © *Luan G, Cai Z, Li Y, and Ma Y.; licensee BioMed Central Ltd Current Status and Prospects of Biodiesel Production from Microalgae.* Biotechnology for Biofuels *6,137 (2013), doi:10.1186/1754-6834-6-137. Licensed under a Creative Commons Attribution 2.0 Generic License, http://creativecommons.org/licenses/by/2.0/.*

machinery engineering [2,11], artificial transcription factors engineering [12,13], and ribosome engineering [14]. All these methods use "Mutagenesis followed-by Selection" as core principle, meaning that firstly introducing genetic diversity by spontaneous mutations, exogenous mutagens, or genetic perturbations, followed by selection of desired phenotypes [6]. Using such methods, iterative rounds of mutagenesis-selection and frequent manual interventions are often required, resulting in discontinuous and inefficient strain improvements, as shown in Figure 1A.

To address the discontinuity of the existing evolutionary engineering approaches and improve the engineering efficiency, we devised a novel method termed as "Genome Replication Engineering Assisted Continuous Evolution, GREACE", which uses "Mutagenesis coupled-with Selection" (Figure 1B) as its core principle. The key element of this novel method is to introduce in vivo continuous mutagenesis mechanisms into microbial cells that are subsequently subjected to continuous selective conditions. Mutagenesis and selection can therefore be coupled to minimize manual interventions, thus providing possibilities to develop a continuous and efficient phenotypes-improving process (Figure 1B). Practically, in vivo continuous mutagenesis can be achieved by introducing genetic perturbations into genome replication machinery so as to trigger inaccurate genome replications. Hypermutable cells with significant genome diversities can therefore be obtained. Offspring cells with mutated genomes that survived the increased selective pressures will be selected during the continuous enrichment processes, and the genomic feature of evolved cells with improved phenotypes can be stably maintained once the elements triggering the inaccurate genome replication are removed from the individually isolated cells.

In this work, we proved the concept of GREACE and tested the efficiency of GREACE using *Escherichia coli* as a model. Genetically modified proofreading elements of the DNA polymerase complex (ϵ subunit encoded by dnaQ gene) were used to trigger perturbations on genome replication for in vivo continuous mutagenesis. We firstly proved the feasibility and intrinsic mechanisms of GREACE by improving kanamycin resistance of *E. coli*. Secondly we show that n-butanol and acetate tolerances of *E. coli*, two important microbial tolerance characteristics, can also be efficiently improved using this method. *E. coli* mutants obtained through the GREACE process showed significantly improved tolerances to n-butanol and acetate, demonstrating the potential of GREACE as an effective and universal approach to improve tough physiological traits required by biofuels production. Furthermore, we discovered a phenomenon that adapting to specific environments calls for specific genetically modified proofreading elements, which may provide insights into understanding and application of evolutionary engineering for strain improvement.

14.2 RESULTS

14.2.1 PROCEDURE OF THE "GENOME REPLICATION ENGINEERING ASSISTED CONTINUOUS EVOLUTION"

The core idea of GREACE is to introduce genetic perturbations into genome replication machinery, so that in vivo continuous mutagenesis can be coupled with simultaneous phenotypes selection. A typical flowchart of GREACE is described in Figure 2. To achieve an inaccurate genome replication, genetic perturbations on genome replication machinery are generated by constructing a mutant library of a proofreading element (PE) gene, designated as PEM-lib. Subsequently, the PEM-lib is transformed into the wildtype strain. Inaccurate genome replication will be triggered in cells containing PEs with decreased proofreading activities, thus continuously generating offspring cells containing different genomic mutations. Notably, the diversity of the PEM-lib will help to generate cells with various genome replication mutation rates and mutation type preferences, ensuring the highest genomic diversity in offspring cells. When cells containing a genetically modified PEs are cultivated under gradually increased selective pressures, only the offsprings with accumulated adaptive mutations can survive and thus to be finally selected. Most importantly, the genome replication will return to regular state once the genetically modified PE is eliminated from the selected cells, so that the genomic features and the evolved phenotypes can be stably maintained.

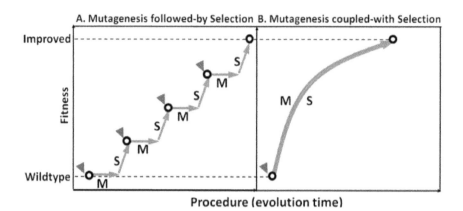

FIGURE 1: Comparisons between the principle of "Mutagenesis followed-by Selection" and the principle of "Mutagenesis coupled-with Selection". (A) Traditional "Mutagenesis followed-by Selection" principle was usually performed by iterative rounds of mutagenesis and selection. Exogenous mutagens or genetic manipulations were required for mutagenesis and the following selection manipulations isolated cells with improved phenotypes, which could be used in next rounds of "mutagenesis-selection". (B) As for "Mutagenesis coupled-with Selection" principle, the two steps are synchronized, so that iterative and lengthy manual interventions are greatly simplified, leading to a continuous and efficient strain improvement process. "S" represents for "selection", "M" represents for "mutagenesis", and red triangles represent for manual interventions such as mutagen treatments or genetic manipulations or selections for improvement phenotypes.

14.2.2 PROOF OF CONCEPT FOR GREACE BY ENGINEERING KANAMYCIN RESISTANCE OF E. COLI

To prove the concept of GREACE, we took *E. coli* as a model strain and selected kanamycin resistance as a phenotype for testing. Among the various cellular proofreading elements including the genes responsible for base selection, exonucleolytic editing, MMR (Methyl-directed Mismatch Repair), and DNA repair [15-18], we chose the dnaQ gene which encodes the ϵ subunit of *E. coli* DNA polymerase III. This subunit is the only one with 3'->5' exonuclease activity in DNA polymerase III, the major DNA polymerase of *E. coli* genome replication, and it was reported to guarantee both the polymerization and error-editing process of DNA replication. Moreover, the strongest mutator phenotypes that have been known were

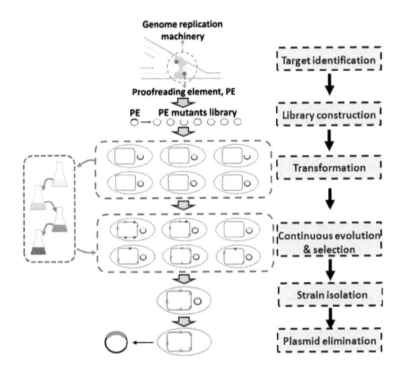

FIGURE 2: Flowchart of the Genome Replication Engineering Assisted Continuous Evolution (GREACE) method. Proofreading element (PE) of the genome replication machinery was selected to generate a mutant library (PEM-lib). For phenotype-improvement, PEM-lib was transformed into host cells, and cultivated in conditions with gradually increased selective strengths (media in flasks with colours from light blue to dark blue). Evolved strains with the most beneficial and adaptive mutation accumulations (red sparks represented for beneficial mutations and black sparks for detrimental mutations) would show the best adaption advantages and dominate the PEM-lib populations. Genetically modified PE mutant would be eliminated from the evolved strains to stabilize the obtained genotypes and phenotypes.

resulted from deficiencies of the dnaQ gene [17,19,20]. Thus, dnaQ was selected as the proofreading element to be engineered.

A mutant library of dnaQ gene (pQ-lib) with the size of 10^6 and average mutation rates of 2–3 amino acids per gene product, was constructed by error-prone PCR. pQ-lib, together with the control vectors pQ-dnaQ (carrying wildtype dnaQ), and the empty vector pUC18, was transformed into

E. coli cells, respectively. All cultures were serially transferred in media containing serial concentrations of kanamycin (15 µg/ml, 100 µg/ml, 200 µg/ml, and 300 µg/ml; X µg/ml kanamycin will be designated as KanX). As shown in Figure 3A, no obvious growth differences were observed among three groups grown in Kan15. Upon increasing the kanamycin concentrations, growth and adaptation advantage of the pQ-lib group became obvious. Finally, the pQ-lib group was able to grow in Kan300, while the pUC18 and pQ-dnaQ groups failed to grow in Kan200 and Kan100, respectively, suggesting the pQ-lib group had a stronger adaptability.

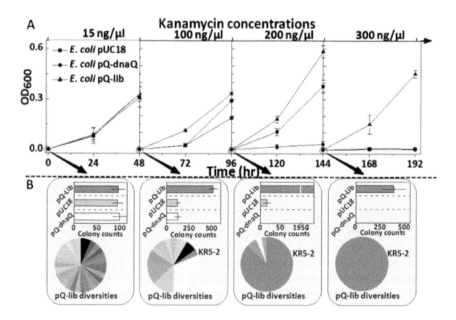

FIGURE 3: Engineering kanamycin resistance of *E. coli* with GREACE. (A)*E. coli* cells transformed with pQ-lib, pUC18, and pQ-dnaQ were pre-cultivated in LB medium and then 1:100 serial diluted in gradually increased kanamycin concentrations, from Kan15 to Kan300. Error bars presented the standard deviation of growth analysis for 3 independent evolution experiments. (B) Cells carrying the above three plasmids were sampled at the end of the kanamycin resistance evolution process. Appropriate 108 cells were spread on plates with kanamycin concentration of the next level, and cultivated for colony counting. Pie charts represent the proportions of dnaQ mutants in 30 randomly selected colonies of *E. coli* cells carrying the pQ-lib at the end of cultivations.

To understand why pQ-lib group showed such adaptation advantages, cultures containing 10^8 cells were collected from pQ-lib, pUC18, and pQ-dnaQ carrying strains grown at each kanamycin concentration and spread onto plates containing higher kanamycin concentrations (cells from Kan0, Kan15, Kan100, and Kan200 liquid culture were spread on plates containing Kan15, Kan100, Kan200, and Kan300, respectively). As shown in Figure 3B, growths of the three groups in Kan0 did not make difference on generation of Kan15 resistant colonies. Growth of the pQ-lib group in Kan15 liquid culture generated more Kan100 resistant colonies than that of the controls, similar results were observed for pQ-lib group grown in Kan100 and Kan200. This suggested that pQ-lib group was able to generate kanamycin-resistant cells more rapidly.

To investigate the dynamic changes of the diversity of pQ-lib during the evolution process, 30 colonies of the pQ-lib group were randomly picked at the end of cultivations at each kanamycin concentration. Plasmids were extracted from these colonies and the dnaQ mutants therein were sequenced. As shown in the pie charts of Figure 3B, the diversity of pQ-lib decreased dramatically, and one mutant termed as dnaQ KR5-2 was enriched, from undetectable level in Kan0 to 100% in Kan200.

To verify whether the observed adaptation advantages were endowed by dnaQ KR5-2, plasmid pQ-dnaQ-KR5-2 was retransformed into fresh *E. coli* cells and cultivated under kanamycin stress. As expected, *E. coli* cells carrying dnaQ KR5-2 indeed exhibited increased evolution speed and adaptability. In Kan15 and Kan100, the dnaQ KR5-2 carrying strain grew to OD_{600} of 1.02 (±0.07) and 0.28 (±0.05) after cultivation for 48 hours, which were 3–4 fold higher than those of the two control strains carrying pQ-dnaQ and pUC18 (Additional file 1: Figure S1). These results confirmed that adaptation advantages of dnaQ KR5-2 under stressful conditions.

To quantitatively evaluate and compare the rates of generating genomic DNA mutations (termed as mutation rates for short) by different dnaQ mutants, we chose the widely used rifamycin resistant colony generation frequency as an indirect indicator [21-23]. Mutation rates determination based on this indicator showed that *E. coli* cells carrying pQ-dnaQ-KR5-2 exhibited a 317-fold increased mutation rate than that of the wildtype control, confirming that introducing genetic perturbation into genome replication machinery indeed triggered inaccurate genome replication.

According to the design principle of GREACE, phenotypes obtained by GREACE should be stable and heritable. To confirm the stability of kanamycin resistance, plasmid pQ-dnaQ-KR5-2 was eliminated from the kanamycin resistant strain *E. coli* KR1. We found that the kanamycin resistance could be well maintained after plasmid elimination (Additional file 1: Figure S2).

14.2.3 GREACE CAN BE SUCCESSFULLY APPLIED TO ENGINEER N-BUTANOL AND ACETATE TOLERANCES

Recently, *E. coli* has been proved to be a promising host for biobutanol production [24-26]. However, n-butanol tolerance of *E. coli* is a bottleneck hampering further increase of n-butanol titer, as n-butanol is highly toxic to microbial cells. Many efforts have been made to improve and analyze butanol tolerance of *E. coli*[13,27-30]. We then wanted to test if GREACE can be used to improve n-butanol tolerance of *E. coli*, a tougher physiological trait comparing with kanamycin resistance. We applied a similar procedure to engineering n-butanol tolerance—*E. coli* cells carrying pQ-lib were serially transferred in media containing gradually increased n-butanol concentrations. Considering that improvement of complex traits such as n-butanol tolerance might require accumulation of much more adaptive mutations [31,32], the pQ-lib carrying cells were transferred in each n-butanol concentrations for three times, upon full growth, before transferring to the next higher butanol concentration.

Figure 4A showed that pQ-lib carrying cells exhibited a better adaptability to the increased n-butanol concentrations. After continuous cultivations and selections, pQ-lib group could grow in medium containing n-butanol concentration of 1.25% (vol/vol, the same below), while n-butanol concentrations of 0.875% and 0.75% were lethal to the *E. coli* cells carrying pUC18 and pQ-dnaQ, respectively. The dnaQ mutant dominating the n-butanol tolerance evolution process was isolated and termed as dnaQ BR1. Further analysis revealed that dnaQ BR1 endowed the host with a 2839-fold increased mutation rate, much higher than that of dnaQ KR5-2 selected under kanamycin stress.

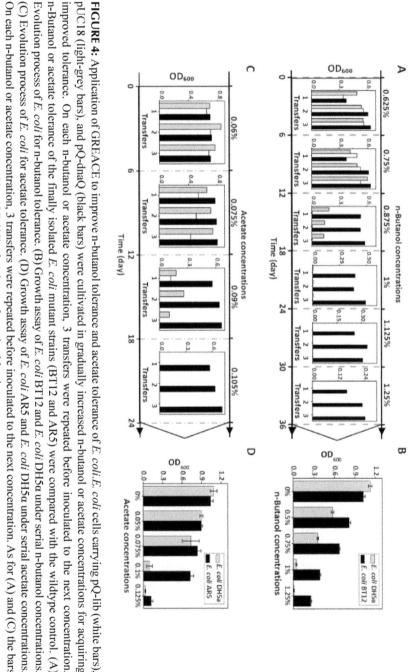

FIGURE 4: Application of GREACE to improve n-butanol tolerance and acetate tolerance of *E. coli*. *E. coli* cells carrying pQ-lib (white bars), pUC18 (light-grey bars), and pQ-dnaQ (black bars) were cultivated in gradually increased n-butanol or acetate concentrations for acquiring improved tolerance. On each n-butanol or acetate concentration, 3 transfers were repeated before inoculated to the next concentration. n-Butanol or acetate tolerance of the finally isolated *E. coli* mutant strains (BT12 and AR5) were compared with the wildtype control. (A) Evolution process of *E. coli* for n-butanol tolerance. (B) Growth assay of *E. coli* BT12 and *E. coli* DH5α under serial n-butanol concentrations. (C) Evolution process of *E. coli* for acetate tolerance. (D) Growth assay of *E. coli* AR5 and *E. coli* DH5α under serial acetate concentrations. On each n-butanol or acetate concentration, 3 transfers were repeated before inoculated to the next concentration. As for (A) and (C) the bars represented for the final optical densities (OD600) of each transfer and cultivation.

After plasmid elimination, an *E. coli* strain, designated as BT12, was obtained. Growth assay revealed that *E. coli* BT12 performed growth advantages upon n-butanol challenge (Figure 4B). At an n-butanol concentration of 0.75%, growth of *E. coli* BT12 reached a doubled OD_{600} value after 48 hours cultivation compared with the wild type strain. When n-butanol concentration reached 1.25%, no growth was detected for the wildtype, while *E. coli* BT12 was still able to grow (25% OD_{600} value of that under no n-butanol stress). n-Butanol shock experiments revealed that *E. coli* BT12 showed a higher cellular stability and tolerance under an extreme lethal n-butanol concentration. After exposure to 2% n-butanol for 1 hour, the survival rate of *E. coli* BT12 was over 100-fold higher than that of the wildtype cells (Additional file 1: Figure S3). We further tested the stability of the n-butanol tolerance of *E. coli* BT12. After frozen at −80°C for 2 weeks and serially transferred for 30 passages, the n-butanol tolerance of *E. coli* BT12 could be well maintained, indicating that good genetic stability and traits heritability can be retained upon GREACE (Additional file 1: Figure S4).

Acetate tolerance is another important trait for biofuels production. Concentrated acetate in hemicelluloses hydrolysates severely inhibited growth and metabolism of microbial cells, thus restricting conversion and production efficiency [33-35]. GREACE was also successfully applied to improve *E. coli* tolerance to acetate. As shown in Figure 4C, the inhibitory effects of acetate on the evolved cells were sharply reduced after transferring *E. coli* cells carrying pQ-lib in gradually increased acetate concentrations for 24 days. pQ-lib carrying cells successfully adapted to an acetate concentration of 0.105% (vol/vol, the same below), while the control could hardly grow in a lower concentration of 0.09%. After plasmid elimination, an *E. coli* mutant AR5 was obtained. AR5 cells showed an 8-fold increased OD_{600} as compared to that of the wildtype strain when grown in the presence of 0.1% acetate (Figure 4D). The dnaQ mutant dominating the acetate tolerance evolution process was isolated and termed as AR6-1, endowing host *E. coli* cells with an 87-fold increased mutation rate.

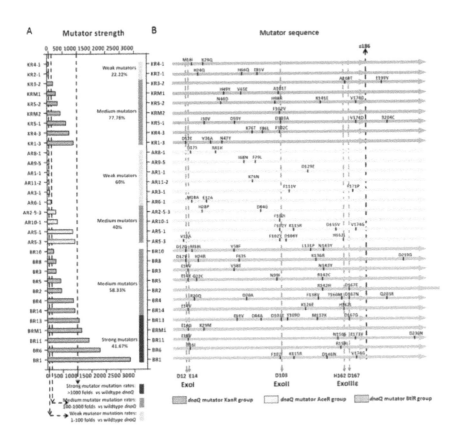

FIGURE 5: Strengths and sequences analysis of the dnaQ mutators selected from kanamycin stress (KanR group), n-butanol stress (BtlR group), and acetate stress (AceR group). (A) Mutator strengths of the dnaQ mutators relative to the wildtype gene, based on the frequency calculation of rifamycin resistant colony generation. For better understanding and comparisons, weak mutators, medium mutators, and strong mutators are distinguished with light blue, marine blue, and dark blue. (B) Summary of the amino acids substitutions on sequences of all the dnaQ mutator genes isolated. The full length of the ε subunit sequence is divided into two parts by the black line, pointing to the ε186. Three red line marked the three conserved and essential regions for the 3'->5' exonuclease activities. Amino acids substitutions are labeled with black bars and characters.

14.2.4 CHARACTERIZATION OF DNAQ MUTANTS SELECTED FROM DIFFERENT CONDITIONS

Three different dnaQ mutants (KR5-2, BR1, and AR6-1), which endowed the host with different mutation rates, were enriched during the evolution process against increased concentrations of kanamycin, n-butanol, and acetate. To investigate whether these enriched dnaQ mutants were the optimal that could always be enriched in independent evolution processes, we performed parallel experiments for acquiring kanamycin resistance, n-butanol tolerance, and acetate tolerance. The finally enriched dnaQ mutants from each process were isolated, sequenced, and analyzed. In the parallel experiments for acquiring kanamycin resistance, nine different dnaQ mutants were isolated from 10 independent evolution processes. This suggests that the evolution process for acquiring kanamycin resistance is not dependent on one preferable dnaQ mutant. Similar results were also observed in parallel processes for acquiring n-butanol tolerance (12 parallel experiments) or acetate tolerance (10 parallel experiments). All dnaQ mutants isolated were different. Sequences of the dnaQ mutants isolated were shown in Figure 5B. As expected, all of the selected mutants endowed the host cells with elevated mutation rates, ranging from 7.7 folds (dnaQ KR4-1) to 2839 folds (dnaQ BR1) (Figure 5A), which indicated the genetically modified dnaQ mutants enriched in GREACE could serve as endogenous mutators to accelerate evolution.

We determined the mutation rates of all 31 dnaQ mutants isolated from parallel experiments, and grouped them into three types based on the increment of the mutation rates over that of the wildtype dnaQ gene (termed as mutator strength for short): weak mutators (1–100 folds), medium mutators (100–1000 folds), and strong mutators (>1000 folds). As shown in Figure 5A, n-butanol stress tended to select strong mutators. A large proportion (5/12) elevated mutation rates of host cells by over 1000 folds, while no weak mutators were found. Acetate stress tended to select weak mutators, as most (6/10) of them falling into the weak group. Kanamycin stress tends to select dnaQ mutants with medium mutation rates, as the majority (7/9) showed medium strengths. This suggested that acquiring desired physiological traits required assistance of mutators with specific mutator strength. Calculation of average strengths of the mutators obtained in acquiring n-butanol tolerance (BtlR group, 959.8-fold increase), kanamycin resistance (KanR group, 375.3-fold increase), and acetate tolerance (AceR group, 249.4-fold increase) also supported this indication. Sequencing analysis revealed a positive relationship between the average mutator strengths and average amino acids substitution numbers in each dnaQ mutant. Among the BtlR group, an average of 3.9 amino acids substitutions were found in each mutant, while the substitutions for KanR and AceR group were 2.9 and 2.1, respectively.

Analysis of the locations of the amino acids substitutions promoted understanding of the decreased proofreading activities of the selected dnaQ mutants. Many efforts have been made to disclose the relationship between sequence and structure characteristics of the ε subunits encoded by dnaQ gene [16,36]. The 3' -> 5' exonuclease activities of the ε subunit was mainly determined by the N-terminal 186 amino acids (designated as ε186) [37]. In this study, most of the mutation (88 out of 94) from all the 31 dnaQ mutants existed within the ε186 region (Figure 5B), demonstrating that ε186 played key roles in proofreading. Three conserved regions essential for proofreading, ExoI, ExoII, and ExoIIIε [38,39], have been recognized in ε186. As expected, a large portion of amino acid substitutions of the selected dnaQ mutants were located in or quite close to these three regions, meaning high possibilities to disturb the natural proofreading function of the respective gene products.

14.3 DISCUSSION

The accurate genetic information transfer guarantees the genetic and phenotypic stability of organism, and it requires complex and precise mechanisms on multiple stages [18]. However, these hierarchal and precise mechanisms turn to be serious barriers for engineering complex phenotypes. To overcome such barriers, we developed GREACE, which implements a novel principle "Mutagenesis coupled-with Selection", and thus enables microbial evolution under stressful conditions in a continuous and efficient way.

As for microbes, expanding offspring genome diversities by elevating replication mutation rates to achieve rapid adaptation and competitive advantages in harsh environments has been discovered and verified both in vivo[40-42] and in silico[43,44]. Various natural mutator genes have been isolated and analyzed. However, some of them failed to work under some harsh conditions, e.g. high temperatures [45,46]. The GREACE method implants a pool of genetically modified proofreading elements into host cells to act as "evolution accelerator", thus provides multiple mutators with diverse characteristics which might guarantee accelerated evolution of host cells under diverse conditions.

Accelerating microbial evolution by mutators with elevated mutation rates during genome replication have been reported previously [47-49]. All these studies used a single specific mutator with a certain mutation rate. However, out work provided the first evidence that adaptation to specific conditions prefers accumulation of specific mutators, especially with specific mutator strengths. This finding was partially supported by Loh et.al [50], who found a narrow range of mutation rates (10–47 folds increase) dominated a designed laboratory survival competition. It has also been predicted by mathematical models that microbial survival and adaptation to different selective pressures might require different amounts and types of adaptive mutations, thus different mutagenesis strengths might be preferred [44]. These results benefit understanding and application of evolutionary engineering strategies for strain improvement, and further support the necessity for using GREACE method, which generated a library of mutators. Characterization of the dnaQ mutants selected from GREACE under different conditions provided interesting clues to understand the intrinsic mechanisms of this novel approach. A universal dnaQ mutant that can effectively improve all physiological traits might not exist, while a group of dnaQ mutants may be helpful, as the most preferable dnaQ mutants will be enriched during the evolution process.

In comparison with other phenotype-improving approaches generating global disturbances by introduction and maintaining of exogenous plasmids [2,11,12], GREACE endows the microbes with stable and heritable phenotypes through scarless manipulations. Genotype stability and phenotype hereditability of the mutants can be maintained at any stage of evolution when the genetically modified proofreading elements are removed from the evolved mutants, and that provides great convenience for further genetic manipulation and application of mutants, e.g. introduction of the metabolic pathway for products synthesis.

Evolutionary engineering has been widely applied to optimize biofuels-producing related characteristics, especially for improvement of substrates utilization and inhibitors tolerant capacities [51-53]. GREACE provides a powerful new tool for evolutionary engineering, as the continuous and exhaustless genetic diversities generated by GREACE can provide nearly every possible solution while the synchronous selection will direct the most advantageous ones. Besides the cellular tolerance phenotypes engineered in this work, GREACE can also be expected to be applied for improving the metabolic capacities of biofuels in combination with the newly arising evolutionary metabolic engineering approach, which establishes the linkage between products synthesis and cell growth [54-56]. In addition, diverse improved phenotypes in mutants evolved by GREACE could be efficiently integrated in a single strain with multiple improved traits by methods like genome shuffling [1,57]. Hence, GREACE can be considered as a promising strain improving approach with wide application potentials.

14.4 CONCLUSIONS

A novel method termed as "Genome Replication Engineering Assisted Continuous Evolution" (GREACE), using "Mutagenesis coupled-with Selection" as core principle was developed to improve microbial tolerance for biofuels production. The GREACE method introduced an in vivo mutagenesis mechanism into microbial cells by introducing a group of genetically modified proofreading elements of the DNA polymerase (ϵ subunit encoded by dnaQ gene) to accelerate the evolution process under stressful conditions. The genotype stability and phenotype heritability can be stably maintained once the genetically modified proofreading element is removed, thus scarless mutants with desired phenotypes can be obtained.

GREACE was successfully applied to engineer n-butanol and acetate tolerances, two important physiological characteristics for biofuels production, demonstrating potentials of the GREACE method for strain improvement in this area. Furthermore, we discovered that adaptation of microbes to specific stresses prefers specific mutagenesis strengths, which may provide new insights on understanding and application of evolutionary engineering for strain improvement.

14.5 METHODS

14.5.1 STRAINS AND CULTURE CONDITIONS

E. coli DH5α (TAKARA) was used for plasmids construction, phenotype evolution, and mutation rates evaluation. *E. coli* cells were grown aerobically in Luria-Bertani medium at 37°C, unless there are special instructions. Antibiotics and kanamycin, n-butanol, and acetate were supplemented as required.

14.5.2 PLASMID AND LIBRARY CONSTRUCTION PROCEDURES

The native dnaQ gene and promoter fragment were both amplified from the *E. coli* DH5α genomic DNA with Phusion DNA polymerase, using primer pairs dnaQ-F/dnaQ-R, and dnaQProm-F/dnaQProm-R, respectively (Additional file 1: Table S1). The dnaQ fragment was cloned into the EcoRI and HindIII sites in pUC18 and the dnaQ promoter sequence was used to replace the lac promoter between the BspQI and EcoRI sites. The final plasmid was named as pQ-dnaQ.

Error-prone PCR was employed to construct a dnaQ mutant library, with primer pairs of dnaQ-EP-F/dnaQ-EP-R (Additional file 1: Table S1). A standard error-prone PCR protocol was taken, and the mutation rate was controlled by adjusting concentrations of manganese and magnesium ions. The products was purified, digested and inserted into EcoRI and NdeI sites of the pQ-dnaQ plasmid to replace the wild type dnaQ gene. The ligation system was transformed into *E. coli* DH5α cells. After cultivation on agar plates, about 106 transformants were obtained and scrapped off, and then the plasmids were extracted to generate a library named as pQ-lib. All enzymes used for DNA manipulation are from NEB.

14.5.3 EVOLUTION AND PHENOTYPE SELECTION

E. coli DH5α cells that were respectively transformed with pQ-lib, pQ-dnaQ, and pUC18 were transferred into fresh LB medium containing 100 μg/ml ampicillin (Amp100 for short) and grown overnight. Then the broth was serially transferred under gradually increased stress conditions. All of the evolution experiments in this part were performed with 10 ml LB media supplemented with Amp100 and serial concentrations of kanamycin, n-butanol or acetate, and cultivated at 37°C, unless there are special instructions. To evaluate cell densities, OD600 was monitored by microplate reader at 600 nm with a sample volume of 200 μl.

(i) Kanamycin resistance: 300 μl of overnight culture broth of the transformants was inoculated into 10 ml LB medium containing Amp100 and Kan15, and cultivated for 2 days. Then 300 μl of the culture broth (with an OD600 value adjusted to 0.2 by fresh LB medium) was transferred to LB medium containing Amp100 and Kan100, and cultivated for 2 days. The transfer and selection processes were repeated to LB media containing Kan200 and Kan300.

On each selection level, about 108 cells from the broth were spread on LB agar plates containing Amp100 and kanamycin concentrations of the next selection level, and cultivated for colony counting. For example, cells from Kan100 broth would be spread on plates containing Kan200.

On each selection level, the culture broth was spread on LB agar plates containing Amp100 and incubated overnight. 30 single colonies were selected randomly and the dnaQ mutants carried were isolated and sequenced.

(ii) n-Butanol tolerance: evolution process for n-butanol tolerance of *E. coli* was similar to that described for kanamycin resistance. After a pre-cultivation in 0.5% of n-butanol, culture broth was stepwise transferred to LB media containing Amp100 and n-butanol with gradually increased concentrations from 0.625% to 1.25%. Inoculation volume and cell densities for the three groups were adjusted to the same. On each n-butanol concentration, three transfers were performed before inoculation into a higher level. Each cultivation was performed for 48 hours. Tubes used in n-butanol tolerance evolution were sealed off with parafilm to avoid n-butanol evaporation.

(iii) Acetate tolerance: evolution process for acetate tolerance of *E. coli* was similar that described for n-butanol tolerance. Cultivations were performed in serial acetate concentrations of 0.06%, 0.075%, 0.09%, and 0.105%.

14.5.4 PLASMID ELIMINATION

The plasmid in the finally evolved strain was eliminated by serial transfer in LB medium without ampicillin but supplemented with the corresponding selective pressure (e.g., high concentrations of kanamycin or butanol or acetate). An appropriate amount of culture was spread on the agar plate with the same concentration of selective pressure before each transfer for colony visualization. Thirty single colonies were streaked on agar plate with ampicillin and those with ampicillin sensitivity were regard as the plasmid-cured ones. Typically 5–10 transfers are sufficient to isolate plasmid-cured strain maintaining the evolved phenotypes.

14.5.5 GROWTH ASSAY

To evaluate tolerance of the GREACE generated cells, growth of the finally isolated mutant strains were analyzed under serial concentrations of n-butanol or acetate and compared with that of the wildtype *E. coli* strain. For growth assay, *E. coli* cells were cultivated overnight at 37°C in LB medium, and then diluted at a ratio of 1:100 into fresh LB media added with serial concentrations of n-butanol or acetate. Cultivations at 37°C, 200 rpm were performed for 48 hours before calculation of OD600 to evaluate the cell densities.

14.5.6 SHOCK EXPERIMENT

To explore tolerance and stability of *E. coli* BT12 strain under extreme lethal n-butanol stress, shock experiments were performed with 2% (vol/vol) of n-butanol. Procedure of the shock experiment was similar with previously introduced [58]. *E. coli* BT12 and *E. coli* DH5α cells were cultivated overnight at 37°C in LB medium, and then diluted at a ratio of 1:100 with fresh LB medium and grown at 37°C to OD600 of 0.3~0.5. The cultures were then diluted with LB to OD600 of 0.3, and n-butanol was added to final concentrations of 2% (vol/vol). After incubation at 37°C for 1 hour, the cultures were serially diluted, plated on LB agar plates and cultivated at 37°C overnight for photographing.

14.5.7 DETERMINATION OF MUTATION RATE

All of the selected dnaQ mutators were retransformed into *E. coli* DH5α strain with clean genetic background to determine the mutation rate. We calculated generation frequency of rifamycin resistant mutant cells to evaluate the mutation rates of the host cells, a method that has been applied widely in types of microbes [59,60]. *E. coli* cells transformed with specific dnaQ mutant were spread on LB agar plates and cultivated overnight in 37°C. Three colonies were inoculated into fresh LB medium containing Amp100 and grown overnight. Approximately 107 cells from the cultivation broth were spread on LB agar containing Amp100 with or without 100 µg/ml of rifamycin, and incubated in dark for 2 days. The rifamycin resistant colonies and total colonies were counted, and CFU data were used for determination of mutation rates.

REFERENCES

1. Patnaik R, Louie S, Gavrilovic V, Perry K, Stemmer WPC, Ryan CM, del Cardayre S: Genome shuffling of Lactobacillus for improved acid tolerance. Nat Biotechnol 2002, 20:707-712.
2. Alper H, Moxley J, Nevoigt E, Fink GR, Stephanopoulos G: Engineering yeast transcription machinery for improved ethanol tolerance and production. Science 2006, 314:1565-1568.
3. Stephanopoulos G, Alper H, Moxley J: Exploiting biological complexity for strain improvement through systems biology. Nat Biotechnol 2004, 22:1261-1267.
4. Brynildsen MP, Liao JC: An integrated network approach identifies the isobutanol response network of Escherichia coli. Mol Syst Biol 2009, 5:277.
5. Nielsen J: Metabolic engineering. Appl Microbiol Biotechnol 2001, 55:263-283.

6. Petri R, Schmidt-Dannert C: Dealing with complexity: evolutionary engineering and genome shuffling. Curr Opin Biotechnol 2004, 15:298-304.

7. Meynial-Salles I, Soucaille P: Creation of new metabolic pathways or improvement of existing metabolic enzymes by in vivo evolution in Escherichia coli. Methods Mol Biol 2012, 834:75-86.

8. Auriol C, Bestel-Corre G, Claude JB, Soucaille P, Meynial-Salles I: Stress-induced evolution of Escherichia coli points to original concepts in respiratory cofactor selectivity. Proc Natl Acad Sci USA 2011, 108:1278-1283.

9. Meynial Salles I, Forchhammer N, Croux C, Girbal L, Soucaille P: Evolution of a Saccharomyces cerevisiae metabolic pathway in Escherichia coli. Metab Eng 2007, 9:152-159.

10. Parekh S, Vinci VA, Strobel RJ: Improvement of microbial strains and fermentation processes. Appl Microbiol Biotechnol 2000, 54:287-301.

11. Alper H, Stephanopoulos G: Global transcription machinery engineering: A new approach for improving cellular phenotype. Metab Eng 2007, 9:258-267.

12. Park KS, Lee DK, Lee H, Lee Y, Jang YS, Kim YH, Yang HY, Lee SI, Seol W, Kim JS: Phenotypic alteration of eukaryotic cells using randomized libraries of artificial transcription factors. Nat Biotechnol 2004, 22:459.

13. Lee JY, Yang KS, Jang SA, Sung BH, Kim SC: Engineering butanol-tolerance in Escherichia coli with artificial transcription factor libraries. Biotechnol Bioeng 2011, 108:742-749.

14. Ochi K: From microbial differentiation to ribosome engineering. Biosci Biotechnol Biochem 2007, 71:1373-1386.

15. Echols H, Goodman MF: Fidelity mechanisms in DNA replication. Annu Rev Biochem 1991, 60:477-511.

16. Maki H, Horiuchi T, Sekiguchi M: Structure and expression of the dnaQ mutator and the RNase H genes of Escherichia coli: Overlap of the promoter regions. Proc Natl Acad Sci USA 1983, 80:7137-7141.

17. Scheuermann R, Tam S, Burgers PMJ, Lu C, Echols H: Identification of the epsilon-subunit of Escherichia coli DNA polymerase III holoenzyme as the dnaQ gene product: a fidelity subunit for DNA replication. Proc Natl Acad Sci USA 1983, 80:7085-7089.

18. Kunkel TA, Bebenek R: DNA replication fidelity. Annu Rev Biochem 2000, 69:497-529.

19. Cox EC, Horner DL: Structure and Coding Properties of a Dominant Escherichia-Coli Mutator Gene, Mutd. P Natl Acad Sci-Biol 1983, 80:2295-2299.

20. Fijalkowska IJ, Schaaper RM: Mutants in the Exo I motif of Escherichia coli dnaQ: Defective proofreading and inviability due to error catastrophe. Proc Natl Acad Sci USA 1996, 93:2856-2861.

21. LeClerc JE, Li B, Payne WL, Cebula TA: High mutation frequencies among Escherichia coli and Salmonella pathogens. Science 1996, 274:1208-1211.

22. Matic I, Radman M, Taddei F, Picard B, Doit C, Bingen E, Denamur E, Elion J: Highly variable mutation rates in commensal and pathogenic Escherichia coli. Science 1997, 277:1833-1834.

23. Schaaper RM: Mechanisms of mutagenesis in the Escherichia coli mutator mutD5: role of DNA mismatch repair. Proc Natl Acad Sci USA 1988, 85:8126-8130.

24. Atsumi S, Hanai T, Liao JC: Non-fermentative pathways for synthesis of branched-chain higher alcohols as biofuels. Nature 2008, 451:86-U13.
25. Dellomonaco C, Clomburg JM, Miller EN, Gonzalez R: Engineered reversal of the beta-oxidation cycle for the synthesis of fuels and chemicals. Nature 2011, 476:355-U131.
26. Shen CR, Lan EI, Dekishima Y, Baez A, Cho KM, Liao JC: Driving Forces Enable High-Titer Anaerobic 1-Butanol Synthesis in Escherichia coli. Appl Environ Microb 2011, 77:2905-2915.
27. Atsumi S, Wu TY, Machado IM, Huang WC, Chen PY, Pellegrini M, Liao JC: Evolution, genomic analysis, and reconstruction of isobutanol tolerance in Escherichia coli. Mol Syst Biol 2010, 6:449.
28. Reyes LH, Almario MP, Winkler J, Orozco MM, Kao KC: Visualizing evolution in real time to determine the molecular mechanisms of n-butanol tolerance in Escherichia coli. Metab Eng 2012, 14:579-590.
29. Zingaro KA, Papoutsakis ET: GroESL overexpression imparts Escherichia coli tolerance to i-, n-, and 2-butanol, 1,2,4-butanetriol and ethanol with complex and unpredictable patterns. Metab Eng 2013, 15:196-205.
30. Zingaro KA, Papoutsakis ET: Toward a Semisynthetic Stress Response System To Engineer Microbial Solvent Tolerance. Mbio 2012, 3:5.
31. Rutherford BJ, Dahl RH, Price RE, Szmidt HL, Benke PI, Mukhopadhyay A, Keasling JD: Functional genomic study of exogenous n-butanol stress in Escherichia coli. Appl Environ Microbiol 2010, 76:1935-1945.
32. Mao SM, Luo YAM, Zhang TR, Li JS, Bao GAH, Zhu Y, Chen ZG, Zhang YP, Li Y, Ma YH: Proteome reference map and comparative proteomic analysis between a wild type Clostridium acetobutylicum DSM 1731 and its mutant with enhanced butanol tolerance and butanol yield. J Proteome Res 2010, 9:3046-3061.
33. Chundawat SPS, Beckham GT, Himmel ME, Dale BE: Deconstruction of lignocellulosic biomass to fuels and chemicals. Annu Rev Chem Biomol Eng 2011, 2:121-145.
34. Mira NP, Palma M, Guerreiro JF, Sa-Correia I: Genome-wide identification of Saccharomyces cerevisiae genes required for tolerance to acetic acid. Microb Cell Fact 2010, 9:79.
35. Pena PV, Glasker S, Srienc F: Genome-wide overexpression screen for sodium acetate resistance in Saccharomyces cerevisiae. J Biotechnol 2013, 164:26-33.
36. Hamdan S, Carr PD, Brown SE, Ollis DL, Dixon NE: Structural basis for proofreading during replication of the Escherichia coli chromosome. Structure 2002, 10:535-546.
37. Perrino FW, Harvey S, McNeill SM: Two functional domains of the epsilon subunit of DNA polymerase III. Biochemistry-Us 1999, 38:16001-16009.
38. Morrison A, Bell JB, Kunkel TA, Sugino A: Eukaryotic DNA polymerase amino acid sequence required for 3'-> 5' exonuclease activity. Proc Natl Acad Sci USA 1991, 88:9473-9477.
39. Blanco L, Bernad A, Salas M: Evidence favoring the hypothesis of a conserved 3'-5' exonuclease active-site in DNA-dependent DNA-polymerases. Gene 1992, 112:139-144.
40. Giraud A, Matic I, Tenaillon O, Clara A, Radman M, Fons M, Taddei F: Costs and benefits of high mutation rates: adaptive evolution of bacteria in the mouse gut. Science 2001, 291:2606-2608.

41. Trobner W, Piechocki R: Competition between isogenic mutS and mut+ populations of Escherichia coli K12 in continuously growing cultures. Mol Gen Genet 1984, 198:175-176.

42. Gibson TC, Scheppe ML, Cox EC: Fitness of an Escherichia coli mutator gene. Science 1970, 169:686-688.

43. Taddei F, Radman M, MaynardSmith J, Toupance B, Gouyon PH, Godelle B: Role of mutator alleles in adaptive evolution. Nature 1997, 387:700-702.

44. Tenaillon O, Toupance B, Le Nagard H, Taddei F, Godelle B: Mutators, population size, adaptive landscape and the adaptation of asexual populations of bacteria. Genetics 1999, 152:485-493.

45. Horiuchi T, Maki H, Sekiguchi M: New conditional lethal mutator (dnaQ49) in Escherichia coli K12. Mol Gen Genet 1978, 163:277-283.

46. Tanabe K, Kondo T, Onodera Y, Furusawa M: A conspicuous adaptability to antibiotics in the Escherichia coli mutator strain, dnaQ49. Fems Microbiol Lett 1999, 176:191-196.

47. Selifonova O, Valle F, Schellenberger V: Rapid evolution of novel traits in microorganisms. Appl Environ Microbiol 2001, 67:3645-3649.

48. Shimoda C, Itadani A, Sugino A, Furusawa M: Isolation of thermotolerant mutants by using proofreading-deficient DNA polymerase delta as an effective mutator in Saccharomyces cerevisiae. Genes Genet Syst 2006, 81:391-397.

49. Abe H, Fujita Y, Takaoka Y, Kurita E, Yano S, Tanaka N, Nakayama K: Ethanoltolerant Saccharomyces cerevisiae strains isolated under selective conditions by over-expression of a proofreading-deficient DNA polymerase delta. J Biosci Bioeng 2009, 108:199-204.

50. Loh E, Salk JJ, Loeb LA: Optimization of DNA polymerase mutation rates during bacterial evolution. Proc Natl Acad Sci USA 2010, 107:1154-1159.

51. Gonzalez-Ramos D, van den Broek M, van Maris AJ, Pronk JT, Daran JM: Genome-scale analyses of butanol tolerance in Saccharomyces cerevisiae reveal an essential role of protein degradation. Biotechnol Biofuels 2013, 6:48.

52. Sanchez RG, Karhumaa K, Fonseca C, Nogue VS, Almeida JRM, Larsson CU, Bengtsson O, Bettiga M, Hahn-Hagerdal B, Gorwa-Grauslund MF: Improved xylose and arabinose utilization by an industrial recombinant Saccharomyces cerevisiae strain using evolutionary engineering. Biotechnol Biofuels 2010, 3:13.

53. Koppram R, Albers E, Olsson L: Evolutionary engineering strategies to enhance tolerance of xylose utilizing recombinant yeast to inhibitors derived from spruce biomass. Biotechnol Biofuels 2012, 5:32.

54. Jarboe LR, Grabar TB, Yomano LP, Shanmugan KT, Ingram LO: Development of ethanologenic bacteria. Adv Biochem Eng Biotechnol 2007, 108:237-261.

55. Jantama K, Zhang X, Moore JC, Shanmugam KT, Svoronos SA, Ingram LO: Eliminating Side Products and Increasing Succinate Yields in Engineered Strains of Escherichia coli C. Biotechnol Bioeng 2008, 101:881-893.

56. Zhang X, Jantama K, Moore JC, Shanmugam KT, Ingram LO: Production of L-alanine by metabolically engineered Escherichia coli. Appl Microbiol Biotechnol 2007, 77:355-366.

57. Stephanopoulos G: Metabolic engineering by genome shuffling - Two reports on whole-genome shuffling demonstrate the application of combinatorial methods for phenotypic improvement in bacteria. Nat Biotechnol 2002, 20:666-668.
58. Chen T, Wang J, Yang R, Li J, Lin M, Lin Z: Laboratory-evolved mutants of an exogenous global regulator, IrrE from Deinococcus radiodurans, enhance stress tolerances of Escherichia coli. Plos One 2011, 6:e16228.
59. Miller JH, Funchain P, Clendenin W, Huang T, Nguyen A, Wolff E, Yeung A, Chiang JH, Garibyan L, Slupska MM, Yang HJ: Escherichia coli strains (ndk) lacking nudeoside diphosphate kinase are powerful mutators for base substitutions and frameshifts in mismatch-repair-deficient strains. Genetics 2002, 162:5-13.
60. Sasaki M, Yonemura Y, Kurusu Y: Genetic analysis of Bacillus subtilis mutator genes. J Gen Appl Microbiol 2000, 46:183-187.

AUTHOR NOTES

CHAPTER 2

Author Information

LM is a PhD candidate in the Department of Wine, Food and Molecular Biosciences, Lincoln University. He holds a B.Sc. degree with First Class Honours in biochemistry from Lincoln University. WV is a senior lecturer in computational biology at the Centre for Advanced Computational Solutions at Lincoln University. He holds one PhD degree in computational systems modelling from Lincoln University and another PhD degree in theoretical physics and was previously a professor of physics at the University of South Africa.

CHAPTER 3

Acknowledgements

The author would like to thank all the collaborators and co-authors of the all manuscripts mentioned in this paper, and Stephanie Seddon-Brown for English proof reading of this manuscript.

CHAPTER 4

Competing Interests

The authors declare that they have no competing interests.

Author Contributions

All authors have contributed equally to this commentary, and all have read and approved the final manuscript.

Acknowledgements

The authors' research on yeast biotechnology is funded by the Federal Ministry of Economy, Family and Youth (BMWFJ), the Federal Ministry of Traffic, Innovation and Technology (bmvit), the Styrian Business Promotion Agency SFG, the Standortagentur Tirol and ZIT – Technology Agency of the City of Vienna through the COMET-Funding Program managed by the Austrian Research Promotion Agency FFG. Additional support by the program "Intelligente Produktion" of FFG, project "Lignoraffinerie" and by the Austrian Science Fund (FWF): Doctoral Program BioToP—Biomolecular Technology of Proteins (FWF W1224) is acknowledged.

CHAPTER 5

Competing Interests
The authors declare that they have no competing interests.

Author Contributions
ZW designed the research and performed the MFC operation. TL carried out the microbial community analysis and prepared the related part of manuscript. ZW prepared the manuscript. BL, CC and JP reviewed the manuscript. All authors read and approved the final manuscript.

Acknowledgements
This work was supported by the National Research Foundation of Korea (NRF) grant funded by the Korea government (MEST) (No. 2012R1A1A2009500). Thanks should be given to Ms Jane Eumie Choi and Dr Euree Choi for help with English editing.

CHAPTER 6

Acknowledgements
We thank E. Korenblum, M.C. Pereira e Silva and C.A. Mallon for their help in the construction of the enriched cultures and for reviewing the manuscript. Further thanks are due to H. Ruijssenaars and R. van Kranenburg (PURAC) for their support. This work was supported by the Netherlands Ministry of Economic Affairs and the BE-Basic partner organizations (http://www.be-basic.org/).

CHAPTER 8

Competing Interests
The authors declare that they have no competing interests.

Author Contributions
JMG carried out most of the research and wrote the manuscript. JMG and JIP designed the study, designed the in vitro expression vector and cloned all cellulase genes to E. coli expression vector, and conducted the in vitro and in vivo screens. JB and VR profiled enzyme activities. PD identified and repaired the 37 cellulase ORFs. BFQ advised JB. KLS advised JIP and VR and helped visualize Topt and IL-tolerance. SWS and BAS advised JG. All authors read, provided edits, and approved the manuscript.

Acknowledgements
This work conducted by the Joint BioEnergy Institute was supported by the Office of Science, Office of Biological and Environmental Research, of the US Department of Energy under Contract No. DE-AC02-05CH11231. JCB was supported by fellowship 9721/11-8 from CAPES Foundation, Ministry of Education of Brazil.

CHAPTER 9

Competing Interests
The authors declare that they have no competing interests.

Author Contributions
BHJ, JAC, HCK, JHH and RAA designed and performed research; BHJ, JAC, HCK, JHH, BAD, JMR, JRK, and RAA analyzed data; and BHJ, JAC, HCK, JHH, BAD, JMR, RAA and JRK wrote the manuscript. All authors reviewed and approved the final manuscript.

Acknowledgements
This work was supported by the Senior Researchers (National Research Foundation of Korea, 2010–0026904) and the Brain Korea-21 (BK-21) programs of the Korea Ministry of Education, Science, and Technology, and the Eco-Innovation project (Global-Top project, 2012001090001) of

the Korea Ministry of Environment. The Korean Basic Science Institute (Chuncheon) is acknowledged for the SEM analysis.

CHAPTER 10

Competing Interests
The authors declare that they have no competing interests.

Author Contributions
DBH, ELH and ZL conceived the work, ZL, CHC, DBH, and ELH wrote the manuscript, CHC performed the pretreatment, composition analysis and enzymatic digestibility analysis. All authors provided input and corrections to the manuscript. All authors read and approved the final manuscript.

Acknowledgments
This work was funded by the DOE Great Lakes Bioenergy Research Center (DOE BER Office of Science DE-FC02-07ER64494).

CHAPTER 11

Acknowledgements
The authors are grateful to Prof. Keith E. Taylor, Department of Chemistry and Biochemistry, University of Windsor, for his helpful comments on the manuscript.

CHAPTER 12

Competing Interests
The authors declare that they have no competing interests.

Author Contributions
JLF and SSJL designed the study and JLF performed the experimental work. JLF and SSJL co-wrote the manuscript. Both authors read and approved the final manuscript.

Acknowledgements

This work was funded by the Competitive Research Program (Grant No. M59002047). Jee Loon Foo acknowledges the support of Nanyang Technological University through the Joint NTU-Berkeley Postdoctoral Program.

CHAPTER 13

Funding

Funding provided through the Irish Department of Agriculture and Food's Research Stimulus Fund (Project Code RSF 07 513; see,http://www.agriculture.gov.ie/research/researchstimulusfundrsf/). The funders had no role in study design, data collection and analysis, decision to publish, or preparation of the manuscript.

Competing Interests

The authors have declared that no competing interests exist.

Acknowledgments

We thank Dr SS Ali (University College Dublin) for his help with the Southern blot analysis and Marcel Ketteniss for help with the secondary screen.

Author Contributions

Conceived and designed the experiments: RH FD EM. Performed the experiments: RH. Analyzed the data: RH FD EM. Contributed reagents/materials/analysis tools: RH FD EM. Wrote the manuscript: RH FD EM.

CHAPTER 14

Competing Interests

The authors declare that they have no competing interests.

Author Contributions

GDL and ZC designed and performed the research. GDL, ZC, and YL analyzed the data. GDL, ZC, YL, and YHM wrote the manuscript. YL and YHM supervised the project. All authors read and approved the final manuscript.

Acknowledgements

This work was supported by National Basic Research Program of China (973 program, 2011CBA00800), and National High Technology Research Program of China (863 program, 2012AA022100).

INDEX